D1222316

STATISTICS AND SOCIETY

DATA COLLECTION AND INTERPRETATION

STATISTICS

Textbooks and Monographs

A SERIES EDITED BY

VOLUME 1: The Generalized Jackknife Statistic, *H. L. Gray and W. R. Schucany*

VOLUME 2: Multivariate Analysis, *Anant M. Kshirsagar*

VOLUME 3: Statistics and Society, *Walter T. Federer*

OTHER VOLUMES IN PREPARATION

STATISTICS AND SOCIETY

DATA COLLECTION AND INTERPRETATION

WALTER T. FEDERER

Biometrics Unit
Cornell University
Ithaca, New York

MARCEL DEKKER, INC. New York 1973

MARCEL DEKKER, INC.

95 Madison Avenue, New York, New York 10016

LIBRARY OF CONGRESS CATALOG CARD NUMBER: 73-79457

ISBN: 0-8247-6083-2

PRINTED IN THE UNITED STATES OF AMERICA

TABLE OF CONTENTS

PREFACE

The initial impetus for this book came from a meeting of the author with T. C. Watkins, past Director of Resident Instruction, and R. P. Murphy, past Head of the Department of Plant Breeding and Biometry, of Cornell University, where it was decided that a different approach of presenting statistics was required. The introductory methods and theory courses in statistics did not appear to be satisfying the needs of all students. A course presenting the ideas, concepts, and philosophies associated with the procurement of numbers with a minimum of emphasis on statistical methodology or on mathematical theory was required; the course was to be concerned with numbers and the use of numbers as related to the many aspects of society. The course was to be complementary to and nonrepetitive of other introductory statistics courses. It was to be an "artsie" type course in statistics in that the student would learn about the subject of Statistics but would not attain competence in the technical aspects of the subject.

Such a course was first attempted in the spring term of 1966 and was repeated during the spring terms of 1967, 1968, 1969, 1970, 1971, and 1972. The course was given at 8:00 a.m. with no announced instructor and as an elective course. Despite the early hour and other adversities, the enrollments in the first seven years were 57, 101, 99, 75, 70, 85, and 100. (In 1972 the time was changed.) In addition to the students registered for credit, several visitors have attended the course each year. The author taught the course during the first three years and the sixth year; Professor B. L. Raktoe, University of Guelph, Professor Carl Marshall, Oklahoma State University, and Professor F. B. Cady, Cornell University, taught the course during the fourth, fifth, and seventh years respectively. These experiences were utilized in the preparation of the present text.

The material presented in the text together with the associated problems, three term quizzes, and a final examination has been considered sufficient for three semester hours of credit at the sophomore level in a university. For the first six years, three lectures were given each week and weekly discussion periods with optional attendance were held. The material has not been difficult for the upper one-fourth of the class but has been for the remaining three-fourths. The concept of orthogonality in experimental designs has been the most difficult concept but without it several students have indicated that the course would not be challenging enough. Student evaluations have been obtained repeatedly but these have produced only minor changes in content or in presentation. Any instructor using this text in a course should make certain that most of the students really understand what has been presented before proceeding to the next topic.

Statistics is defined to be the development and application of techniques involved with (1) the design of investigation, (2) the summarization of results from an investigation, and (3) inference from the results of the investigation to the population from which the results were derived. Hence, a prime concern should be with the origins of data and how to obtain reliable information in any investigation; unfortunately, methodology and mathematical statistics books define away these important problems related to real world situations. The emphasis in this text is on the relationship of sample and population, statistic and parameter, and on the underlying variability of all observations in the real world. The universality of variability and the various sources of variation represent the underlying theme throughout the book. One designs to take account of certain types of variation. Consequently, this text emphasizes design concepts and techniques, mostly with some attention being devoted to summarization concepts and techniques, and with little attention being devoted to inference. This is the reverse of the emphasis in current statistics texts.

A search was made for a suitable text for this course. In the course of this search, the author was amazed to discover that methods books contain mostly computational procedures and techniques of computation with little or no space being devoted to methods of procuring meaningful and accurate data; it appears that many statistics books are concerned mainly with the manipulation of numbers and algebraic symbols or with mathematical manipulations. Although many students are interested in these manipulations, many are not. This latter group would form the nucleus of students for the material covered in this text. An instructor or student using the present text will find the books by E. B. Wilson, "An Introduction to Scientific Research", by D. Huff, "How to Lie With Statistics", and by M. J. Slonim, "Sampling in a Nutshell", useful for supplementary reading material, especially with regard to chapters I to V.

Several types of courses may be taught from this text by changing the emphasis and sequence of chapters and by adding material in selected areas.

During the first four presentations of the material, the chapter sequence followed was I to IV, VI, VII, V, VIII, X, XI, and XIII; most of the material in chapters IX, XII, and XIV was omitted. The present sequence of chapters is considered to be better pedagogically because the sample survey concepts presented in chapter V are easier to comprehend than the experiment design and treatment design concepts in chapters VI and VII. During the fifth and sixth presentations of the course the chapter sequence was essentially that presented in the text except that some of the material in chapters IX to XII and XIV was included with the material in chapters V to VII; also, sample survey designs received more emphasis and experimental and treatment designs less emphasis than given in the text. A third type of course that could be taught from the text would emphasize sample survey design and analysis concepts using the material in chapters I to V, parts of VIII to XII, XIV, with either more emphasis on summary statistics or with a considerably expanded version of chapter XIII on the type of statistics gathered and the methods used to procure the data for the United States, for the United Nations, or for some selected country like India. This text is designed for a one-semester course but could easily be expanded into a two-semester course by supplementing, for example, the material in chapters V, VIII, XIV, IX, and/or XI.

I am indebted to the Literary Executor of the late Sir Ronald A. Fisher, F. R. S., and to Oliver & Boyd, Edinburgh for their permission to reprint tables VIII.5, XI.1, XI.2, of t-values in section XI.5, and of chi-square values in section XIV.5. from their book _Statistical Methods for Research Workers_.

The author wishes to acknowledge the interest shown by and the comments received from Prof. B. L. Raktoe, Prof. Carl Marshall, Prof. Foster B. Cady, and his other colleagues at Cornell University in the preparation of this text. Prof. Oscar Kempthorne devoted a considerable amount of time and effort and offered valuable suggestions; his efforts are greatly appreciated. Sincere gratitude is expressed for the editorial assistance of my wife, Lillian. Also, gratitude and thanks are acknowledged to Mrs. Donna Van Order, Mrs. Anne White, Mrs. Norma Phalen, and Mrs. Helen Seamon for their efforts and skills exhibited in typing the manuscript.

Copies of the supplement to the text are available from the publisher upon request.

Walter T. Federer

STATISTICS AND SOCIETY

DATA COLLECTION AND INTERPRETATION

CHAPTER I. INTRODUCTION AND DEFINITIONS

Numbers and conclusions drawn from numbers are commonplace in the printed media of our society. In order to evaluate the soundness of the conclusions it is necessary for the reader to develop a constructively critical and questioning attitude toward the manner in which the numbers were obtained and toward the manner in which they were used. Consideration of these topics forms an important part of this book. If valid and meaningful conclusions are to be drawn from numbers, the numbers must be obtained from a carefully planned and carefully executed investigation. The reader should be able to know when to consider a printed report seriously and when not to. He should have little difficulty in making a decision about the conclusions cited in the following article, reprinted from The Journal of Irreproducible Results, volume 15, number 1, page 18.[1]

PICKLES AND HUMBUG

(A bit of comparative logic)

> Pickles will kill you! Every pickle you eat brings you nearer to death. Amazingly, "the thinking man" has failed to grasp the terrifying significance of the term "in a pickle". Although leading horticulturists have long known that *Cucumis sativus* possesses indehiscent pepo, the pickle industry continues to expand.
>
> Pickles are associated with all the major diseases of the body. Eating them breeds wars and Communism. They can be related to most airline tragedies. Auto accidents are caused by pickles. There exists a positive relationship between crime waves and consumption of this fruit of the cucurbit family. For example:
>
> 1. Nearly all sick people have eaten pickles. The effects are obviously cumulative.
> 2. 99.9% of all people who die from cancer have eaten pickles.
> 3. 100% of all soldiers have eaten pickles.
> 4. 96.8% of all Red sympathizers have eaten pickles.
> 5. 99.7% of the people involved in air and auto accidents ate pickles within 14 days preceding the accident.

[1] Reprinted with the permission of G. H. Scherr, Publisher, Journal of Irreproducible Results, P.O. Box 234, Chicago Heights, Illinois, 60411.

6. 93.1% of juvenile delinquents come from homes where pickles are served frequently.

Evidence points to the long-term effects of pickle-eating:

Of the people born in 1839 who later dined on pickles, there has been a 100% mortality.

All pickle eaters born between 1849 and 1859 have wrinkled skin, have lost most of their teeth, have brittle bones and failing eye-sight, — if the ills of eating pickles have not already caused their death.

Even more convincing is the report of a noted team of medical specialists: rats force-fed with 20 pounds of pickles per day for 30 days developed bulging abdomens. Their appetites for wholesome food were destroyed.

The only way to avoid the deleterious effects of pickle-eating is to change the eating habits. Eat orchid petal soup. Practically no one has any problems from eating orchid petal soup.

(Anonymous)

Reading such a report, one would probably be amused. However, there are situations where the value of a report cannot immediately be assessed, and it is here that some aids such as those described in the following chapters would be useful.

Instead of being faced with a report such as the above, one might want to open a business and conduct a market research study prior to selecting a place for said business. This is exactly what Fred Finn did before he and his wife Mickie opened their fabulously successful "Mickie Finn's" in San Diego (see Ithaca Journal, April 16-22, 1966, page 11). The odds against opening a nightclub and having it be a success are quoted as being on the order of 50,000 to one. The odds became somewhat different after the market research study indicated that San Diego was the best location "because of its growth potential, mild climate, and tourist advantages". Studies of this nature may be utilized in many ways.

All books must have objectives and goals. Those for this introductory text are:

1. to develop ideas and concepts related to measurement, data collection, and experimentation,
2. to define and describe various aspects of the subject of statistics,

3. to study some uses and abuses of data,

4. to develop a constructively critical attitude toward numbers and users of numbers in our society.

Toward this end, a number of topics will be covered. In order to understand and orient the material presented, it will be helpful to read the associated material listed at the end of each chapter. To understand the concepts and techniques of statistics more fully, it is highly desirable for the reader to work the problems at the end of the various chapters.

In order to have numbers that have a meaning for us, we shall obtain the height, weight, eye color, hair color, and age of each student in the class, as described in the next chapter. This survey, or collection of numbers, represents characteristics of the group of individuals composing this class. If the aggregate of individuals in this class represents all the individuals of interest in the survey, they constitute the universe, or population, of interest. If, however, the universe is composed of a wider aggregate or group of individuals, say the full-time undergraduate students of this university, then the members of this class represent only a part, or sample, of the universe or population. Complete enumeration or survey of a characteristic in the population is defined as a census, whereas enumeration on only a part of the population is known as a sample or as a sample survey.

Populations or universes may consist of characteristics of people, of acres in farms, of urban and nonurban dwellings, of commercial deep-freeze storage facilities, of the various genotypes obtainable in the F_2 generation of a cross of two parents, of the various strains of Neurospora (bread mold), of the various commercial pesticides available in 1966, of all 25-watt light bulbs, of the number of dandelions in my lawn at time X, of the dresses in a wearing-apparel shop at time X, or of any other defined aggregate or group in a specified place, area, or context. Populations are classified and described by numbers. The more developed and the more studied populations are associated with rather complete and full numerical characterizations. For example, United States agriculture is characterized by numerous annual publications describing all commercial crops and livestock produced in the United States every year in terms of number of acres, production, amount commercially sold, amount stored, prices received, time when sold, etc. In addition to annual publications, reports on many of these items are made monthly or weekly, and the newspapers contain daily reports on prices paid and prices received for many agricultural commodities. The rapid advances in high-speed computing will add many more numbers and consequently reports to this list.

In the United States we are almost engulfed in numbers such as those mentioned above. In addition, we have social security numbers, credit card numbers, telephone numbers, zip code numbers, batting averages, bowling averages, etc. When will a university student be known as a number? That day is fast approaching unless the advances in high-speed computers tend toward the use of alphabets rather than of numbers. It is a fact, however, that the more developed a society, the more that society is characterized by numbers. Knowledge of the characteristics of a society allows intelligent action to be taken in order to further develop the society.

Numbers used to describe a characteristic of a population and which are derived from all members of the universe are called parameters. Parameters represent facts about the population. Numbers derived from a sample and which may be used to "guesstimate", to estimate, or to approximate the value of the parameter are called statistics. Thus a grade average of a student in any given semester is an estimate of the student's grade average for his four years at college; in this example, the one semester (specified) grade average is a statistic, and it estimates the grade average for a total of eight semesters. The latter grade average is the parameter; the population is the totality of grades and credit hours from all courses taken by the student in eight semesters.

As opposed to the above use of the word "statistics" as represented by columns of numbers, averages, percentages, ratios, and the like, there is a science of Statistics which deals with the development and application of statistical procedures. Courses are taught under the heading of Statistics; it is one of the areas of specialization in many colleges; and there are people known as statisticians who work and make their livelihood in Statistics. In addition to undergraduate degrees, graduate degrees are awarded by over 100 universities in the United States (see The American Statistician, December, 1971), and the subject is rapidly advancing owing to the vigorous research programs at many of these graduate training and statistical research centers. The subject of Statistics is concerned mainly with the following items:

1. to design or to plan experimental investigations (experiments) and sample surveys,
2. to summarize the information from the numbers collected from experiments and sample surveys,

3. to relate or to infer facts about the population utilizing facts from the experiment or the sample.

The above serves well as a definition of Statistics. The Civil Service Commission says that "Statistics means the collection, classification, and measured evaluation of facts as a basis for inference ". This is elaborated as "a body of techniques for acquiring accurate knowledge from incomplete information, a scientific system for the collection, organization, analysis, interpretation, and presentation of information which can be stated in numerical form". The two definitions are equivalent, though different words are used.

Since all people are involved with Statistics in one way or another, they are, or should be, interested in the statistical techniques or tools useful to them in the context of their interest. Most people are interested in understanding and/or using the tools of the statistician; they are not interested in developing new statistical tools, as this is the job of the statistician. Often a user of a tool does not understand or know how to use the statistical tool. In this case, consultation with a statistician conversant with the use of the tool or with an appropriate publication on its use should suffice. It should be emphasized that not all statisticians are conversant with the use of all statistical tools. For example, an expert on inference may be of little or no use in designing an experimental investigation and *vice versa*. Despite this, the training of consulting statisticians often equips them to answer many or most of the problems arising in design, in summarization of numbers, and in making inferences from the sample to the population. The fact that the statistician consults with and works with others on the use of statistical procedures in all fields of investigation illustrates the usefulness and importance of the various aspects of Statistics.

A brochure entitled "Careers in Statistics"[2], and published by the three professional societies of statisticians in the United States, The American Statistical Association, The Biometric Society (ENAR and WNAR), and The Institute of Mathematical Statistics, describes what statisticians do, where they work, some concepts of Statistics, the education of statisticians, the future of Statistics, and the salaries that statisticians may expect to receive.

---

[2]Copies of this brochure may be obtained by writing E. M. Bisgyer, The American Statistical Association, Suite 640, 806 15th Street, N.W., Washington, D. C. 20005.

The advent of fast and new electronic calculating machines has touched off a frantic search for people who can write instructions to make the machines work. For example, an article in the March, 1967, issue of *Fortune* entitled "Help Wanted: 50,000 Programmers" indicates one aspect of this search. A person who knows Statistics and can write statistical programs is a relatively rare individual and therefore in great demand. This area is expanding, and the demand for people trained in these areas is on the increase with a corresponding increase in starting salaries.

Statistics as a science and subject unto itself is a branch of applied mathematics and probability. As such, it is rigorous and well defined within a framework of definitions and assumptions. Whenever a statistical procedure is applied to a real-life situation the assumptions may or may not be justified. This means then that the application of statistical procedures *always* involves a degree of subjectiveness. The degree of subjectiveness should be constantly questioned and evaluated in order to make proper use of the statistical procedure under consideration.

In any continuing program involving sample survey and experimental investigations two purposes should always be served. These are:

1. to describe or characterize the population from the facts obtained from the investigation,
2. to utilize the facts from the investigation to design better sample surveys and experiments in future investigations.

The latter item is often forgotten by experimenters, sample surveyors, and statisticians. What is called subjective under (1) may not be under (2) above. For example, in making laboratory determinations on the chemical composition of plant samples, the samples were numbered 1A, 1B, 2A, 2B, 3A, 3B, etc. 1A and 1B and so forth were duplicate samples and the chemist was so informed. The results of the duplicate samples were strikingly close; in fact, much closer than was believed possible for the chemical and sampling procedures used. To check on this, numbers were assigned to the duplicate samples in an arbitrary manner, resulting in a random allocation by chance of the numbers to the samples. The chemist was not told which samples were duplicates. The results for duplicate samples were no longer close. This example illustrates a method of changing the investigational procedure in such a manner as to treat duplicate samples "fairly" with respect to nonduplicate samples. More such procedures will be shown in the forthcoming chapters and in the problems associated with each chapter.

I.2. Problems

I.1. (Do not spend long periods of time thinking about this problem. It will require some ingenuity in setting up an investigational procedure to obtain the desired results. Many measurements will need to be taken in order to obtain the desired accuracy.) A table with unknown length is to be measured to within .001 of a foot. A measuring instrument is available which is calibrated only in feet. The calibration marks at the one-foot intervals are correct to .0001 of a foot. Design an investigational procedure (experimental design or design of the experiment) for utilizing the measuring instrument to obtain the length of the table to the desired accuracy. Do not devise schemes for calibrating the measuring instrument, even though this might be the realistic thing to do in certain cases.

Instead of a measuring instrument calibrated in feet, suppose that a rod of length 30.500 ± .001 inches is available. Devise a procedure for measuring the above-mentioned table accurate to .01 inch.

I.2. Describe ten different universes or populations with which you are associated or come in contact in everyday life. Are you in contact with all elements of the universe or only a sample of them?

I.3. Obtain three examples of everyday usage of statistics from the news media. Describe the use as you envisage it.

CHAPTER II. MEASUREMENT

II.1. Introduction

Man is continually attempting to quantify more and more characteristics about himself and his environment. This quantification is arrived at by measuring characteristics of an aggregate of individuals or items and it is expressed on a number-, letter-, symbol-, or word-scale; our present number system was devised for this purpose. Measures describing certain characteristics about an individual or group of individuals are the number of stone axes owned, the number of fish caught last season, the average quantity of milk consumed per meal by 10-year old boys, the percent of all protest rallies attended last year, the number of redheads dated during one's senior year in high school, a grade in a freshman chemistry course, etc.

In order to obtain meaningful and consistent numbers to characterize certain aspects of a population an appropriate measuring device is necessary. To have repeatable or reproducible measurements (numbers obtained by the measuring device), it is necessary to have a measuring device with a prescribed or measurable margin of error. Note that we do not say that the measuring instrument or device must be error-free, but only that the error of measurement must fall within prescribed limits. Having established the limits of error of measurement, we are then in a position to determine whether or not we can measure a characteristic of the individuals in the sample or in the population with the desired accuracy.

There are situations wherein it has not been possible to devise a method of determining the limits of error; this is true for certain areas of the social sciences as well as for various areas of all sciences utilizing experimentation. To overcome this deficiency, investigators may need to exhibit considerable ingenuity in devising procedures which yield repeatable measurements and a measure of the error associated with the measuring device. An example of such a situation is illustrated in problem I.1.

II.2. Measuring Devices

Many diverse measuring devices are utilized by experimenters, teachers, administrators, merchandisers, consumers, etc.. We shall list some of the more common measuring devices with some comments on their use.

II.2.1. Counting Measurements

Through the centuries man has utilized counting to measure his wealth and worldly goods and other items of interest to him. From past records we note that, at first, man could only count to two. The concept of one unit meant "one", and anything not one was "more than one". This probably arose from the fact that man has two arms, two legs, two eyes, etc. Even today, many people think in terms of two categories such as all-or-none, for-or-against, dead-or-alive, ill-or-well, liberal-or-conservative, racist-or-nonracist, student-or-nonstudent, mature-or-immature, drunk-or-sober, male-or-female, etc. Sometime later man counted to three on the joints of his finger. Then, man learned to count to ten using the fingers of his hands. This method is still utilized by man today! As time went on various number systems evolved including our present number system, and man was able to count his money, wives, children, herds of cattle, sheep and goats, fleets of ships, members in an organization, men in an army, etc. Even today man still measures characteristics about himself and his environment by counts such as pollen count, number of dresses or suits, number of courses or credit hours, number of confrontations, number of boyfriends, etc.

II.2.2. Ordering or Ranking Measurements

The ranking of individuals, just as counting, is a form of measurement with a long and useful history and it is still relevant in many present-day contexts. For example, the girl who has a number, say $b \geq 2$, of boyfriends may, and usually does, have preferences. She may rank the set of b boys she considers to be her boyfriends with respect to dancing ability, conversational ability, masculine appearance, good looks, or general overall likeability, where these characteristics have definite meaning in her mind. For any given trait the b boys are ranked from 1 to b with allowance for ties whenever she does not rank two or more boys differently. Such rankings are useful to her in making decisions as to which boy to date for a particular event.

In other situations, beauty contestants are ranked with respect to "beauty", entries in a pie-baking contest are ranked for appearance or palatability, contestants in a track event are ranked with respect to order of placement, entries in a horse race are ranked with respect to order of finish, athletic teams in a tournament are ranked with respect to their win-loss record, etc. The ordering or ranking of a group of individuals represents a form of measurement with which we come in daily contact. We should note that ordering does more than merely "pick-a-winner", it ranks all individuals relative to each other.

The measuring instrument is generally the person performing the ranking. The ranking by a particular individual is subjective, and is not always repeatable by another individual but can usually be repeated by the same person. The degree of arbitrariness or subjectiveness varies with the trait being ranked. Most placements of contestants in a track event are repeatable, despite the occasional "rhubarbs" encountered, whereas the ranking of two boyfriends with respect to "likeability" certainly varies with each girl, fortunately for the survival of the species *Homo sapiens*. Despite the arbitrariness of some rankings, ranking is and will continue to be a useful form of measurement for all of us.

As an alternative to ranking a number of individuals or items, actual measurements might be taken on their performance. For example, suppose that one is comparing ten light bulbs representing ten different brands for length of life; suppose that the ten bulbs are inserted in ten electrical sockets, and suppose that the electricity is alternatively turned on for five minutes and off for five minutes until all bulbs have been burned out (fail to light again) or until the electricity has been turned off 1000 times. In the latter case, some bulbs may not yet have burned out. For those that have, the investigator could rank them as first, second, third, etc. to burn out, *or* he could count the number of times electricity had been turned off before a bulb has failed to light again. The latter form of measurement is known as order-statistics; statistical theory is available to estimate the average length of life for all ten light bulbs even though some have not burned out at the termination of the measurements. The topic of order-statistics is beyond the scope of this book.

II.2.3. *Lineal Measurements*

Perhaps the first measuring device that comes to one's mind is a ruler, a yardstick, or a meterstick used for lineal measurements. These instruments may be calibrated in feet, inches, 1/16 inches, centimeters, millimeters, or other units of measurement. It has been implied from our kindergarten days that these units are fixed units that never vary, although this implication is never made explicit until high school and college physics courses are encountered, and even here it receives only limited attention. Do we ever stop to question how much variation there is between the calibration marks on the rulers manufactured as brand X? Our experience has told us that commercially available rulers, yardsticks, and metersticks are calibrated closely enough that we need not worry about this kind of error in everyday life. Unfortunately, when this sense of security is carried over into scientific research requiring very precise measurements it may sometimes have less happy results. Would any of us recognize the fact that brand Y rulers actually measure only 11 $\frac{15}{16}$ inches even though they are calibrated as 12 inch rulers? A yardstick that was 35 inches long could add considerable profit to a merchandiser of cloth goods sold by the yard. How many of us ever bother to check the lineal measurements represented to us as a specified length? Probably none, because we trust that our State and Federal Bureaus of Weights and Measures will ferret out unscrupulous merchandisers and penalize them, thereby freeing us of this worry.

Do any of you know how tall you are to the nearest centimeter when you first arise in the morning, and how tall you are to the nearest one-half inch when you retire at night? Do you know that height measurement can vary by two inches or more from morning to evening? (An acquaintance of mine has height measurements of 6'2" and 6'4" officially recorded on United States Air Force records.) Has your height been measured with or without shoes and stockings? Such questions lead us to the idea that the height of a person must be defined in precise terms or we shall be unable to determine what is meant by it except in general terms. Even so, these measurements are more descriptive than categories such as midget, short, medium, tall, and "how's the weather up there you pro-basketball prospect"?

What is an inch, and was an inch always an inch? We note from the dictionary that the word inch is derived from the Latin <u>uncia</u> meaning twelfth part. Historical records show that the inch has been used to measure items in the Anglo-Saxon world since the days of William the Conqueror. It was first defined as the width of a man's thumb, and later as the length of the thumb from the joint to the tip, both rather variable measurements. The first real effort to establish a precise basis for the inch was made under Edward I during the last part of the thirteenth century, when it was defined as being equal to three round, dry barley corns laid end to end. Thus, if one knew what barley corns were and had three of the desired type, one would have a fairly precise standard for calibrating a ruler into inches. Only recently has the inch been standardized in the precise form needed to manufacture and use present-day high-precision instruments.

In addition to the inch, other units of distance have been used through time. Once upon a time a "short piece" and a "far piece" sufficed, as did "one day's journey". Today, "one day's journey" on foot is quite different from one by a jet or missile. Later, a cubit (elbow to fingertip), a foot length, a stride or pace, etc. were used to measure distances, but these measures were too variable for any type of precise work and greatly bothered the early map-makers.

Today we do have very precise units to measure distances. Not only do we have the English system of inch, foot, yard, rod, mile, and league but also the metric system of micron, millimeter, centimeter, decimeter, meter, decameter, hectometer, and kilometer, and also a nautical measure system of fathom, cable, nautical mile, marine league, and degree of a great circle of the earth. Systems of square and land measures, cubic and volume measures, and circular or angular measure have been devised to describe characteristics in man's universe. As civilization becomes more complex, man requires more diverse and precise units of measure and measuring instruments.

II.2.4. Volume Measurements

Culinary artists to this day tend to utilize very imprecise units for both dry and liquid measure: a "pinch of salt", a "whisker of paprika", a "dash of salt", a "sprig of mint", a "dab of flour", a "few drops of vanilla", etc. with such directions as "stir until the right consistency is achieved" or until it "feels right" all tend to make certain that no one can reproduce the epicureal masterpieces of great chefs! But, we do have a precise set of dry and liquid measures in the English system and volume and capacity measures in the metric system, as well as an apothecaries' liquid measuring system, to allow determination of measurements to the degree of repeatability desired in present day technology. The units in the system have been precisely defined and precision measuring devices developed.

In the English system of liquid measure we have the teaspoon, tablespoon, gill, cup, pint, quart, gallon, barrel, and hogshead, as well as special measurements such as a "fifth", a number 2 can, etc. In dry measure, the English system has the pint, quart, peck, and bushel, as well as the

British dry quart. In the metric system, the cubic millimeter, cubic centimeter, cubic decimeter, and cubic meter are units of volume measure; in the capacity measure the units are the milliliter, centiliter, deciliter, liter, decaliter, hectoliter, and kiloliter. Measuring devices for dry and liquid measures consist of containers graduated in the desired unit.

II.2.5. Time Measurements

Early man may have been satisfied with the interval of sunset-to-sunset or of the beginning of a season to the end of a season. The sundial, the graduated candle, and the hourglass were early timepieces for measuring shorter units of time, and units such as the second, minute, hour, day, week, month, year, decade, century, and eventually the light-year were devised to enable man to record events in time. Elaborate and diverse timing devices have emerged to keep pace with present day technology. We have calendar watches and clocks, transistorized timepieces, and atomic clocks. Sports events are timed in tenths of a second, whereas distances to far-off celestial bodies are measured in the number of years it takes light to travel from that body to earth. Airplane flights keep a tight time schedule in order that passengers do not have to wait. We are very dependent and subservient to the timepiece in our everyday lives as well as in certain forms of scientific inquiry.

II.2.6. Weighing Measurements

Another measuring device very much on the minds of weight-conscious Americans is the bathroom scale, or more generally the scale, calibrated in pounds and ounces. For scientific investigations, the scale is calibrated according to the metric system in kilograms, grams, centigrams, and milligrams. We have spring and balance scales with all degrees of accuracy for both types. Do we ever bother to ascertain the accuracy of the scales used? Why do we *always* weigh more on the doctor's scale than on our home bathroom scale? Is the difference in the scales, in the way in which the scales are read, or in the amount of clothes and time of day? Which scale is "incorrect"? Can we even talk about incorrectness until we have defined what is meant by the weight of an individual? We could carry our bathroom scale to the doctor's office and calibrate our scale with his in the range of our particular weight, but "it's too much bother, and anyway I weigh less on my scale".

A few years ago, a research organization checked a scale used for weighing heavy objects; the scale was found to weigh high for relatively light objects and to weigh low for relatively heavy objects. This meant that the difference in weight between heavy and medium, heavy and light, and medium and light objects was smaller than it should have been. The error in measurement of weights could have led to erroneous conclusions. A simple check would have revealed this error which had gone undetected for an unknown length of time. If a scale is utilized for precise weights with important consequences, as for instance, in certain research investigations, it should be calibrated against a known standard throughout the total range of weights employed on the scale.

When you buy a package of meat in a grocery store, and the label states that the weight is two pounds four ounces, what is your concept of this weight? Do you take this to be 2 lbs. 4 oz. of meat at this moment, of meat at the time of packaging, of the meat and paper, or of meat, paper and the butcher's heavy thumb? Do you ever bother to weigh the meat at home or to have the butcher weigh the meat at the store? It was reported recently that a particular butcher had sold over 30 pounds of paper per week at the price of steak, and the public was warned to have the meat weighed at the store. One can be fairly certain that the number of overweight packages is smaller than the number that are underweight.

In addition to avoirdupois weight, we have apothecaries' weight and troy weight systems; the latter system is used by jewelers, but how many of us ever use it to check whether our one-carat diamond actually weighs 3.086 grains = 200 milligrams?

II.2.7. Measurements by a Judge

Another very common measuring device is the human judge; he serves as the measuring device for sports events, beauty contests, taste panels, reading other measuring devices, scoring plant strains for disease infection, etc. One of the key criteria for a judge is the ability to discriminate and to differentiate between varying levels of the characteristics under consideration. For example, if all pies within a specified range taste the same to a person, he will be unable to discriminate between the small differences that a researcher in home economics is studying; or if all the girls involved in a beauty contest seem equally beautiful to a person, this person is useless as a judge who has to pick a winner; a referee unable to observe fouls and violations is not suitable to referee contests; if all plant strains appear to be equally infected with a disease to the judge when in fact they were not, this person's scores are useless in differentiating between the strains.

The ability to discriminate can often be sharpened in many cases with adequate training. However, some individuals may never be able to attain a high level of discrimination with regard to a particular characteristic despite considerable training. One of the key characteristics of outstanding research personnel is their ability to observe and to discriminate among the various types of evidence encountered and then to organize and to sort out the pertinent facts. Successful researchers and judges are keen observers.

II.2.8. The Questionnaire Measurements

Another very common type of measuring instrument is the questionnaire. It has many and diverse forms, but mainly the common goal is to elicit information from or about people, their activities, and their attitudes. A widely known form is the ordinary examination, with as many forms as there are instructors or persons giving the test. There may be true-false, multiple-choice, completion, matching, discussion, etc. types of questions and various combinations of these on any given examination.

Another form of the questionnaire is the income tax form which we fill out every year for federal, state, and some city govermnents. This questionnaire seeks the answer to only one question, "How much income tax do you owe for last year?".

Another type of questionnaire associated with surveys and censuses seeks to obtain information on type of dwelling, occupancy and ownership of dwelling, income and expenditures of occupants, attitudes of occupants toward various items ranging from prejudice to choice of political opponents, preference of automobiles and applicances, use of foods and products, attitudes about children or pets, ownership of cars and boats, attitudes toward government policies, etc. A problem with these questionnaires is that they are often constructed by people who forget one simple fact: if the person being interviewed does not understand the question but some answer is given, the answer is as meaningless as one generated by a random or chance process. For example, suppose that we ask people to state which of three medical policies they favor. If they know little or nothing about the three policies, we might expect about $\frac{1}{3}$ of the people to favor each policy; if we toss a common six-sided die and record the proportion of times the events (1 or 6), (2 or 5), and (3 or 4) (or any other pairing) occur, we would expect approximately $\frac{1}{3}$ in each group. Unfortunately, many people conclude from such results that the policies are equally favored. This could be true or it could be that the people did not understand the question. This situation may be of frequent occurrence and should be born in mind when one studies the results from a survey.

Application forms may attempt to obtain information about an individual for university admission, for credit rating, for security clearance, admission to graduate school, or job application. Often these forms are very brief, but occasionally the inventor of forms becomes a little too enthusiastic. Does the age of a person's wife have anything to do with his being a national security risk? Probably not, because security may be granted even when this space is left blank.

Forms of questionnaires are varied, and we complete one form or another almost daily. Many of you will be involved with developing questionnaires. Please be precise, exact, and unambiguous.

II.2.9. Chemical Measurements

Another type of measuring device is the chemical determination. Large laboratories are constructed for the sole purpose of performing chemical determinations on plant, animal, and mineral samples. The results are utilized in many ways. For example, the Food and Drug Administration checks on the contents and quality of foods and drugs. Limits of variation are set for individual items, and the manufacturer must conform to these standards. Since not all people are "good guys" large state and federal organizations have been set up to ascertain that John Q. Public obtains what is stated on the contents of his purchase.

Other chemical laboratories check soil samples for fertility content, milk samples for butterfat content, food samples for pesticide residues, concentration and content of drugs, concentration and content of alcoholic beverages, cigars and cigarettes for concentration of tars, resins, nicotine, etc., content of cosmetics, concentration and identity of weed seeds in crop or lawn seed, etc. An important statistical problem in connection with all these is the design of the sampling procedure and the establishment of limits of variation that will be tolerated in the samples. For many items the statistical standards have not yet been established; there are too many items and too few statisticians. In other cases, standards have been set arbitrarily; many of these may be shown to be relatively impossible to attain when studied statistically. For example, for certain certified seed requirements, the presence of one specified noxious seed in a sample makes the entire lot of seed unsuitable for sale. In order to find such a seed in a lot and to be certain that it will not appear in the sample, it would be necessary to inspect the entire lot seed by seed. This is too expensive and time consuming for commercial seed production. Some other means such as field inspection must be used to eliminate the specified noxious weed seeds from the seed lot.

II.3. A Class Survey in Measurement

In order to obtain some experience with measuring and measuring devices, and in order for us to have a set of measurements for class use in various ways, let us perform a class survey. The object of the survey is to measure some characteristics of members of this class. As a part of this survey, each member of the class should supply the measurements requested in problem II.4. Here we have defined eye color to be one of five categories (blue, green, grey, brown, or black) and hair color to be one of four categories (red, blond, brown, or black). Age is requested in units of years and months. Class standing is defined in terms of year in college such as freshman, sophomore, junior, senior, graduate, or other. If the student is unable to fit himself or herself into one of the above categories (for example, hair color might be purple, orange, or mouse-colored), the instructor should be so notified. Perhaps a category such as "other" or "miscellaneous" should be added, but one needs to define which items make up this category.

As a second part of the survey, let us take a series of measurements in class. First of all, suppose that we wish to obtain the following measurements on each student present:

Name ____________________

Height to nearest millimeter ____________ mm.

Height to nearest foot ____________ ft.

Height to nearest yard ____________ yd.

Weight to nearest pound ____________ lbs.

Eye color (circle one) blue green grey brown black

Hair color (circle one) red blond brown black

At each of two stations (A and B) in the class, select a team of three students. Members of the team at station A should be the three shortest class members, and at station B the three tallest. The reason for height differences will be explained later. Have one member of each team be the recorder of the measurements on one of two forms, A and B, on which a student has written his name. Form A is to be used for station A measurements and form B for station B. A second member of each team is to measure height measurements in millimeters and to determine eye and hair color or to read weights on a bathroom scale. The third member of each team is to measure the height of a student to the nearest foot and nearest yard. The measurements on the six members of the two teams should be obtained first in order that they are not missed; then the remaining students are measured as they file past at each of the two stations. Students should complete problem II.5 during the time that measurements are being obtained on the other students.

The measuring devices required are two straight eight-foot dowels one inch in diameter and calibrated in yards, two eight-foot dowels calibrated in feet, two eight-foot one inch by two inch boards on each of which two metersticks have been mounted end-to-end, and one bathroom scale.

Before proceeding with the measurements, we must decide how we are going to take height and weight measurements. Shall we measure students with or without shoes? If without shoes, it is suggested that the student stand on sheets of paper. With regard to the height measurement in yards, all persons in most classes should have been recorded as two yards tall unless a recording or judgement error was made or unless very short or very tall people are in the class. People under $4\frac{1}{2}$ feet in height should be classified as one yard and people over $7\frac{1}{2}$ feet should be classified as three yards tall. No one in the class measurements given in chapter VIII was in either of these categories. There should be no variation in the heights of most classes measured to the nearest yard.

For most classes, which usually do not contain a dwarf or a giant, there should be two groups of measurements recorded in feet; these should be five and six feet. If any other measurement appears we would suspect a recording error or an incorrect reading of the measuring device. Thus, measured in feet, all members of a class are either five or six feet tall. There is variation in heights when measured in feet but there is none when measured in yards. Also, we should not expect the different people performing the measurements to agree on all heights, as individuals near five feet six inches tall may be classified differently by the two measurers.

In problem II.4 the student is to record height in inches; we shall not check on the student's version. Instead, we turn to the heights obtained in millimeters. Without summarizing the data obtained, one would guess that all heights should fall between 1450 and 1950 mm. There will be many categories of height. Very few individuals now are recorded as having the same stature, and the variation in units of height is considerable. Furthermore, if we compared two heights recorded by the same individual and two heights by the two different individuals, we note that they would not always agree. In fact they probably will agree less frequently than they disagree. This is due to the fine calibration and to the fact that the measured heights of an individual will vary as the subject's posture varies from measurement to measurement; also, the individual performing the measurement will not always read the measuring instrument in the same manner. There is a degree of subjectiveness in reading the meterstick when measuring the height of an individual. We could do better and have less variation in duplicate measurements if a flat surface such as the length of a table were being measured. With a little care, there should be little variation in duplicate measurements or between people performing the measurement. However, if the measurements were taken to the nearest one-half, one-fourth, or one-tenth of a millimeter variation would again occur. There is a limit of repeatability for any measurement; if the unit is fine enough there will be variation in measurements by the same and by different individuals.

The same problem will be encountered with the measurement of weights. The weights given by individual students are obtained from a variety of scales and a variety of conditions. The weights recorded in class are obtained on a single bathroom scale, read by a single individual, and can be checked by a second individual to minimize reading and recording errors. The variation or measurement error in the weight of an individual is influenced by such things as time of day in relation to eating, amount of clothes worn, method and correctness of reading a scale, the particular scale used, and the correctness and legibility in recording the weight. An individual's weight may vary by five pounds or more from the time just before breakfast until just after the last meal of the day even though the same clothes are worn and the same scale used.

Probably the most subjective measurement recorded will be eye and hair color; despite the few categories involved, the variation between the human measuring devices involved in the classification will be noticeable. Were the student's age and weight guessed, there would be even more variation and subjectiveness in these recorded characteristics.

We will utilize the data obtained in various ways as we proceed through the term.

II.4. Standardization of Measuring Devices

The need for standardization is illustrated by an example. Four metersticks were purchased and arrived in a box which originally contained 12 metersticks. Two of the metersticks differed from the other two by more than one millimeter in the calibration marks. The metersticks carried the same brand name and lot number. This points up the fact that whenever a new measuring device is utilized it should be checked against a standard; the standard should have known accuracy. A measuring device with unknown accuracy may be useless for the purposes at hand. If a standardized meterstick were available, the newly purchased metersticks could be checked against it.

As was pointed out earlier, an inch was not always an inch. The United States Congress in 1866 pronounced the international prototype meter bar in Paris as equal to 39.37 inches. A few years later, a platinum-iridium meter bar became the official standard for all United States measures of length. By 1940, the National Bureau of Standards was able to define an inch accurately to one-millionth of its length, but even this did not prove accurate enough; there was a difference in the British and the United States inch in that the former was equal to 2.539995 cm and the latter to 2.540005 cm. This difference of 0.00001 cm greatly affected the manufacture and use of precision instruments during World War II. In 1960, the official inch for both Britain and the United States was set equal to 2.54 cm exactly, and an international agreement made the meter "equal to 1,650,763.73 wave lengths in vacuum of the radiation corresponding to the transition between the energy levels of 2_{p10} and 5_{d5} of the atom Krypton 86". Thus, much of the confusion in linear measurements appears to be taken care of for the time being.

Standards have also been developed for gravity, time, temperature, voltage, weight, angular, and frequency measurements, and prototypes of almost all standards are kept in France, Britain, or the United States; it is therefore possible to obtain measuring devices of the desired calibration and accuracy. Copies of the prototypes have been widely dispersed for use in making measuring devices.

Measuring devices should be checked against a standard when first used and occasionally thereafter, since it is difficult or impossible to do accurate work with an incorrect device. Scales should be checked for accuracy throughout their usable range. Human judges should be checked for discriminatory power and for level of discrimination. Questionnaires should be pre-tested in order to eliminate ambiguities and lack of clarity. Unfortunately, instructors do not find it convenient to pre-test quizzes and examinations prior to use; occasionally some poorly designed questions appear on an examination. Chemical and physical procedures should be checked when first initiated and occasionally thereafter, in order to ascertain that the process remains accurate within the prescribed levels; procedures that are usefully accurate for one type of material may be inaccurate for a second kind of material.

Whenever possible, measuring devices should be calibrated against known calibrated standards and they should be recalibrated at intervals. Other types of devices can be checked by including duplicate samples and samples of known content along with unknown samples. However, one must be careful not to let this checking take an inordinate amount of one's time. This can be a gross waste of effort and can result in very inefficient use of resources.

II.5. Variability

We have noted in section II.3 that variability is present in our measurements. When the unit of measurement is fine enough, we find variation between two duplicate measurements of height of a single individual even though the measurements might be made only minutes apart and by the same person. Variation occurs between the height measurements made by two different individuals. Even more variation is found between heights of the members of this class, as there is considerable variation due to causes other than measurement errors in our recorded heights. In fact, we need only look around us to note that two individuals who are "alike as two peas in a pod" are very difficult to find, expecially since there are no identical twins in most classes. Variation is universal in characteristics of all populations. We live in a variable world. This is good because it has allowed the human race and individuals of other populations to survive. Since variability is universal we must learn to live with it and to design experimental and survey investigations in such a way as to overcome its effects. There are several ways of accomplishing this, as will be described in chapters V, VI, VII, and XII. Some types of variability occur in an organized fashion (chapter XI); as a result it is possible to devise statistical procedures for summarizing the sample facts and for making inferences about the population facts. The procedures are so many, varied, and detailed that numerous statistical books have been and are being written on the subject. The more one knows about the nature of the variability, the more useful are the sample facts in making inferences about population facts.

II.6. Bias in Measurement

An unstated tenet in the collection of numbers obtained from utilizing a measuring device is that the sum of errors in a positive direction is about equal to the sum of the negative errors of measurement; over a large number of trials one would expect the errors to sum to zero or nearly so. Suppose that this is not the case, and that the magnitude of inaccuracies in one direction exceeded those in the other. The nature of this type of discrepancy is termed a _systematic error_ or more commonly a _bias_. The reason for selecting a short person and a tall person for making the measurements described in section II.3 was to introduce an inaccuracy of this nature. Because the short person has to look up it might be assumed that he or she would record heights higher than would a taller person. Then, if on the average, short investigator A _always_ reads the measuring device one unit higher than does tall investigator B the bias of A compared to B is +1, and the bias of B compared to A is -1. Note we did not state which, if either, of the two investigators takes correct measurements in the sense that if he measured all individuals in the population he would obtain the population parameter for the characteristic measured.

II.7. Error in Measurement

As we have seen, the causes of variation in measurement are many and varied. Barry [1964], chapter 2, and Wilson [1952], sections 9.1, 9.5, 9.8, describe some of these. There are systematic errors or biases, personal errors, mistakes, and errors due to assignable causes. In addition, variation in measurements may be caused by unassignable causes due to the cumulative effect of a number of uncontrolled and often unknown minor variables. If the magnitude and sequence of these variations are completely unpredictable and nonsystematic, that is, they form a _random_ sequence, we denote them as _random variations_ or _random errors_; the sum of the random errors over all individuals in a population should be zero.

The total variation in measurements excluding mistakes or blunders may be written as

total variation = assignable causes + bias + random error.

The error of measurement is often defined as

error of measurement = bias + random error.

Quite often the bias term is ignored when in fact it may be the larger factor in the error of measurement.

If differences between individuals in the population rather than the individual measurements are utilized a constant bias term will disappear. Thus, the method of computation may affect the effect of the bias term; the investigator must be aware of this in choosing his method. In order for a measuring device to be useful, some measure of its reliability and of its accuracy should be available.

II.8. Accuracy Versus Precision

In statistical terminology, the term precision refers to the repeatability of a measurement. Low precision means that there is wide variation in repeated measurements on the same object, whereas high precision means that there is little variation between repeated measurements. Accuracy, on the other hand, refers to the size of the bias term plus the precision. Thus, low accuracy could result from a large bias term with either low or high precision, or from a zero bias term coupled with low precision, as illustrated in figure II.1. High accuracy results when the bias term is small or zero and precision is high.

Perhaps the best illustration is to consider the drawings in figure II.1 as targets for darts, arrows, air pellets, or bullets. A rifle producing a pattern like the one on the upper right hand target when fired from a rigid fixed support would be considered highly accurate. If a similar pattern were obtained by a person holding and firing the rifle we would say that this person is an excellent (highly accurate) shot and that the rifle is also highly accurate, provided, of course, that the person was not correcting for a bias in the rifle.

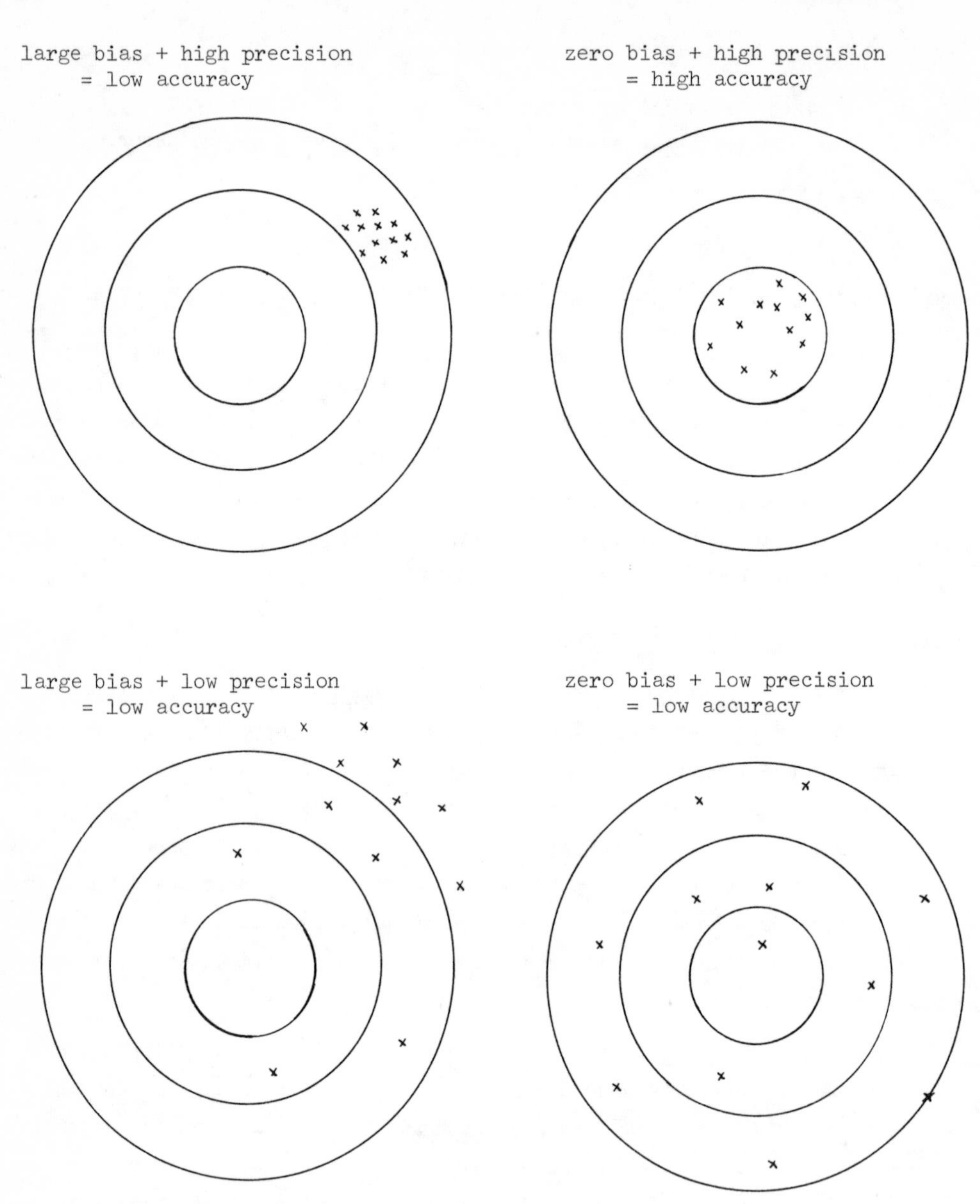

Figure II.1. Graphical representation of precision and accuracy.

II.9. Problems

II.1. Select a measuring device used in one of your laboratory classes and a set of measurements made on it; describe the types of errors of measurement and methods of eliminating or controlling these errors.

II.2. Construct a measuring device and calibrate it. Determine the limits of error. (See Youden [1962] pp. 89, 114.)

II.3. Using an appropriate measuring device such as a micrometer or vernier caliper, measure the thickness of pp. 1-40, 41-80, 81-120, 121-160, and 161-200 of this book, each 10 times. How would one estimate the thickness of a single page? Would the above method be more precise in estimating average page thickness than to measure one page 2000 times? Why or why not?

II.4. Supply the following data for yourself:

Name ____

Height to nearest inch ______________________(in)

Weight to nearest pound ______________________(lbs)

Age to nearest month (as of Feb. 1 of the current year) _____(yrs)

_____(mos)

Eye color (circle one) blue green grey brown black

Natural hair color (circle one) red blond brown black

Class standing (circle one) Freshman Sophomore Junior Senior Graduate Other

II.5. Comment on the class survey with respect to the efficiency and conduct of the investigation; comment on the accuracy of taking the measurements and on ways in which the accuracy might have been affected.

II.10. References and Suggested Reading

The three books listed below are easy to read and elementary in orientation; they are recommended reading:

Barry, B. A. [1964]. Engineering Measurements. John Wiley and Sons, Inc., New York, London, and Sydney. pp. x + 136.

Wilson, E. B. [1952]. An Introduction to Scientific Research. McGraw-Hill Book Company, Inc., New York, Toronto, and London. (Sections 4.3, 4.5, 4.6, 6.4, 7.1, 7.2, 7.7, 9.1, 9.5 and chapter 5.) pp. x + 373.

Youden, W. J. [1962]. Experimentation and Measurement. National Science Teachers Association, Washington, D.C. pp. 127.

The following two references are for those who wish to pursue further the philosophy of measurement:

Campbell, N. R. [1928]. An Account of the Principles of Measurement and Calculations. Longmans, Green, London.

Ellis, B. [1966]. Basic Concepts of Measurement. Cambridge University Press, London, pp. 220.

An intriguing article on psychological testing and measurement is:

Anastasi, A. [1967]. Psychology, psychologists, and psychological testing. American Psychologist 22:297-306.

CHAPTER III. DATA COLLECTION

III.1. Introduction

Suppose that one were to collect the phone numbers of all college coeds in a given university. This would represent a lot of numbers. If the sole purpose was to obtain a large array of numbers, such a collection would appear to be useless. If, furthermore, the telephone numbers were five years old, they would be even more useless!

Suppose that in addition to these five-year-old phone numbers we had height and weight measurements associated with each one. With no purpose or usefulness, the array of numbers would still be nothing more than just a lot of numbers. If, on the other hand, some use were to be made of the height and weight measurements, the numbers come to represent facts, and they are designated as data. A datum is defined as a fact from which a conclusion can be drawn. It is important to differentiate between data and numbers from which no conclusions are to be drawn. If one only wants numbers and not data, these can be quickly generated by the "side-foot" with a high speed computer where a side-foot is defined to be a stack of computer sheet paper one foot high; it does not refer to length of paper.

Several aspects should be considered in data collection; some of these are discussed informally below, while the precise steps in scientific investigation will be discussed in the next chapter.

III.2. Why Collect Data?

As indicated in the previous section, there should be a reason for collecting data on a given item. For example, in the survey we conducted in class, data were collected on various characteristics of members. The reasons for collecting these data were to introduce the idea of measuring devices and measurements, to illustrate errors, variability, and bias in measurements, and to obtain a set of data for illustrative purposes later.

Although it is imperative to collect the data required for drawing conclusions on the items of interest, it usually is not a good idea to collect data just because it is easy or because someone, somewhere, sometime may use them, even though data of this nature occasionally turn out to be exactly what is needed at the end of the experimental investigation or survey. The latter possibility is of infrequent occurrence in well-thought-out investigations.

III.3. What Data are to be Collected?

Before starting an investigation, one must determine the nature and characteristics of the data to be collected. For the survey described in section II.3 it was decided to collect height, age, weight, eye color, hair color, and class standing measurements for each student. The classes or categories for eye and hair color and class standing were designated. Although the definition of these categories or classes was simple for our survey, this is not always the case. Classes must be made as distinct as possible and free from ambiguity. Wilson [1952, sections 7.1 and 7.2] discusses class definitions and gives some examples.

III.4. How are Data to be Collected?

In determining the manner of collecting data, it is necessary to consider the conditions under which an investigation is performed. For example, if our class survey had been performed out-of-doors during a snowstorm, it would have been difficult to get students to take off their shoes. Since there should be no such problem in a lecture room, height measurements could be standardized to some extent by taking all measurements without shoes. The unit of measurement, yards, feet, or millimeters, was predetermined. Measuring devices, which were 8 foot dowels marked in yards or in feet or which were 8' x 1" x 2" boards on which two metersticks were screwed end to end were selected, as were the persons doing the measuring and recording. The form on which the data were to be recorded and the order of the data were determined. The results of one class survey are ordered and presented in chapter VIII; the data will be utilized for various purposes.

Wilson [1952, section 6.2] discusses and gives examples of the form and manner of recording data. Careless recordkeeping can considerably decrease the value of data or even make them useless. Permanent, strongly-bound notebooks are recommended for recording laboratory measurements. Data should be entered in the notebook at the time of observation, preferably in permanent ink. Recopying of data allows yet another chance for recording errors to appear. It is preferable to use ink in the event that the notebook is used in legal cases as evidence. This is also necessary if the notebook is used extensively, as penciled figures tend to smear and rub off. The name of the observer and the period covered for the observations should also be recorded in permanent ink. Each set of observations should be dated and described. If more than one person uses a book, each person should initial the material entered by him. Do not crowd data on the page; remember that paper is cheap! Above all, be neat and legible. Also, a safe and permanent place should be selected for storing the notebook when not in use. If the data are extremely valuable and hard to replace, it may be desirable to store in a fireproof safe a photocopy or ozalid copy of the properly identified data sheet.

As stated in the previous section, the observer must first determine what observations are to be made. Having made this sometimes difficult decision, he is in a position to determine the order of appearance of the characteristics in the notebook and the order of taking the observations, which may be very important in certain investigations. Time can be saved and accuracy of recording can be improved by consideration of the order of taking the observations and proper orientation of the form for recording. For example, let us suppose that a florist has 20 pots of flowers of variety A and on these he is counting the number of saleable flowers produced each day. The 20 pots are arranged in two rows, one row on each side of a greenhouse bench. Suppose a notebook has been prepared with the numbers appearing in serial order from 1 to 20. Consider the following three numbering systems for the pots:

Numbering system 1

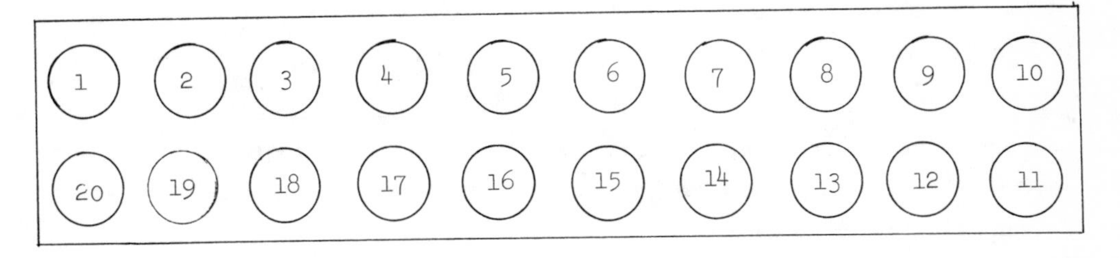

Numbering system 2

Numbering system 3

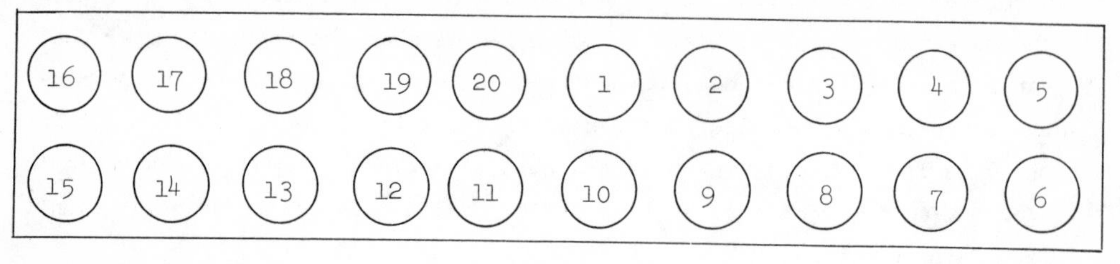

With numbering system 2, it is necessary to walk back to the beginning of the bench each time observation 11 is to be recorded. For small investigations this may not be important, but for investigations covering a relatively large area, the proper numbering system could become quite important as a time-saving device. Numbering system 3 could be more efficient in saving time than system 1 if the point of starting the observations were in the middle of the bench rather than at the end of the bench.

In questionnaire construction, it is highly desirable to order the questions and to categorize the answers in such a way that summarization of the data from a questionnaire is a relatively simple task. For example, true-false, completion, and matching questions are easy to grade on an examination. Discussion questions are difficult and time-consuming to grade, but they may be the only way to obtain the information desired. If the data are to be put on punched cards, it is wise to pre-code the questionnaire rather than to code each question individually after completion of the questionnaire. This saves an enormous amount of time and reduces the chances for error.

Significant digits represent the number of figures retained in a measurement or observation. For example, the numbers 137, 137000, .137, .0000137, etc. have the same three significant digits. The number of significant figures to be retained for observations should be determined prior to conducting the investigation, as difficulties in summarization arise when different numbers of significant digits have been retained. Usually, sufficient significant digits should be retained to have at least a range of 30 or more units in the sample in order that rounding errors have small effects on a number of statistics computed from the data.

III.5. Where and When are the Data to be Collected?

In some investigations there is only one possible place where the investigation can be conducted; consequently, there is no problem involved in making the choice between places. However, there will still be a necessity for determining the time for the investigation to take place: other classes or other investigators may need to utilize the space for other purposes. For our class survey, we determined that the survey would take place on a given day of a given month and year, and at a specified time, in a specified classroom. At any other hour, it would be difficult to assemble as many people as there are in a class in one place.

In other surveys, there may often be vast numbers of places where the survey could take place. For example, consider that we wished to take a survey of 500 of the striking transit workers in the city of New York on January 3, 1966; workers could be interviewed at home, on picket lines, in meeting halls, or a variety of other places. If the decision were made to interview them in their homes between the hours of 10:00 a.m. and 8:00 p.m., it would further need to be determined whether other members of their family or other occupants of the building will be allowed to be present during the interview and what is to be done about refusals, about not-at-homes, about another occupant of the building who insists on being present during the interview, etc.

In sampling foods for determining pesticide residue, we need to determine the time, place, and material to be sampled and the laboratory or laboratories making the chemical determinations. In judging food products, as, for instance, cherry pies, the time and place, as well as the conditions, need to be explicitly stated in order to avoid difficulties in scheduling the investigation.

A record of the time and place of the investigation and a complete description of the items involved in the investigation are necessary for future use of the data. These should be recorded immediately and in an acceptable form. It is fatal to trust to memory because there are lapses for all humans.

III.6. Who Collects the Data?

If a number of people are involved in the collection of data, it is necessary to determine precisely which person is to record each specific observation, so that important items are not omitted. If possible all measurements or observations in one group should be made by one person and/or measuring device, in order to standardize the results. Note that this was done in the class survey. There are differences, sometimes small and sometimes large, between measuring devices; it is important to eliminate this source of variation from our data if at all possible, as the more variation that can be eliminated due to assignable causes the less will be the variation among observations.

Clarity and preciseness of directions to technicians and associates are of utmost importance in minimizing confusion and in ascertaining that the desired data will be collected. It is best to have written copies of directions prepared well in advance of the investigation. The launching of a missile is an illustration of teamwork resulting from carefully prepared and rehearsed directions. In order for the launch to be successful, all members of the team must perform their work. If a technician fails to remove a plastic cap in the fuel lines, the missile will not be launched successfully. Likewise, if the designated person fails to water the plants in a greenhouse or to feed laboratory animals over a long week-end, the plants and animals may be dead or in very poor condition. Even when a reliable person has been designated to perform an important task, it is well to have a means of checking on him in most investigations.

III.7. Complete Description of Collected Data

As we have indicated throughout the preceding sections, it is absolutely necessary to have a complete description of the data available in a permanent and available form. If one had the five-year-old telephone numbers with associated height and weight measurements, but these numbers were unlabeled, they would be nothing but an aggregate of numbers. Any set of data can be rendered useless if they are not completely labeled and described.

Record sheets must be sufficiently well identified to reorder them properly should they be placed in the wrong order. The entire set of data may be labeled and described adequately, but if the individual record sheets are not, this could cause difficulty. For example, if an investigator is conducting two surveys simultaneously, unlabeled pages of individual questionnaires could become mixed, and it may be impossible to separate them if they are alike in all other respects; this could render the results of both surveys useless for the purpose at hand.

If the names of students were removed or omitted from examinations and problems, the class average would stay the same; however, if we allotted grades in an arbitrary manner to the student names, the grade associated with an individual would most likely change. This procedure would not lead to happy students because some students deserving high grades would receive low grades and *vice versa*.

Examinations and problems should be identified with one's name on every sheet; if the sheets are stapled together, a name on the front or back page would suffice.

III.8. Disposal and Storage of Data

After data have been collected, it is necessary to plan for their storage until the next process in handling them is started. Also, after the data have been utilized and the study is complete, it has to be decided what is to be done with the set of data. If all information has been gleaned from the study, the data should be destroyed. Data which have no further use should not be allowed to occupy valuable space.

If the data are to be saved for future use, one should be certain that they are stored in a safe, dry, fireproof place. IBM cards should be tightly packed to avoid curling and crinkling. Notebooks should be stored correctly to keep them from warping.

III.9. Conclusions Drawn with no Data and Phony Statistics

In the 1960 presidential election, John F. Kennedy stated that 25 million people in the United States go to bed hungry. In commenting on his statement, *Time* magazine stated that in their opinion 44 million people go to bed hungry every night because they were all dieting! This facetious remark illustrates the fact that such statistics have no basis of fact and are phony. Our world is full of phony statistics. For example, when the little boy told of seeing a truck 40 times as large as a house, he was remonstrated for exaggeration by his teacher who said, "Johnny, if I told you once, I told you a million times not to exaggerate!". No data were involved in obtaining these phony statistics.

The first paper in Sielaff's [1963] book is one written by Daniel Seligman entitled "We're drowning in phony statistics", and it contains a wealth of phony statistics examples; it first appeared in the November, 1961 issue of <u>Fortune</u>. Mr. Seligman divides phony statistics into two categories, one which he calls <u>Meaningless Statistics</u> and the second <u>Unknowable Statistics</u>. He defines a Meaningless Statistic as one wherein a number or phrase is associated with an undefined term or ambiguous thought in such a way that it is unclear what is being added together to arrive at the figure. For example (from Seligman), New York's Mayor Robert F. Wagner announced in his annual report to the City Council that "over-all cleanliness of the streets had risen to 85 percent in 1960", that it was up from 56 percent in 1955, and that this figure was based on "personal inspections and evaluation of 4,493 city blocks". Mr. Seligman tried to ascertain what objective criteria were used to determine whether a street was clean or not. He met with no success, and one wonders whether the "personal inspection" of 4,493 city blocks would not have kept the good mayor so busy that he would have found no time to do anything else!

An Unknowable Statistic is one whose meaning may be perfectly clear, but is derived from an alleged fact that no one could possibly know. Mr. Seligman gives the following as an example: in the October, 1958 issue of <u>This Week</u> magazine, in a contribution by Dr. Joyce Brothers, it was stated that "the American girl kisses an average of seventy-nine men before getting married". Again Mr. Seligman attempted to obtain the source of this information, but his letter was never answered. Although the above statistic is of the Unknowable variety, it could also be of the Meaningless type in that it is not quite clear what is meant by men kissed.

Mr. Seligman, in giving examples of various loss statistics, finds that one quarter of the Gross National Product (G.N.P.) is being frittered away! He states that enough loss statistics could easily be accumulated "to show" that losses in the United States exceed the G.N.P., but that anyone compiling such a list would run a considerable risk, since "there are a lot of people around who would take it seriously"!

The following example is in the same vein as Mr. Seligman's interesting examples. It was found by a student answering problem III.1; the original author is unknown.

1967 POPULATION BALANCE SHEET

OR

WHO'S TO DO THE WORK???

Population of United States	198,000,000
People 65 years or over	55,000,000
Balance left to do the work	143,000,000
People 21 years or under	58,000,000
Balance left to do the work	85,000,000
People working for the government	35,000,000
Balance left to do the work	50,000,000
People on relief and Appalachian Program	24,000,000
Balance left to do the work	26,000,000
People in the Armed Forces	11,000,000
Balance left to do the work	15,000,000
People in City or State Government	12,800,000
Balance left to do the work	2,200,000
Bums and others who never work	2,000,000
Balance left to do the work	200,000
People in hospitals or asylums	126,000
Balance left to do the work	74,000
People in jail	73,998
* Balance left to do the work	2

* Two? Why, that's you and me! Say!! Then you'd better get a wiggle on 'cause I'm getting awfully tired of running this country alone!!

-Anonymous

III.10. Problems

III.1. Give an example of an "Unknowable Statistic" and of a "Meaningless Statistic" as defined by D. Seligman in the paper "We're drowning in phony statistics". Use an example from printed material you have read recently. Describe why it fits the category given.

III.2. An experimenter used 1000 ripe peaches (from Georgia) in an experiment. He wished to determine which quarter of a peach was the most tender. (We shall refer to this as the Georgia peach squeezing experiment.) He designated the peach fruit by quarters starting with the suture, or indentation, on the fruit, as follows: (1) left front, (2) left back, (3) right back, and (4) right front. (He did not cut the peach.) A device was utilized which measured the amount of pressure required to penetrate the peach skin. On each peach, the experimenter measured the quarters in the order numbered above. He reached the conclusion that the least amount of pressure was required to penetrate the right front quarter, the next lowest was for the right back quarter, the next in amount of pressure was for the left back, and the left front quarter required the greatest pressure for penetration. Comment on various aspects of this experimental investigation. How should the investigation have been designed to determine which quarter of the peach was most tender?

III.11. References and Suggested Reading

Sielaff, T. J. (Editor) [1963]. Statistics in Action. Readings in Business and Economic Statistics. The Lansford Press, San Jose, California. (First paper). pp. vii + 251.

Wilson, E. B. [1952]. An Introduction to Scientific Research. McGraw-Hill Book Company, Inc., New York, London, and Sydney. (Chapter 1-7, 9) pp. x + 373.

CHAPTER IV. PRINCIPLES OF SCIENTIFIC INVESTIGATION

IV.1. Introduction

The class survey described in chapter II was carried out in an inefficient manner the first year it was taken; it was also incomplete, because not all individuals were present. Consequently, we did not obtain all the data desired during the specified class period, and, unless a part of another class period were relegated to securing the missing data, the investigation would necessarily remain incomplete. However, the data collection had served several purposes, and we had sufficient data for our needs. Hence, we did not take time to complete the first-year survey. This is an illustration of the fact that careful, precise planning is necessary for all investigations. For example, if a team taking the measurements goes from seat to seat as they did in that first year, we would note that finding the name and recording the data would take the most time. If all students had been lined up and if they had come to the recorder and measurer to be measured, in the order in which their names appeared on the sheets, we would have obtained the measurements more quickly. Alternatively, we could have followed the plan described in section II.3, and all measurements could have been obtained in one class period. Better organization allowed the data to be taken during one class period the second and third years that the class survey was conducted.

There are many steps involved in planning scientific investigations, and we shall discuss a number of these. In general, we follow the material in the references cited in presenting the principles of scientific investigation. First, however, a number of definitions and concepts will be required.

To illustrate the necessity of following basic principles for all investigations, we shall cite several examples. During a visit to the offices of a national survey and market research organization, the author was invited to participate in a discussion of a proposed survey on medical needs of the elderly. The survey leaders gave the investigator a thorough cross-examination on all aspects of the survey from beginning to end. In particular, they questioned him on the basis of the principles of scientific investigation as listed below in section headings. They asked about the purpose, usefulness, possible uses, and interpretation of the survey data. After several hours of intense cross-examination, the investigator said, "Gentlemen and lady (a lady psychologist was present), I have never been subjected to such treatment as this in my whole life! But, I can tell you that your procedure truly impresses me as an excellent way to proceed! Let's continue." Afterward, the head of the organization told me that they had learned from sad experience that every client must be subjected to such a thorough cross-examination in order to prevent legal proceedings later because of a misunderstanding about the information desired and the segment of the population to be involved. Every scientist should likewise subject himself to such an examination to determine whether or not it is a worthwhile problem and what is the best way to obtain an answer.

As a second example, the same organization was approached by a large publishing company and asked to conduct a national survey, costing over $200,000. Near the end of their "cross-examination period", the members of the survey organization found that the publishing company wanted only a small segment of the information that would have been obtained from the proposed survey. The desired information was made available to the publishing company for about 1/10 of the cost of the proposed survey. This example illustrates that careful planning and a thorough understanding of what is wanted can greatly lower the cost of an investigation, while in other areas, time saved can be the important item. Careful, precise, and rigorous planning of investigations often saves time and money and increases the value of the results. A person following the procedures of scientific experimentation will make more rapid progress than one who does not, he will have more time for recreational activities, and he will be more productive than his poorly organized colleagues.

IV.2. Preliminary Concepts and Definitions

Science is the term applied to a body of systematized knowledge. Systematization of knowledge begins with the organization of everyday observations; these are then classified, their sequences become known, and their importance is evaluated through systematic investigation. Scientific method or scientific inquiry is the procedure whereby knowledge is acquired and is an attempt to extend our range of knowledge. The evaluated knowledge is then used to formulate a hypothesis: a tentative or a postulated explanation of a phenomenon and a statement about a population characteristic. The hypothesis is tested by experimental investigations to ascertain its plausibility. A hypothesis that is relatively well verified and possesses some degree of generality is a theory. A theory that has been verified beyond all reasonable doubt at the moment is designated as a law. We say "at the moment" because new knowledge may later indicate that a law of the past was only a theory; for example, Newton's laws of motion were replaced by the laws of quantum mechanics; his laws are limiting forms of the new laws for heavy bodies (see Wilson [1952], pages 29 and 30 for this and other examples). Also, Mendel's laws of inheritance, the exponential growth law, the growth and decay law, rate of increase law, etc., are laws used in biology today. Frequently we find it expedient to use mathematics, the language of science, to express a law: thus, the exponential growth law is written as $W = Ae^{bt}$ = weight at time t; the rate of increase of growth at a specified time, t_0 say, $\frac{dW}{dt} = Abe^{bt}$ evaluated at $t = t_0$.

A hypothesis, a theory, and a law differ in degree of plausibility and generality. For example, suppose that we were to obtain the following sequence of sample observations:

$$0, 3, 6, 9, 12, 15, 18, 21 .$$

A reasonable hypothesis would be that these sample observations fit the sequence 3n for $n=0,1,2,3,\cdots$. But just because these data fit the 3n sequence is not proof that the next number in the sequence will be 24; there are several sequences which these numbers could fit. Once such sequence would be

$$3n + (n-1)(n-2)(n-3)(n-4)(n-5)(n-6)(n-7) .$$

The use of the latter sequence to predict the next number in the sequence would give a vastly different number (5,064) from 24. This illustrates the fact that sample data often support a number of hypotheses. While some of these may be ruled out because they conflict with known laws or theories, a series of investigations will be necessary to differentiate among the remaining hypotheses. Thus, support of one hypothesis by the data does not necessarily establish the truth of the hypothesis. This is a fact that is often overlooked.

The object of a scientific investigation is not to prove the scientist correct, but to establish the truth. In science, the investigator is seeking the truth and the true facts about a phenomenon. It ultimately does not matter what his dogmas, beliefs, or superstitions are because someone, somewhere, sometime will find the truth. The scientific world is completely impersonal and in the long run, is impervious to personal biases inserted in works. A great scientist said he believed that 95% (a phony statistic, of course) of the people were working on the wrong things (an undefined term) at any one point in time. Even if this statement resembles some of those in Seligman's article (Fortune, November, 1961) discussed in chapter III, there is some truth in it. Has the reader ever encountered articles on statistics written by a famous heart specialist or articles on heart surgery written by a famous statistician? When heart surgeons follow the heart surgeon's statistics and when the statisticians follow the statistician's heart surgery, both groups could be following erroneous information Misplaced self-confidence introduces misconceptions which may lead investigators astray. Egotism often goes too far in science as it does in everyday life. Don't we, without any political experience, feel that we know how to run our government, or anybody else's government, better than the political experts who have made this their life's work? Aren't we then like the statistician writing articles on heart surgery? Do we attempt to force our point of view on others, or do we attempt to set up a logical, reasonable basis for our point of view and let the listener believe what he will? Although personal bias is often difficult to overcome, the investigation must be conducted in such a manner as to minimize or to eliminate all possible biases. The investigator must have self-esteem, but he should not indulge in self-worship; it may be necessary to point out to ourselves or to our colleagues that self-esteem is not a necessary and sufficient condition for a proof of a theorem in statistics or in mathematics. One's opinions are irrelevant: the only item of relevance is whether or not the stated theorem is true. Often when the problem becomes difficult it is human nature to resort to "appeal to intuition or good-will or accept me as an authority". Unfortunately, this type of reasoning crops up continuously in everyday life as well as in the social and political and other sciences. When hypotheses are difficult to state and much more difficult to prove or disprove, people often skip all steps of scientific experimentation and

1. accept the word of an authority, friend, or idol as "law",
2. say something often enough so that they have propagandized themselves and perhaps others into believing that a "law" exists, or
3. camouflage the whole topic with half-truths, phony statistics, and vague meaningless words and then derive a "law".

For example, we often hear such statements as "hippies (all) have a message", "hippies (all) have brilliant minds", "artists (all) tell the story of social progress", "Negroes (all) are excellent athletes", "he is an odd-ball and therefore is intelligent", "a generation gap exists between parents and children", "a pietist-secularist gap exists", "change (all) represents progress for society and therefore is good", "he is a brilliant mathematician and therefore he is brilliant in political science", and so on. (The word "all" has been inserted in common statements to illustrate the falsity of these sweeping statements.) We definitely know that some of the hippies are stupid, that there are nonathletic Negroes, that no generation gap exists in many families, that a brilliant mathematician may be a complete ignoramus when it comes to politics. We know that some of these problems are not new; they have always existed in one form or another. We know also that the human being is very susceptible to propaganda, especially if it allows him to do what he wants to do and he will often go to great lengths to justify his position. Knowing this, we should always question statements involving sweeping generalities, and currently popular adjectives and phrases. Do you really know what "doing his thing" means? Do you know precisely what your neighbor means when he uses terms such as "liberal" and "conservative"? What is a conservative revolutionary radical? What is a liberal arch-right-winger? What do the words "modern" and "relevant" mean? When people speak of "civil rights" do they mean rights for themselves or for everybody? We must examine carefully and critically what we hear, see, and read, if we are to follow the principles of scientific investigation in everyday life as well as in scientific inquiry.

Furthermore, we should note that the scientific method builds on previous knowledge. Ignoring or destroying the knowledge and "know-how" of a previous "establishment" has been demonstrated time and time again to be a foolhardy method for making progress. The destruction of an organized society has in the past set civilizations back several hundred years in a social and technological sense. Radical changes in society often have disruptive effects. Therefore, the scientific method, an orderly, continuous quest for knowledge, making use of all previous knowledge, leads to the most rapid advance.

When a truth has been established (not assumed), we have a law. The law is often phrased in the form of a mathematical equation. As soon as the law is established, investigation ceases in the direction that produced the law, and proceeds in another direction toward establishing another law. Several laws later, the investigation may turn attention to a generalization of the current laws, if this is possible, as was the case with a generalization of Newton's laws of motion.

The orderly quest or pursuit of new knowledge is known as <u>research</u>. Webster defines research as "(1) Careful search: a close searching. (2) Studious inquiry; usually critical and exhaustive investigation or experimentation having for its aim the revisions of accepted conclusions in light of newly discovered facts." Research should yield new knowledge. The term research is not to be confused with <u>re-search</u> which means a search or refinding of already established and known facts, or merely follow-the-leader investigation. Scientific research results in an advancement of the state of knowledge and, therefore, in an advance of science.

It is useful to differentiate between <u>empirical research</u> and <u>analytic research</u>. The former deals with investigations involving measurements; the latter deals with laws, axioms, postulates, and definitions in the field of inquiry. In mathematics, philosophy, and theoretical physics research is generally analytic in nature, whereas in experimental physics, biology, social sciences and business, much of the research is empirical in that it involves measurements and observations on various characteristics. The mathematician states a theorem within a framework of definitions and axioms. He then sets out to prove the theorem in a mathematical sense, utilizing the axioms and definitions. He does not collect data from observations to prove the theorem. The biologist, on the other hand, states a hypothesis and then conducts an investigation to collect data on the plausibility of the hypothesis. He may or may not accept the hypothesis on the basis of the facts obtained, but seldom, if ever, does this prove that the hypothesis is true or false. The empirical facts only substantiate the claim for the hypothesis; they do not prove it.

A necessary part of research is <u>inference</u>, which is a process of reasoning whereby the mind begins with one or more suppositions or propositions and proceeds to another proposition. It is a psychological process. Inference may be <u>deductive</u> or <u>inductive</u>. <u>Deductive inference</u> is the process of determining the implications inherent in a set of propositions. For example, in plane geometry consider the statement: "_If the three sides of one triangle are equal, respectively, to the three sides of a second triangle, then the triangles are congruent_", where the term "congruent" means that they have the same size and shape. The process of reasoning and the conclusion "the triangles are congruent" is the deduction and the remaining part of the statement is the proposition. The conclusion is reached through the process of deductive inference. This type of inference is associated with analytic research; it is also used in empirical research in various ways, such as in the constructing of hypotheses, theories, and laws.

Aristotle was one of the first to stress the systematic nature of science and to teach the use of reasoning in the development of science. Syllogism is one form of logical deduction that was used to a large extent; this method of deductive inference begins with two premises, usually a major premise and a minor premise, or with propositions so related in thought that a person is able to infer a third proposition from them. The following examples illustrate this method of deduction.

Major Premise: All living plants absorb water (_induction_).
Minor Premise: This tree is a living plant (_observation_).
Conclusion: Therefore, this tree absorbs water (_deduction_).
Major Premise: Human beings are men and women (_induction_).
Minor Premise: This person is a man (_observation_).
Conclusion: Therefore, this person is a human being (_deduction_).

The following excerpts and examples were taken from Professor K. Choi's (formerly at Cornell University) introductory lecture in a course on model building on the subject of deductive inference:

Major Premise: Every M is P
Minor Premise: S is M
Conclusion: Therefore, S is P.

Or, to cite a particular example

Every virtue is laudable,
Kindness is a virtue,
Therefore, kindness is laudable.

It is really remarkable that from this kind of start, which seems so innocently simple, there has been developed a grand array of powerful procedures by means of which reasoning can be kept tidy and dependable.

Lewis Carroll, the author of Alice in Wonderland and a mathematician at Christ Church, Oxford, wrote several books and pamphlets explaining the application of the rules of logic. He showed that one could, from the three premises,

1. babies are illogical,
2. nobody is despised who can manage a crocodile,
3. illogical persons are despised,

derive the conclusion:

Babies cannot manage crocodiles.

This example of logical reasoning indicates, incidentally, that a logical conclusion is no more valuable than the premises on which it is based. The above deduction is not difficult. But take a look at another of Lewis Carroll's examples. What is the conclusion from the following nine interlocked premises?

1. All who neither dance on tightropes nor eat penny buns are old.
2. Pigs that are liable to giddiness are treated with respect.
3. A wise balloonist takes an umbrella with him.
4. No one ought to lunch in public who looks ridiculous and eats penny buns.
5. Young creatures who go up in balloons are liable to dizziness.
6. Fat creatures who look ridiculous may lunch in public provided they do not dance on tightropes.
7. No wise creatures dance on tightropes if liable to giddiness.
8. A pig looks ridiculous carrying an umbrella.
9. All who do not dance on tightropes and who are treated with respect are fat.

This one takes a bit of thinking, but it is straightforward (using the procedure of logic) to discover that these nine premises yield the conclusion:

No wise young pigs go up in balloons.

One may not be overwhelmed by the significance of these examples, but it is rather impressive when one realizes that involved and complicated premises do in fact lead unambiguously to certain definite conclusions and that clearly formulated procedures of logical thinking can unravel a curious mass of interlocked statements of this kind. Knowledge of the past furnishes the major premise; a particular problem or situation supplies the minor premise. The deduction obtained by the psychological process of reasoning constitutes the conclusion.

Inductive inference forms a large share of the third part of the definition of the subject of Statistics. This type of inference is characterized by the fact that from the sample data, we draw conclusions concerning population facts. For example, from tasting a small piece of cake we draw conclusions concerning the entire cake; by sticking our toe in the water at one spot we draw conclusions about the relative temperature of all the water in a swimming pool; a polling organization obtains answers from a small representative sample relative to preference of two candidates, A and B, for election to an office and from the sample proportion draws conclusions about the proportion of registered voters preferring candidates A and B; from a sample selected entirely by chance from two lots of light bulbs the experimenter sets up an experimental investigation to determine the average length of life of light bulbs and from the investigation draws conclusions about the average length of life of all bulbs in each of the two lots; from a sample of the weather early in the day we draw inferences, dangerously of course, about the weather for the entire day, and based on these observations we decide the type of clothes to be worn for the day; and so forth for many other situations requiring decisions or conclusions in investigations and in everyday life experiences.

We have noted that the scientific method or inquiry involves the following:

1. use of hypotheses, theories, and laws in the investigation,
2. use of a scientific attitude which involves critical and searching observational ability,
3. use of deductive and inductive inference,
4. use of orderly and organized quest of knowledge.

We also may note that the above are followed very rigorously by detectives and successful prosecuting attorneys in solving fictional cases in novels and on television. Such people as Sherlock Holmes, Maigret, and detectives on television shows are exceptionally keen observers detecting the tiniest bits of evidence and organizing them into an impregnable case for or against a character.

In research, a scientific attitude is required. The scientist must have a passion for facts and truths, he must be cautious in his statements, he must have clarity of vision, he must be discerning, he must be persevering, he must be thorough, and he must have a good sense of organization in bringing together related facts. The person who disregards pertinent information in an interpretation of a scientific endeavor does not have the proper scientific attitude. The person who knows what he believes without wanting to be bothered by facts cannot fit into the scientific community without discord. In science one is seeking the truth, and someone will find it sometime. A scientist must be discerning and must work on important and general problems if he is to make significant advances in science. As E. B. Wilson [1952] states in the opening sentence of his book, "Many scientists owe their greatness not to their skill in solving problems but to their wisdom in choosing them". The scientist must be persevering, but he should also know when he is defeated and is unable to solve the problem. At the same time he must be absolutely thorough in his investigations. Adequate controls, observations, and measuring instruments are a necessity. In discussing the need for adequate controls, Wilson [1952], page 41, illustrates this by citing the following example: "It has been conclusively demonstrated by hundreds of experiments that the beating of tom-toms will restore the sun after an eclipse". On the question of adequate numbers of observations, two examples from Wilson [1952], pages 34 and 46, are pertinent. A nutritionist published an article on an experiment with a surprising result. When a visitor to his laboratory requested to see more of his evidence he replied, "Sure, there's the rat". The use of one rat in a nutritional study is obviously insufficient for drawing conclusions. In the second example, an experimenter in medical research reported that of the animals used in the study, 1/3 recovered, 1/3 died, and no conclusion could be drawn about the remaining subject because that one ran away!

By its very nature, research investigation requires inference. Inductive inference, as we have seen, proceeds from the results obtained from a sample, a part, or an experiment to characteristics of the entire population or of all the members of a class. This type of induction is based on partial evidence, and as such it is characterized by a degree of uncertainty. The evidence for inductive inference may often be stated on a probability basis. A large part of the science of Statistics is devoted to procedures relating to inductive inference.

In model building, which is a process for finding a mathematical formula explaining a phenomenon, the process of deduction and induction is utilized over and over again, as illustrated in figure IV.1. We postulate a model, construct an experiment to test the adequacy of the model, conduct the experiment, compare the data with the postulated model, and then start the cycle over again. This process of investigation continues until a mathematical formulation is found which explains the phenomenon under study. We then have John Doe's law.

In all of scientific investigation there are certain principles that must be observed if the investigation is to be a success rather than just an experience. Before listing these principles we define some additional terms.

Two additional terms requiring definition are _sample survey_ and _experiment_. A _sample survey_ is an investigation of what is present in the population. When the sample is a 100% sample, it is called a _census_. Any observation that appears in the population could appear in the sample; any condition not represented in the population will not be observed in a sample or a census.

In many investigations, however, it is desired to investigate conditions which do not appear in a population, but could appear in an _experimental investigation_ or _experiment_, where an _experiment_ is defined as the planning and collection of measurements or observations according to a prearranged plan, for the purpose of obtaining factual evidence supporting or not supporting a stated theory or hypothesis. The experimenter may, and often does, introduce conditions which do not exist in any naturally occurring population. In the _experiment_, the investigator controls the conditions;

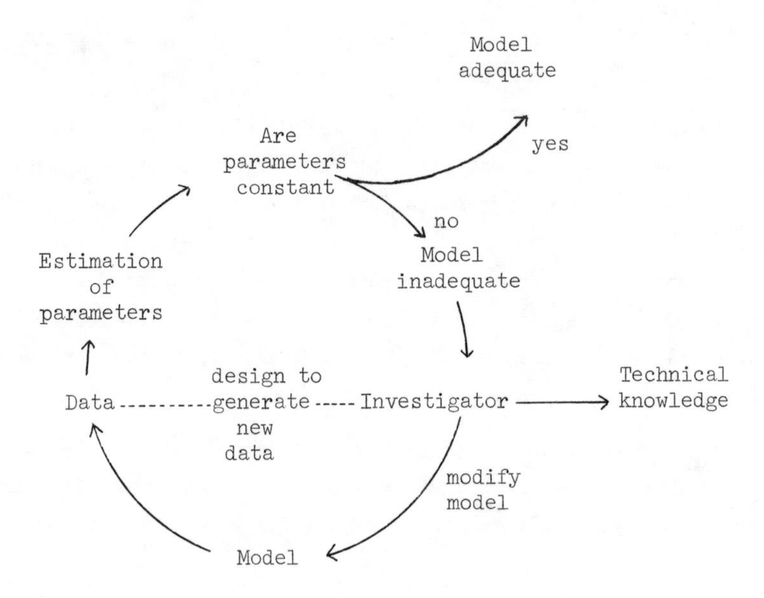

Figure IV.1. Model-building (adapted from Box and Hunter [1962]).

in a survey, the conditions are those that prevail in the population. For example, in a survey of the types of floodlights used for night lighting slopes at ski centers in the Northeast in the 1964-65 season, we would find that mostly mercury vapor bulbs were used and that General Electric Lucalox sodium vapor light bulbs were not. In order to test the effectiveness of the latter bulbs for lighting, an experiment would need to be designed as was done at the Grosstal Ski Area (Rochester Democrat and Chronicle, 2/18/66, page 10D). This sodium vapor bulb would not have been found in a survey of bulbs used to light ski slopes but it was used in the experiment being conducted at Grosstal.

Some additional terms are defined below for use in the following sections:

-- A treatment is a single entity or phenomenon under study in an experiment. (E.g. a mercury vapor bulb or a sodium vapor bulb could be a treatment.)

-- An absolute or single phenomenon experiment is one which contains a single treatment.

-- A comparative experiment is one designed specifically to compare two or more treatments.

-- A treatment design represents the arrangement and selection of treatments for comparative purposes or for ascertaining responses to several levels of a number of variables. It could, and often does, include controls and standards for comparison.

-- An experimental design (or experiment design) is the arrangement of treatments in an experiment; it is used in all types of experimental investigations.

IV.3. First Principle: Clear Statement of Problem Requiring Solution

In order to make progress in a research investigation, it is essential that the problem to be solved and the question to be answered are explicitly and clearly stated in written form. Explicitness and clarity are necessary in order to make certain that the resources of the survey or experiment are expended in a fruitful and efficient manner. The collection of unneeded data or the omission of essential data because of vagueness in stating a problem leads to inefficient investigations. Incompleteness of statement of the problem and of the questions to which answers are sought is wasteful of progress and of resources. Because of the fallibility of human memory, it is necessary to state the problem precisely in written form; since many items are forgotten during a term, one has evidence of this in every course. Then, why should one attempt to remember the important items associated with a research investigation? It is a relatively simple matter to follow the first principle of scientific inquiry, but it is surprising how often it is disregarded.

As one outstanding biologist, A. C. Hildreth, once said, "One can't imagine any more of a fool's paradise than to experiment or investigate in whatever direction the mind may wander, without any purpose or objective. Such meandering investigations lead to naught but a wastage of funds and resources".

In analytic research, the framework of definitions and axioms as well as the conclusions to be drawn must be explicitly and rigorously stated if one is to have any hope of effecting a solution and drawing a conclusion.

IV.4. Second Principle: Formulation of Trial Hypotheses

As defined previously, a hypothesis is a statement of fact about some characteristic of the population; the statement may or may not be true. From the statement of the problem, we formulate a statement of the hypothesis or hypotheses; we should record these in written form. Some of the essential conditions for the adequacy of a hypothesis as it related to the problem under consideration are:

1. It must possess sufficient clarity to permit a decision relative to sample or experimental facts.
2. It must lend itself to testing by investigation.
3. It must be adequate to explain the phenomena under consideration.
4. It must allow a reliable means of predicting unknown facts.
5. It should be as simple as possible.

In a proposed research project, a team of psychiatrists and psychologists was requested to submit hypotheses in written form in order that others could understand the goals and objectives of the study. The result was a 35 page jumble of words full of ambiguities and violating every one of the above essentials of an adequate hypothesis! After some time and after sufficient prodding, it was encouraging to note that these same individuals were stating hypotheses in one or two-line lucid and simple statements.

One type of hypothesis often used in statistics is the null hypothesis or the hypothesis of no difference. For example, suppose that two samples of light bulbs were drawn by a random process from the same lot, and the average length of life was determined for each of the samples. The null hypothesis, which may be stated as "the average length of life, say μ_1, of light bulbs for the population from which the first sample was drawn is equal to the average length of life, say μ_2, of those in the population from which the second sample was drawn, or alternatively as $\mu_1 - \mu_2 = 0$", would be true since $\mu_1 = \mu_2$ when both samples are drawn from the same population of light bulbs. In other cases, we do not know whether the null hypothesis is true or not; for example, in the case of the effectiveness of a new vaccine compared to the effectiveness of a standard known vaccine. We postulate that there is no difference, and then we set up an investigation to ascertain whether or not the sample or experimental facts are in agreement with the null hypothesis. Statistical procedures have been devised to determine whether or not the sample facts agree with the null or any other specified hypothesis, where the term "agree" has been defined.

IV.5. Third Principle: A Careful, Logical, and Critical Evaluation of the Hypotheses.

Once the hypothesis has been critically evaluated for adequacy, the next step in scientific inquiry is to search the literature to determine whether or not the proposed problem and formulated hypotheses will add new knowledge, whether or not they are plausible, and whether or not the problem is worthwhile. The statement that "Many scientists owe their greatness not to their skill in solving problems but to their wisdom in choosing them" may be applied to individuals in all walks of life including business and politics as well as to research investigators.

The problem of searching the literature can be systematized and made efficient; Wilson [1952] discusses this subject in chapter 2 of his book, and he gives reasons for a literature search, the structure of scientific literature, methods of searching, and methods of recording the desired information.

After completing the literature search, one may decide that the problem is not worthwhile or that it has no solution; in this case, the first two principles listed above would have to be reformulated and one would proceed in another direction.

The first three principles are very important to the success of any investigation, scientific or otherwise. It is here that the experimenter has a "go/no-go" situation. He must decide whether to carry out an experiment or to wait until further information and/or resources are available.

IV.6. Fourth Principle: Design of the Investigation.

Any empirical investigation involves a procedure for collecting the data. The observations may be collected in a haphazard manner to yield little information on the formulated hypotheses, or they may be obtained from a well-planned and well-designed procedure to yield the desired information. In planning the investigation, items such as those discussed in sections III.3 to III.7 must be considered. The data collected must be such that evidence for or against the hypotheses is obtained. The lack of coordination of this step with the first three can, and often does, lead to serious deficiencies in the investigation.

In research investigations, consideration of the following items is required:

1. The characteristics to be observed on the different entities or treatments in the experiment or survey.
2. The design of the investigation, that is, the design of the experiment or of the sample survey (chapters V and VI).
3. The treatment design for the experiment and selection of adequate controls and standards (chapter VII).
4. The number and kind of observations to be made (chapter XII).
5. The selection of procedures and measuring instruments resulting in meaningful data and controlling extraneous influences.
6. A detailed outline of costs, equipment, and personnel required to carry out the experiment or survey.
7. An outline of the statistical procedures to be used in the summarization of the data.
8. An outline of the summary tables, graphs, charts, pictures, and figures desired for the investigation.
9. Safeguards against accidents and carelessness.

After a careful and critical review of all the above items, the investigator may decide that he requires more facilities than are available or that he will need to reformulate the problem and the hypotheses in order to conduct an investigation within his resources.

For example, if the problem requires the purchase of an instrument costing $50,000, if the entire budget for the investigation is $25,000, and if it is not possible to rent the instrument or to use another measuring device, the problem will need to be reformulated and the investigation carried out within the available budgetary limitations. As a second example, suppose that the investigator wishes to conduct a sample survey using personal interviews from the population of individuals over 18 years of age in a preselected geographical area, say within the city limits of Syracuse, New York. Suppose that the average cost of a personal interview is $5.00, and that 500 interviews are desired. If his total budget is $500, he will have to utilize other procedures such as mailed questionnaires, telephone interviews, a combination of the above, or a reduction in the number to 100 interviews. He may decide that none of these alternatives is satisfactory and that the survey should be delayed until additional funds of $2,000 are made available for the investigation.

After preparing the pertinent summary tables, graphs, and charts and the statistical analyses in outline form, the investigator should compare these with the stated problem and formulated hypotheses. If they do not conform, he should change the procedure to obtain the information related to the objectives of the investigation.

In preparing for an extension survey a number of years ago, the investigators stated the problem, the objectives, and the questions to be answered in precise form. When the questionnaire had been prepared, a comparison of the questionnaire with the objectives of the survey indicated that there were no questions on the questionnaire concerning five of the desired objectives. Luckily, the questionnaire had not been distributed, and it could be reformulated to conform with the goals of the survey.

The entities or treatments in the experiment must be selected with care. The success of the experiment, to a large extent, depends upon the selection of the treatments in an experiment. The importance of treatment selection varies with the investigation, but in general it is a very crucial item. We devote an entire chapter (VII) to the subject of treatment design.

IV.7. Fifth Principle: Selection of Appropriate Measuring Instruments, Reduction of Personal Biases to a Minimum, and Establishment of Rigorous and Exact Procedures for Conducting the Investigation

As was evident from our classroom survey, the selection of an appropriate measuring device was important for a differentiation between the heights of individual students. A measuring device calibrated in yards gave units too coarse to differentiate and one calibrated in feet only separated the class into two groups, that is, five feet tall and six feet tall. The measuring device calibrated in millimeters was certainly calibrated finely enough for our purposes. Presumably, calibration in centimeters would have been sufficient.

For all types of investigation, the selection of the measuring device and of the unit of measurement is very important. In order to accelerate progress in economic, sociological, psychological, educational, and certain biological investigations, it will be necessary to devise appropriate and precise measuring instruments for the study of various phenomena. For example, how many people living in the world in 1958 could have predicted the interim state of affairs in Cuba or in Czechoslovakia? Could knowledgeable people in social and political sciences devise a scheme which would eliminate deadly physical conflict between countries or even between juvenile and/or adult groups in the large cities of the world? The author believes that the greatest advances to date have been in the physical, medical, and biological sciences, but that *by far* the greatest need for progress is in the social and political sciences. We may learn the exact chemical and physical nature of the animal *Homo sapiens*, but unless we also learn how to control him the law of the jungle soon may prevail. The sociological and psychological problems of welfare recipients, of children from broken homes or temporary common law marriages, of migrating workers, and of a very fluid and mobile labor force considered necessary for a low percentage of unemployed as well as the problems of teaching more material in the same amount of time of increased pressures on children, of mass transportation, of food production for the future, of pollution, and so on are with us now. Do we have measuring devices to obtain factual and exact information leading to hypotheses and theories which will allow appropriate action to be taken by politicians, by private citizens, and by government officials? As a professor of sociology once said, "Our State Department asks the sociologist what will happen in country X if we institute policy A, policy B, or policy C. We sociologists only shrug our shoulders and profess ignorance. But the State Department has to have a policy, so the politicians have taken over the job of the sociologist by default". Perhaps, _and the author emphasizes "perhaps" because of his incompetence in this area_, the social scientist should spend more time devising measuring devices and less time conducting investigations, in order to accelerate progress in this important area. Perhaps more laboratories such as the Primate Laboratory in Madison, Wisconsin, are needed. Perhaps startling new techniques which as yet have not been envisaged are necessary. The problems are known in this area and hypotheses can be stated. But can we design an investigation to obtain the relevant facts? If we cannot, then our time should be spent on developing and calibrating measuring devices and in designing procedures to obtain the desired information rather than on obtaining irrelevant data.

Even though finely calibrated measuring devices are available, precautions are necessary to ascertain that the instruments will be used correctly and the measurements will be free from bias. For example, one could take three readings and discard the most discrepant one. Discrepant from what? One could use an "intelligent placement" of treatments in an experiment which could consistently allow one treatment to appear in the most favorable position. A "favorite" treatment or method could be scored high whenever the investigator knew its identity. In grading examinations, the grader should be unaware of the student's identity to avoid possible favoritism. The instructor may be biased toward neatness and clarity of exposition, and he may have to be constantly on guard against this bias in grading examination questions.

In taste-testing experiments, it is wise to keep each taster's score secret, since a known score of one taster may affect the score of the second. In judging diving events, the judges are required to show their scores simultaneously in an attempt to make the scores independent. Judges should be trained in a standard manner and should know the procedures of judging the material being considered. It benefits no one to allot the ribbons in a haphazard manner. Measuring devices and judges should not be changed in the middle of an investigation and they should not be affected by environmental conditions.

Rigorous and exact procedures have to be devised for discarding or for utilizing "far out" observations. For example, suppose that one observes the following data:

Treatment									Arithmetic average
Control	9.8	17.8	10.1	18.2	10.2	10.3	9.7	9.9	12.0
New treatment	12.0	12.1	11.8	12.1	11.8	12.0	12.3	11.9	12.0

What should one do with data of this type? Are the numbers 17. and 18.2 discrepant observations ("wild things") or are they representative of this type of data? We note that excessively large or excessively small observations have considerable effect on the arithmetic average. If one has no criteria for discarding "wild things", there are statistical procedures for reducing their effect. One could, for instance, utilize a different type of average such as the median or the middle observation. The medians are 10.15 and 12.00, respectively, for the two treatments.

IV.8. Sixth Principle: A Rigorous Conduct of the Investigation

Having proceeded according to the five principles above, the experimenter has again to decide whether or not he should perform the investigation. If he decides to proceed, then the experiment or survey must be conducted in an exact and rigorous manner, making certain that the data obtained from the investigation are free of all but random measurement errors and ambiguities. Prior to the execution of the investigation, it may be well for the investigator to reconsider his method for conducting the investigation. During the conduct of the investigation, he must be keenly aware of possible difficulties and be able to spot unexpected results that may come up.

IV.9. Seventh Principle: A Complete Statistical Analysis of the Data and Interpretation of the Results in Light of the Experimental Conditions and Hypotheses

While statistical procedures are valuable aids in reducing data to summary form, the object of an investigation should not be the application of statistical procedures. Instead, it should be the relationship and interpretation of the results in light of the stated problem and hypotheses.

The results of statistical analyses should provide evidence for or against the stated hypotheses.

In addition to providing factual information concerning the various hypotheses, the results of the investigation and the statistical analysis should produce facts for the design of future sample surveys or experiments. This is the second purpose of any investigation, as listed in chapter I.

All statistical computations should be checked and rechecked for correctness. The arithmetic and the algebra should be free from computational, algebraic, and copying errors. The statistical procedure should be checked for appropriateness; use of an incorrect procedure can be as misleading as an arithmetical error. Unfortunately, articles published in scientific journals contain all too frequently one or more of the above types of errors. For example, one investigator professed to support view A over view B with a small number of observations, 15; unfortunately, he made a mistake in arithmetic which, when corrected, supported view B. The fact that <u>eight</u> pages of a well-known scientific journal had been used to publish his mistakes is appalling!

In another instance, a famous psychologist presented the results of a research study in his presidential address to a psychological society. The results had appeared in reverse order and contradictory to him, and he requested a recheck by the computing center. He was assured that the results were correct. He interpreted the results as best he could. Some time later he needed an additional table to be included in the printed address: the computations had to be performed on a desk calculator, as the high speed computer was being repaired. He found that the items had been coded 1,2,⋯,20 when they should have been coded 20,19,⋯,2,1. He was now faced with a complete reversal in interpretation. He did this by emphasizing the need for complete checking on all phases of statistical analyses and of processing the data.

IV.10. Eighth Principle: Preparation of a Complete, Accurate, and Readable Report of the Investigation

No investigation is complete until a comprehensive, correct, and readable written report of the investigation has been made available in the required form. In some instances, the written report only has to be made available in handwritten or typewritten form to one or more individuals such as a teacher, administrative head, a graduate committee, fellow investigators in one or more laboratories, or personnel of a company; in other cases, it is printed and published as an article in a professional journal, newspaper, or book, and it is made available to the general public. The conclusions drawn should be substantiated and in agreement with the facts obtained from the investigation. In the written report, the description of the experiment or survey and of the statistical procedures utilized should be complete enough to allow the reader to decide whether or not the conclusions reached are in agreement with the facts of the investigation and the description of the data and of the method of collecting the data should be detailed enough to allow the reader to decide whether or not they bear on the hypothesis under investigation or are so mixed up with other effects that the data do not or may not allow testing of the hypotheses. The mixing up of effects sometimes occurs in investigations, and the resulting data may support several hypotheses in addition to the one of interest.

A possible misconception among investigators is that the experimental results should indicate differences, that is, the null hypothesis should be rejected and that this represents "positive results" from an investigation. However, correct and informative data <u>always represent positive results whether or not the null hypothesis is rejected</u>. There are <u>no</u> "negative results" if the principles of scientific investigation are followed. It is just as important to know that the null hypothesis is probably true as to know that it is probably false. Improper procedures in an investigation leading to meaningless results are negative, but reliable data on a phenomenon are never negative. Whether or not differences are found, a written report should be prepared and submitted to the appropriate public.

A principle could appear to be established were investigators only to report the differences found and to ignore those cases when differences were not found. For example, in an examination of the effect of drug A on patients, suppose that only cases of recovery were reported while cases of nonrecovery were not. These reports would erroneously indicate 100% effectiveness for drug A when in fact it could be that drug A had a detrimental effect on certain groups of individuals. If these were never reported by investigators, one would only find this out by conducting a crucial experiment of one's own. It has happened in some fields of inquiry that the reporting of "positive results" and the omission of "negative results" in scientific publications has lead to misconceptions.

In this connection, one should not confuse the failure to find a solution to a problem with the fact that the results of an investigation indicate no differences among the treatments or variables. The confusion of these two ideas could be one reason for the attitude developed toward "negative results" in certain areas of investigation.

IV.11. Discussion

Close adherence to the principles of scientific inquiry will lead to progress and efficient utilization of resources. The investigator who follows these principles will make considerably faster progress than his counterpart who just investigates or who plans poorly. The planning stage of an investigation is probably the most important step in an investigation, but all steps are important. The adage that "a chain is no stronger than its weakest link" is definitely true of any investigation. Failure to observe any one of the principles of scientific experimentation may doom, or at least weaken, the entire investigation by leading to difficulties in the design and in the analysis and interpretation of the data.

Many of the steps in scientific investigation do not involve statistics, but failure to observe any one often leads to difficulties in statistical analyses The fourth and seventh principles listed above are highly statistical in nature, the fifth is partly statistical, and the remaining principles are mostly nonstatistical in nature.

Figure IV.2 shows the relationship between the topic headings in sections III-2 to III-9 and the eight principles of scientific investigation. There are other interrelations, but the major ones are indicated in the diagram.

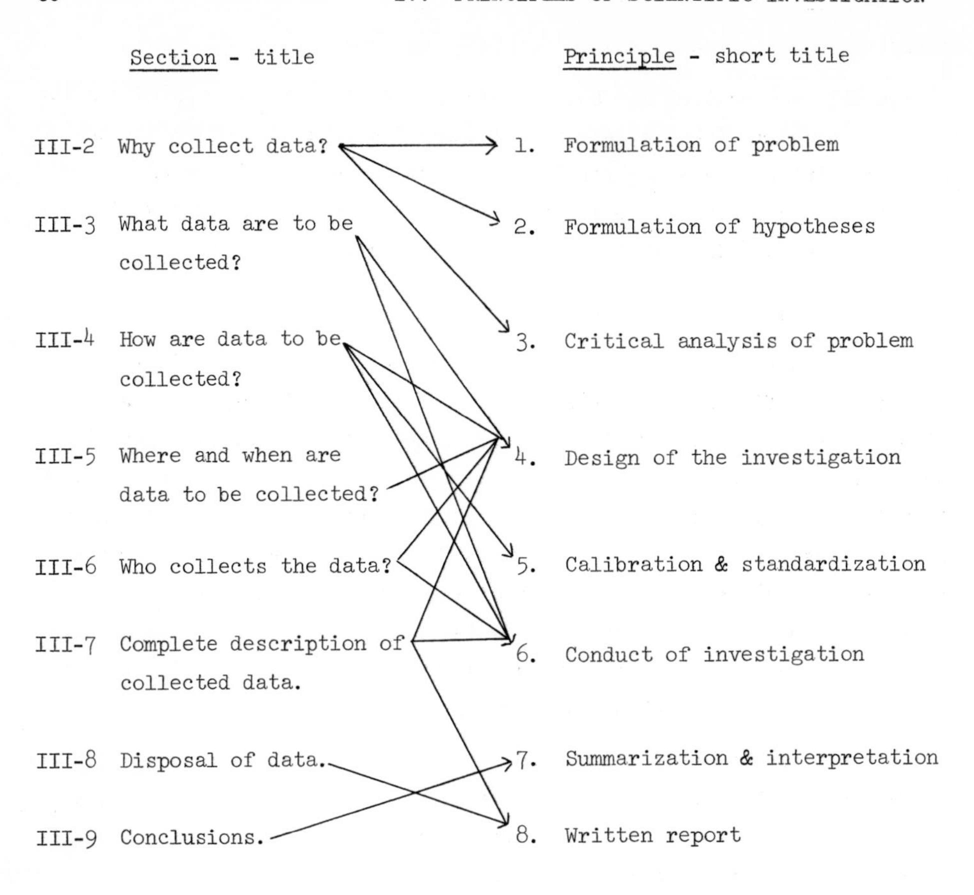

Figure IV.2. Interrelations between sections in sections III.2 to III.9 and the principles of scientific investigation.

IV.12. Problems

IV.1. This problem is designed to illustrate the selection of treatments or kinds of observations bearing on the hypotheses being considered. In all cases, select the minimum number of observations (readings on the hour) differentiating among the five hypotheses listed below. Our clock has only an hour hand and we can obtain only observations on the hour. One of the observations is to be at 12:00 noon. The five hypotheses are:

1. The clock has stopped.
2. The clock keeps perfect time.
3. The clock revolves forward three complete revolutions every 12 hours.
4. The clock is running backward at the rate of one revolution every 12 hours.
5. The clock is running backward at the rate of three revolutions every 12 hours.

Given that readings can be observed only at 3:00, 6:00, 9:00, and 12:00 o'clock, what observations are necessary to differentiate between the above hypotheses? Suppose we hypothesize that the clock is running either forward or backward 1, 2, 3, or 4 revolutions per 12 hours, or that it could have stopped. If readings are available only on the hour, what readings are necessary to differentiate among these nine hypotheses? Would it be helpful if an observation at 12:00 noon on 1000 successive days were available? Why, or why not?

IV.2. Given 12 objects 11 of which weigh the same and the 12th is either lighter or heavier and a balance scale which will accommodate up to 12 objects on each side, design a procedure of four or fewer weighings such that the odd object can be identified, and that it can be determined whether it is lighter or heavier than the others. (Note: Obviously seven weighings would determine the relative weight of the odd object if the 12 objects were first divided into six pairs and each pair weighed to give six weighings; one of the six pairs would not balance. By noting the direction of the balance and by weighing one of the objects from the pair not balancing against one object from a pair that balanced, one can determine the odd object and whether it is lighter or heavier on the seventh weighing. It is possible to find a solution in three weighings but a four weighing design will suffice for this class.)

IV.3. The dictator of a "People's Republic" has 10 workers making gold bars weighing 10 kilograms each. Each worker is to make gold bars of identical size and place 10 of them in a box each day as his quota. The dictator, being an efficient person and knowing that some or all of the workers are materialistic and would filch gold whenever possible, wants to check on the workers as simply as possible and wants to detect any worker who does not make gold bars weighing 10 kilograms. Furthermore, the process is such that if any amount is taken from a bar it would be precisely x units. How would he do this with one weighing?

IV.4. A man buys a newspaper and a bag of peanuts. He becomes engrossed in reading the newspaper while absentmindedly munching on the peanuts from the bag. He has eaten all but two peanuts from the bag. In observing the remaining two, he notes that they are full of worms. What inferences might he make from the sample of peanuts consisting of the two remaining peanuts? What is the type of inference used? What elements would tend to make the sample of two peanuts representative or nonrepresentative of the population of peanuts in the sack with respect to worminess?

IV.13. References and Suggested Reading

Box, G. E. P. and Hunter, W. G. [1962]. A useful method for model building. Technometrics 4:301-318.

Federer, W. T. [1955]. Experimental Design. The Macmillan Co., New York, pp. xix + 544 + 47. (Ch. 1)

LeClerg, E. L., Leonard, W. H. and Clark, A. G. [1962]. Field Plot Technique. Burgess Publishing Co., Minneapolis, Minn. (Ch. 2)

Wilson, E. B., Jr. [1952]. An Introduction to Scientific Research. McGraw-Hill Book Co., New York, pp. ix + 373. (The reader may omit chapters 10, 11, and 12 if he does not have time to read all of the book.)

In addition to many books on scientific experimentation, the reader will find it valuable to read:

Bennett, J. H. (Editor) [1965]. Experiments in Plant Hybridization, Gregor Mendel (with introduction and comments by R. A Fisher). Oliver and Boyd, Edinburgh and London, pp. ix + 95.

CHAPTER V. SURVEYS AND SURVEY DESIGNS

V.1. Introduction

A survey is an investigation of what is present in a particular population. If every member of the population is investigated or measured relative to the characteristic being considered, this is called a census or 100% sample survey. If only a fraction of the population is investigated, this is called a sample survey. Whenever one collects data in a survey, a population and a characteristic of the population are automatically implied. Therefore, whenever a survey is contemplated, it is absolutely essential that the population and the characteristic to be measured are precisely defined. Failure to do this can lead to a set of meaningless numbers and endless difficulty in interpretation.

In the class survey described in chapter II, our universe or population was defined to be those students registered in the course for credit. Visitors were not a part of the population. Furthermore, we decided to measure the heights of class members in yards, in feet, and in millimeters, to measure weights of individuals in pounds, and to ascertain hair and eye colors. But was this a sample survey or a census? We had attempted to obtain a census, but some students were absent and some refused to have their weight taken. Hence we ended up with a sample survey of those attending class on the given day and of those who cooperated in the survey.

All surveys should have a goal. Ours was to acquaint students with measurements and with a survey and to provide a set of data for future class use.

There are many ways to conduct a survey. Some methods are better than others, some lead to erroneous conclusions, and some lead to naught. The sampling procedure should not be selected "to prove one's ideas or predjudices" but rather to seek the truth as a true scientist should. Mrs. Science is an extremely hard taskmaster; if one does not find the truth, someone, sometime will.

Although very few of us have ever thought of ourselves as samplers, we are continually taking sample surveys. One of the first surveys one takes every day is of the weather. From the sample of weather facts observed, we decide on wearing apparel and perhaps on activities for the day.

Then some of us sit down to breakfast to continue our sampling. If the first taste of hot cereal is satisfactory, we continue eating. If the cereal is burned or is too heavily salted, we may discontinue eating. This process of sampling continues all day long, in one form or another, from sipping coffee to answering questions on an examination. Each of us has his own peculiar method of sampling and in drawing inferences from the sample facts about the population parameters. We shall discuss various methods of taking surveys, that is, sample designs that are useful in survey work.

Another question that arises is whether a sample or a census should be taken. Since we know that samples may vary considerably, why not always take a census in order not to have sampling errors affecting our results? In certain situations it may be impossible to take a census. For example, would it be possible for a girl to date <u>all</u> boys in the city in which she lives in order to pick a husband? The answer is obviously "no"; she samples until she finds the "right one" and obtains a proposal. Likewise, in sampling firecracker stocks for exploding, ammunition for firing, wines for tasting, light bulbs for length of life, pies for suitability for human consumption, and so forth, the sampling is destructive; in order to have some material left to sell or to use, a census is out of the question. In heavily populated countries such as India and China, a census of people and some of their characteristics is not feasible because of the logistics problem of training and supervising so many census reporters. Also, it would be almost impossible to take a census of a species of fish in a large lake such as Cayuga Lake. Even if the logistics problem could be solved, the cost of taking a census would become astronomical.

Thus, we see that not only are we all samplers but that some situations require that we take a sample of a very small fraction of the total population. In order to ascertain the temperature of a cup of coffee, a small sip is recommended over a mouthful simply because of the possible consequences. Likewise, one should not dive into a pool of water until one knows the depth and the temperature of the water, and it is fatal to let the doctor take too large a sample of your blood.

The accuracy of a result may even be inversely related to the amount of sampling. For example, it has been found in taking inventory on liquor stocks that the reliability of the results and legibility of writing decreased and the "shrinkage" in stocks increased as the size of the sample was increased!

Also, suppose we wish to know the weight and length of 1000 fish caught on a given day by 200 fishermen. We could ask the fishermen, but we all know about "fish stories"! Alternatively, we would obtain a more reliable estimate by actually weighing and measuring a randomly selected sample of 100 of the fish caught by the fishermen. As another example of the necessity of taking a sample, the Iowa State Soil Conservation Service needed to know how much erosion of soil had been caused by the 13-inch rainfall in the month of June. After the rains subsided, they had only a short period of time before the farmers started working their fields and covering up the evidences of erosion. The survey was planned and executed during a period of four days (and nights). Obviously, a census would have been impossible in this situation. (These last examples bring up questions on sample size; these will be considered in a later chapter.)

So far we have considered only situations in which sampling is possible; there are cases where sampling is impossible. For example, students are not allowed to take only a sample of tests, and a teacher is not allowed to grade only a sample of his students. The Internal Revenue Service will not let us estimate deductions from only a sample of deductions. The banks cannot determine whether or not we are overdrawn by considering only a sample of the checks we have written. Thus, many situations require a census or complete enumeration.

For additional comments, the reader is referred to M. J. Slonim's book entitled "Sampling." The book is almost devoid of mathematical results and contains a wealth of examples from all walks of life. Several examples used herein were taken from this book.

V.2. Steps in a Survey

The principles of scientific investigation given in chapter IV could be used in any survey, or one could list the steps in a different manner; an alternate form specifically related to a survey is given below.

The first step in planning a survey is to define the population or universe to be surveyed as related to the objectives or goals of the survey. For example, if we decide to study selected characteristics of dental patients who patronize Clinic HURT staffed by Drs. Filmore, Payne, and Toothacher, we must decide what is meant by "patronize" and by "dental patients" and whether or not we wish to stratify the patients of the three dentists. If we select those individuals who enter the clinic, we might find that the girl who is very interested in young Dr. Filmore may appear quite frequently in our survey, but that no dental work was performed on her teeth. She might be "patronizing" but still not fit our definition.

After determining precisely what population we wish to study, the second step in planning a survey is to develop a list or a description of all individuals in our population. This is called the sampling frame. From the patients' records, we could obtain a complete list of names of all patients of Clinic HURT during a specified time period. From registration records, we could obtain a listing of all members of this class; from voter registration lists, we could obtain a listing of voters in a precinct, and so on. In another case, we could describe our population to be all goldfish in a pond which did not slip through a net with holes 1/2-inch in diameter. It is imperative to describe the population and the individuals in a population.

The third step in planning a survey is to describe precisely the information being sought. In a medical study, the information desired may be average time to recurrence of tuberculosis or cancer rather than the incidence of recurrence of the disease. In a food preference study, it may be important to know what food a person would eat if all types were available; an individual may be eating what he does simply because his local supermarket does not stock his favorite foods. In a study on clothing sizes in an army, it is essential to define whether we wish information on sizes of clothing issued, on actual sizes that an individual would wear if he or she had a choice, or on actual sizes that an individual should wear as determined by a panel of "experts." The type and amount of information desired should be carefully and thoughtfully considered.

The fourth step in planning a survey is to determine whether or not the information is already available. It is much cheaper and less time-consuming not to take a survey than to take one. Nothing is to be gained by obtaining a second, third, or even a fourth survey on the same item. In fact, in surveying people there is much to lose in that people quickly become tired, and sometimes angry, if they are interviewed too often and quite often this means more than once. When this happens, the nonresponse problem increases in later surveys. Someone who has gone through three long interviews in one week is quite likely to become a nonrespondent.

The fifth step pertains to solicitation of only essential information. For example, many of you would become quite perturbed, and rightly so, if on each quiz you were required to write, in addition to your name, your parents' names and ages and their educational background, and your social habits during the week. This information, with the exception of your name, is useless for purposes of obtaining information on your comprehension of subject matter in a given course. Anyone will occasionally, or even frequently, be guilty of obtaining useless information or information that is readily available elsewhere, and hence one needs to be constantly on guard. As M. J. Slonim says, "With the exception, perhaps, of stilts for a serpent, there is nothing more useless or ridiculous than a mass of figures collected at great travail, added, multiplied, divided by the cube root of π, and converted (by a $350-an-hour electronic computer) to homogenized index numbers - that have no bearing on the problem".

The sixth step concerns sample size which in turn is related to precision desired for the sample survey, to the available funds, and to the method of taking the survey (for example, personal interview versus mailed interview). When one defines the population, sample size, and other information about the universe, one must define the observational unit and the sampling unit. The observational unit is the smallest unit on which information is obtained, while the sampling unit is the smallest unit used in the selection process in a survey design. For example, suppose that a list of houses in a city is available and that a random selection of houses is obtained; the house is the sampling unit. Then, suppose that information on income is obtained for each person living in the house; the person is the observational unit. In some situations the observational and the sampling units are identical. If, in the above example, information on the type of roof of the house is required, then the observational and the sampling units would be identical for this characteristic.

The question concerning the desired number of sampling units for a survey is considered in chapter XII. However, in order to obtain an idea of the effect of increasing sample size on the variability of a statistical estimate, let us consider the following example. Suppose that a sample survey is designed as a simple random sample as described in section V.8, suppose that n = 100 individuals are polled in a city as to their preference for one of two mayoral candidates A and B, and suppose that 60 favor A and 40 favor B. The proportion favoring A is $\hat{p} = 60/100 = 0.6$; the proportion favoring B is $(1 - \hat{p}) = 40/100 = 0.4$. A measure of the variability in $\hat{p}$, called the estimated standard deviation of $\hat{p}$ or of $(1 - \hat{p})$, is $\sqrt{\hat{p}(1 - \hat{p})/n} = \sqrt{0.6(0.4)/100} = \sqrt{0.0024} = 0.049$. (This statistic is described in chapter X.) We may note that as n increases, the estimated standard deviation of a specified $\hat{p}$ decreases; for example, for $n = 1000$, $\sqrt{0.6(0.4)/1000} = \sqrt{0.00024} = 0.0155$. The value for $n = 1000$ is about 1/3 of the estimated standard deviation for $n = 100$ and $\hat{p} = 0.6$. Writing the estimated standard deviation in the form $\sqrt{\hat{p}(1 - \hat{p})}/\sqrt{n}$, we note that this ratio decreases as the square root of the sample size n increases. As the sample size n approaches infinity, the ratio $\hat{p}(1 - \hat{p})/n$ approaches zero for any $\hat{p} \neq 0$.

In the above example, A appears to be leading, but how does one know whether or not the proportion of people favoring A in the sample is the same as in the entire city? One cannot know this without taking a census, but one can obtain an interval estimate on the true proportion preferring A (or B). To do this, we may (as explained in chapter XI) construct an interval estimate of the true proportion as $\hat{p} \pm 2\sqrt{\hat{p}(1 - \hat{p})/n}$ which for the above example is $0.600 \pm 2\sqrt{0.6(0.4)/100} = 0.600 \pm 2(0.049)$, or from 50.2% to 69.8% who prefer A. (The reason for the 2 is explained in chapter XI.) If candidate A desires more confidence in the result, he might wish to take an additional 900 randomly selected individuals resulting in a sample size of $n = 100 + 900 = 1000$. Then, if the value of $\hat{p}$ remains equal to 0.6, the interval estimate of the true proportion will now become $0.6 \pm 2\sqrt{0.6(0.4)/1000} = 0.6 \pm 2(.0155)$ or from 56.9% to 63.1% who prefer A. If such were the situation, A would have more confidence in his victory than previously.

Sample survey design is a seventh step in the planning of a survey. Although we shall discuss several sample survey designs and several additional ones are discussed in Slonim [1960], McCarthy [1957, 1970], Cochran [1963], Sukhatme [1954], Yates [1960], and Hansen, Hurwitz, and Madow [1953], it is always a good idea to consult with a statistician on a survey design.

As an eighth step in planning a survey, one must construct the reporting form or questionnaire. Simplicity, brevity, and clarity are three very important ingredients for any questionnaire. The form of the question is also very important. Beware of ambiguous questions such as the "wife-beater" question type: "Do you still beat your wife?" If the answer is "no", we don't know whether the person has stopped beating his wife or has never beaten her. Some steps to consider in questionnaire construction are:

1. comparing the list of objectives of the survey with the list of questions to determine whether they coincide,
2. making certain that only relevant questions are included,
3. using as short a questionnaire as possible to achieve the objectives,
4. critically examining each question for ambiguity, clarity, and pertinence,
5. pretesting the questionnaire on a small sample and revising, if necessary, and
6. consulting an expert on questionnaire construction.

If one carefully considers all of the above, useful information should be obtainable. Also, it is necessary to decide whether the interviews are to be conducted by mail, by telephone, from a panel, by personal interview, or by some other means. It is essential to realize that the method of obtaining an answer to a question may, and often does, influence a respondent's answer. One's answer to a question of age may be quite different over the telephone than in a personal interview. Answers by telephone may not be considered as binding as answers written in response to a mailed inquiry. The nonresponse problem is often quite high with mail questionnaires. What are we to do if only 5% of our mailed questionnaires are returned? Such questions as these must be carefully considered and decisions reached before proceeding with a survey.

As a ninth step in planning a survey, it is necessary to train the individuals who obtain the interviews and record the results. It is essential that interviewers do not influence the answers given by interviewees. The method of asking a question can influence the answer, and the interviewer must (unless the respondent does) write down the answer correctly. A method of checking on the interviewers is necessary, since they might be tempted to fill out questionnaires without even contacting the interviewees. One method of training individuals to obtain and to record answers correctly is to have them interview each other during the training sessions and to illustrate how to obtain unbiased answers.

The tenth step in planning a survey is to determine what will be done with "not-at-homes" or "refuse-to-answer" interviewees, how the returns are to be audited, what information is desired from the survey, and what safeguards are to be utilized to guard against reporting and processing errors. It is a well-known fact that individuals often report less income than they actually receive and that the age or weight of women is often given as less than it actually is.

To stress further the importance of adequate training of enumerators and of adequate checks, let me relate two incidents that happened to a fellow statistician, Mr. F, when he was enumerated in two different ways.

In the first instance, a young lady came to Mr. F's home and told him that she was taking a survey for a laxative company. After the proper introductions, she asked a number of questions. One question was, "How often do you use laxatives?" (A loaded question.) Mr. F answered that he never had. Then, to his horror, he observed that she wrote down that he used laxatives 4 to 5 times a week. Mr. F asked why she had deliberately written down a falsehood. She replied that the company would not like it if she were to write down his answer. Mrs. F answered in the same manner as her husband, but the enumerator indicated on the schedule that Mrs. F used laxatives every day of the week. In a second survey in which Mr. F happened to be one of the "unlucky" individuals drawn, he was asked why he did not shop in Ithaca (Another loaded question). Mr. F, being a very serious individual, explained in detail why he did not. Again the enumerator did not write down his answers but something else. When asked by the indignant Mr. F why he did not write down the correct answer, the enumerator replied that it really did not matter because everything would average out anyway!

After a careful consideration of each of the above ten steps in planning a survey, the surveyor conducts the survey, summarizes the results from the survey, and interprets and reports on the results of the survey in a careful, precise, and scientific manner.

V.3. Types of Sample Survey Designs

Although we cannot hope to cover all possible types of sample survey designs, we shall present some of the simpler and more commonly used ones. Both nonprobability and probability sample survey designs will be discussed in the following sections. When the chance of selecting any one of the possible samples of size n is known, the sample survey is denoted as a probability sample survey design. A format similar to that followed in presenting experimental designs in chapter VI will be used in this chapter. Again it should be emphasized that it is wise to consult with a person well-versed in sample design before conducting the survey.

V.4. The Purposely Biased Sample

In certain instances, an individual wants to obtain some numbers which he hopes to pass off as data. Extremists, politicians, organizations, advertisers, labor unions, and so forth, often obtain information only from people who respond in the desired manner. They may take a sample of those people who attend one of their meetings and then use the results as if they represented the entire population of people. Statements such as "two to one prefer Morning Cough Cigarettes", "nine out of ten dentists prefer Evening Cough Cigarettes", and "every person interviewed stated that All Day Cough Cigarettes were milder than those of their own brands", have always appeared unrealistic and incorrect to this writer, especially when it is reported that the two-to-oners did their survey in "several large cities" as follows. Four large cities were selected. Two interviewers went to a busy area of the city. One member of the team gave away free samples of Morning Cough Cigarettes. He walked along with an individual who accepted a free cigarette and lit up. The other member of the team was stationed a short distance down the block; he interviewed two of those his colleague had walked with to one of those that he did not walk with but who was smoking a cigarette. Thus, for every three individuals interviewed, two were found to be smoking Morning Cough to one who was not.

For the "9-out-of-10" group, it is reported that free samples of Evening Cough Cigarettes were distributed in a cafeteria line and that "9-out-of-10" had smoked the free cigarette. This was reported as a preference for Evening Cough.

With regard to the statement that All Day Cough Cigarettes were *always* milder than a smoker's own brand, if a cigarette is lit and if the smoker takes a puff and after a short time takes a second puff, the first puff always appears milder! This was the procedure used to compare the cigarettes, that is, All Day Cough was always smoked first.

V.5. Convenience Sample

A convenience sample is defined to be one which is convenient to take without any regard to bias or representativeness of a universe. For example, one might decide to interview 100 people on some selected subject; the 100 interviews might be taken at Sloe's Bar and Grill because the interviewer frequents the place, at a National League Football game because the interviewer wished to see that game, at the 100 households nearest his home, or the 100 interviews may be limited to the interviewer's girl friends and their girl friends. As examples in other areas, the interviewer may take 100 lumps of coal from the top of a carload, the interviewer may select his sample solely from the members of his class, a comrade may interview his fellow comrades attending a given meeting, or the doctor may select the interviews from patients treated at the hospital where he works.

One could continue giving many such examples of convenience samples, since man is naturally a lazy creature and often sacrifices quality to do less work. The big question about convenience samples is that we have little or no idea about their representativeness of any population and no idea about their reliability.

V.6. Judgment Sample

The judgment sample is one obtained by a self-styled expert in a subject matter area wherein the "expert" reasons that his intimate knowledge of all sampling units in an area allows him to select a sample which represents specified characteristics of the population. For example, suppose a School Board member, Mr. I. M. Smart, uses Mr. and Mrs. I. Dontno, Mr. and Mrs. I. M. Rich, Mr. and Mrs. I. M. Fashionplate, and Mr. and Mrs. U. R. Average to represent the opinion of the members of a school district. Mr. Smart may firmly believe that these four couples represent the opinion of all members, but his confidence in his great discerning abilities by no means makes the sample representative. Individuals do exist who are able to take the pulse of a population with regard to various issues, but the ordinary person using judgement samples is usually only fooling himself (and perhaps others). Even the true experts can go astray when something unusual happens which they do not know how to take into account. Judgment samples are to be used with extreme caution but even when so used, they provide no measure of reliability as do the sample designs discussed in sections V.8 to V.12.

V.7. Quota Sample

The population may be subdivided into several subpopulations or strata representing the various categories. In this sense the sample is representative. Suppose then that the interviewer or surveyor is allowed to select individuals or units within each stratum or subpopulation in any manner he desires, until he obtains the specified number or quota of individuals. This type of sampling is called quota sampling. To illustrate this, suppose that our population is stratified into men over and under forty, women over and under forty, and persons with or without a college degree to give eight strata. The surveyor then selects 50 individuals in each category to obtain the desired information. There are many ways in which the sample could be selected. Many quota samplers simply obtain a convenience sample or a judgment sample in each stratum. More than likely, any difficult-to-obtain individuals are omitted. The representativeness of such a sample within each stratum is highly suspect, since there is no way of ascertaining the reliability of samples taken in this manner, as there is with probability samples.

V.8. The Simple Random Sample

The preceding types of sample designs are nonprobability samples; the chance of selecting a sampling unit is unknown. When the chance of selecting a sampling unit and of selecting a sample is known, then the sample is a probability sample. A simple random sample of size n may be defined to be one in which all elements in the population have an equal and independent chance of being selected in the sample, where independent means that the selection of one element in the sample does not affect the selection of another element. Alternatively, if all possible samples of n elements have an equal chance of being selected, the sample is said to be a simple random sample of size n.

To illustrate, suppose that the population consists of the elements a,b,c,d,e, and f; suppose that a sample of size n=3 is to be selected. The possible samples of size 3 are abc, abd, abe, abf, acd, ace, acf, ade, adf, aef, bcd, bce, bcf, bde, bdf, bef, cde, cdf, cef, and def, resulting in 20 samples of size n=3. If the samples are numbered, and if a number between 1 and 20 is randomly selected such that any number has an equal chance of being selected, then a simple random sample results. Alternatively, we could randomly draw one member from the population of six letters, then randomly draw one member from the remaining five letters (sampling without replacement), and then randomly draw one member from the remaining four letters. The latter procedure would give an equal and independent chance of selecting any element in the population, resulting in a random sample of size n=3.

It is not enough to give an equal chance of selection to each sample member. The chance of selection must be equal and independent. Suppose that we decide to have two samples, abc and def, of size n=3 of the 6 letters, and suppose that we flip an unbiased coin to determine whether we take the sample abc or the sample def. Now every element in the population has an equal chance of 1/2 of being selected, but, every time a is selected so are b and c. Thus, the chances of selection are not independent, even though they are equal.

In a simple random sample, the observational equation of the ith observation from the population may be expressed as:

i^{th} observation = true population mean + bias + deviation.

When an observation is for a randomly selected individual in the population, the term deviation is denoted as random error. Symbolically, the above equation may be written in the form:

$$Y_i = \mu + \text{bias} + e_i$$

where the symbols and the words in the two equations are pairwise equivalent, and i = 1,2, ... , N = the number of observations in the population. In the sample of n observations, the ith observation may be expressed as:

$$Y_i = \text{sample arithmetic mean} + \text{deviation from mean}$$

$$= \bar{y} + (Y_i - \bar{y}) = \bar{y} + \hat{e}_i \ .$$

The statistic $\bar{y}$ estimates the true population mean μ plus the bias. If the bias is zero, then $\bar{y}$ is an unbiased estimate of μ. The statistic $N\bar{y}$ is an estimate of the population total.

V.9. Stratified Simple Random Sample

If the sample units are first grouped, blocked, or stratified according to some criteria and if a simple random sample of size n_i is selected in the ith stratum for all i (i.e. in every stratum), this type of sample design is defined to be a stratified simple random sample. Suppose that our population consists of the first nine letters of the alphabet and that we block or stratify the nine letters into two strata such that vowels are in stratum 1 and consonants are in stratum 2, resulting in

stratum 1	stratum 2
a	b
e	c
i	d
	f
	g
	h

Now suppose that we take a simple random sample of one item from each stratum. The 18 possible stratified random samples are:

	1	2	3	4	5	6	7	8	9	10	11	12	13	14	15	16	17	18
stratum 1	a	a	a	a	a	a	e	e	e	e	e	e	i	i	i	i	i	i
stratum 2	b	c	d	f	g	h	b	c	d	f	g	h	b	c	d	f	g	h

Taking the same number of sampling units from each stratum is denoted as equal allocation.

Suppose on the other hand that we wish the sample size, n_i, in each stratum to be proportional to the stratum size, say N_i. Since stratum 2 is twice as large as stratum 1, we should select twice as many observations in stratum 2 as in stratum 1. This type of allocation of sample sizes is denoted as proportional allocation. The latter form has several desirable characteristics and is frequently used in sample survey design. In this case, the 45 possible stratified random samples are:

	1	2	3	4	5	6	7	8	9	10	11	12	13	14	15	16	17	···	43	44	45
stratum 1	a	a	a	a	a	a	a	a	a	a	a	a	a	a	a	e	e		i	i	i
stratum 2	b	b	b	b	b	c	c	c	c	d	d	d	f	f	g	b	b		f	f	g
	c	d	f	g	h	d	f	g	h	f	g	h	g	h	h	c	d		g	h	h

Other types of allocation are possible. For example, the variability in stratum 2 might be twice as large as in stratum 1, and hence we might wish to take four times as many individuals in stratum 2 as in stratum 1: twice as many because of size and twice as many because of greater variability. Cost considerations and other factors may also be involved. In certain situations, one item or a few items may make up a relatively large proportion of the population; for example, one large farm may produce 10%, say, of a commodity in a given county, a second farm may produce 5%, and a third farm 4%. The surveyor may, and often does, decide to enumerate these three farms and to sample the remaining farms in the county. The three large farms would make up one stratum in which a 100% sample would be taken.

In the absence of bias, the observational equation for the ij^th^ observation in a population containing k strata of size $N_1, N_2, \ldots, N_k$, for $j = 1, 2, \ldots, N_i$ and $N = (N_1 + N_2 \cdots N_k)$ = total population size may be expressed as Y_{ij} = true population mean + deviation of true stratum from true population mean + deviation of observation from true stratum mean. Symbolically, this may be expressed as:

$$Y_{ij} = \mu + (\mu_{i.} - \mu) + (Y_{ij} - \mu_{i.})$$

$$= \mu + d_i + e_{ij}$$

where $\mu = (N_1\mu_{1.} + N_2\mu_{2.} + \cdots + N_k\mu_{k.})/N$ and $\mu_{i.}$ is the true stratum mean. For a randomly selected observation, e_{ij} is the random error deviation. In the sample of size $n = (n_1 + n_2 + \cdots + n_k)$, where n_i individuals are randomly selected from the ith stratum, the observational equation may be written as:

$$Y_{ij} = \bar{y} + (\bar{y}_{i.} - \bar{y}) + (Y_{ij} - \bar{y}_{i.})$$

$$= \bar{y} + \hat{d}_i + \hat{e}_{ij}$$

where $\bar{y}_{i.} = (Y_{i1} + Y_{i2} + \cdots + Y_{in_i})/n_i$, $\bar{y} = (n_1\bar{y}_{1.} + n_2\bar{y}_{2.} + \cdots + n_k\bar{y}_{k.})/n$, $\hat{d}_i = (\bar{y}_{i.} - \bar{y})$ is the estimated stratum effect for the ith stratum which is $(\mu_{i.} - \mu)$, $\bar{y}_{i.}$ is an estimate of the true stratum mean $\mu_{i.}$, and $\hat{e}_{ij} = (Y_{ij} - \bar{y}_{i.})$ is the estimated random error component for the ij^{th}

observation. If n_i/N_i is a constant for all i, then the allocation is proportional; $\bar{y}_{i.}$ is an estimate of $\mu_{i.}$ and $(N_1\bar{y}_{1.} + N_2\bar{y}_{2.} + \cdots + N_k\bar{y}_{k.})$ is an estimate of the population total $N\mu$.

V.10. Cluster and Area Samples

If the population is divided, naturally or not, into subgroups or clusters and if a simple random sample of clusters is selected, this is denoted as a cluster sample design. If the clusters form areas, then this is defined to be area sampling. A 100% sample of the clusters results in a stratified sample where the clusters are the strata.

In certain areas of the United States, the land is blocked into sections which are one mile square, and the cluster is the land which falls within the section lines. One of the early area samples in the United States was a random selection of sections of land in Wyoming. Apple trees are clustered together in an orchard; a random selection of orchards results in a cluster sample. The group of individuals within a household represent a cluster; a simple random sample of households in a given city results in a cluster sample.

Nothing has been said so far about how many individuals are selected within each cluster. One could enumerate all sampling units or individuals within each cluster, or, alternatively one could obtain a simple random sample of n_i individuals within each of the selected clusters. Likewise, one could utilize equal or proportional allocation in the clusters. Proportional allocation is frequently used in cluster sampling. The statistical analyses are simpler for proportional sampling than they are for disporportional sampling. The observational equation for a cluster sample is of the same form as for a stratified simple random sample.

Because natural phenomena frequently lead to clusters of sampling units, a cluster sample design is frequently utilized in survey work.

V.11. Stratified-Cluster Sample Design

As might be suspected, the sample design could become more and more complicated as more and more levels of stratification are used. Moe [1952] conducted a survey on farmers' opinions of agricultural programs. He stratified New York by counties; then each county was divided into areas of approximately five farms each. A simple random selection of clusters or areas of five farms was made within each county; the number of clusters selected being proportional to the number of clusters in a county. All full-time farmers within an area or cluster were sampled. We would denote such a sample design as a stratified-cluster (or area) sample with a 100% sampling in each cluster and with proportional sampling of the clusters. This design proved effective in obtaining results and the statistical analysis remained simple. This survey of about 1500 farmers was planned, conducted, and summarized in a two-month period. The teamwork necessary to conduct such a survey is described by Moe [1952]; the results are reported in an easily understandable manner since Mr. Moe uses ratios such as 6-out-of-10 or 9-out-of-10 to report the opinions of farmers on various topics.

V.12. Every k^{th} Item With a Random Start Sample Design

Another type of probability sample design often used when a listing or an ordering of all sampling units is possible is to select an interval, say k, to select a random number between 1 and k, say c, (see section V.14) and then to take the cth, $k+c^{th}$, $2k+c^{th}$, etc. item on the list. Such a sample design is denoted as an every k^{th} item with a random start sample design. For example, suppose that we have a list of 5000 houses in a city, and we wish our sample to be 100 houses. We select a random number between 1 and 50, say 13, and then take the houses numbered 13, 63, 113, 163, 213, ..., 4913, and 4963, as our sample. There are 50 such samples possible; hence, the chance of selecting any given sample is one in fifty.

We may wish to take several such samples in order to ascertain reliability. Suppose that we wish to take five such samples, and still to have a total sample size of 100. We would then randomly select five numbers between 1 and 250, and for each sample the sampling interval would be 250. Thus, suppose our first randomly selected number were 90, the first sample then would consist of the 20 items numbered 90, 340, 590, 840, 1090, 1340, 1590, 1840, 2090, 2340, 2590, 2840, 3090, 3340, 3590, 3840, 4090, 4340, 4590, and 4840. The remaining four samples would be obtained similarly.

This type of sampling often results in a considerable saving of travel time in planning the sample and in taking the survey. However, caution must be exercised against cyclical variations with sample designs of this type. For example, a sample of every k^{th} (say 50th) business from a list of businesses in New York was obtained by a surveyor. The results appeared absurd until he noted that every business appearing in the sample was located on or near Fifth Avenue!

V.13. Some General Comments

As we have noted, most people are surveyors and some even make their living taking surveys. Most of us have heard of Gallup, Roper, Nielsen, National Analysts, and so forth, and there are many survey firms of which we have never heard. We could rightly presume that these firms use a variety of procedures to obtain results. This writer was rather surprised by an article appearing in The Ithaca Journal dated March 16, 1967 and entitled "Survey Firm Apologizes." The survey firm was called the "Voice of the Voter." In the article, the survey procedure was described as follows: "Some 1500 questionnaires were distributed in the county the weekend of Feb. 4-5, but only 35 were completed and returned..... 'Our organization considered that at least 100 replies back would be necessary before we could effectively go ahead with further action'." This firm appeared to imply that 100 responses out of 1500 would have been satisfactory; however, the firm did use only 35 returns out of 1500. What about the opinions of the 1465 out of 1500 who did not respond? Even if 100 returns had been available, what could one really conclude about the opinions of the 1400 nonrespondents? Better survey procedures would have eliminated the need for an apology from the survey firm.

Another mistake often made by surveyors is that of taking a prejudiced sample of size one and then drawing conclusions from said survey. The proud husband often talks about his wife's shopping habits, cooking procedures, and so on. Likewise, the biased parent believes his child to be unusually bright and thinks of all children's reactions in light of his child's reaction. This idea of taking a sample of size one goes further than within the family. Slonim [1960], pages 86-7, describes a situation where the owner of a large hosiery factory believed that 100,000 dozen pairs of nylon hose worth about one megabuck were disappearing from his plant every year. They used all means of ferreting out the culprit who might be appropriating them. As everyone and everything was searched, it developed that the company estimated the annual output of nylons based on the performance of one operator. It turned out that this one operator was the best in the plant and she used less yarn than any other operator; the 1,200,000 pairs of nylons that were presumed to have been filched had never been made!

In another situation, a hybrid seed corn company in Iowa estimated the amount of seed corn for sale in the spring from the corn that had been harvested in the fall. The latter often contained 40% or more of moisture before the moisture content was reduced by drying to 15% moisture for storage. A chart was used to determine the amount of corn at 15% moisture content that would be obtained from X pounds of corn at 45%, say, of moisture. Every spring, the company appeared to have lost several thousand pounds of valuable seed corn. They suspected piracy, but from all appearances they had reliable and trustworthy employees. After considerable investigation, attention was turned to the chart itself. It turned out that the chart had been developed for moisture contents of less than 35% and that a straight line prediction was used. After additional investigation, it was found that both the chart and the machine used for determining moisture content were biased and inaccurate for samples of corn with more than 35% moisture content. No one had taken any valuable seed corn. It had simply never been there in the first place.

The above are examples one would not like to experience personally. With proper survey procedures one can avoid most, if not all, of the pitfalls of inadequate or biased sampling. Close adherence to the ten steps for survey work as listed in section V.2 will result in representativeness of samples and reliability of results.

V.14. Random Selection, Chance Allotment, and Equal Opportunity

We hear a great deal about equal opportunity these days, and it is a concept that we shall utilize in order to obtain unbiased estimates of parameters. To assure unbiasedness we are forced to use probability samples, which give an equal opportunity or chance of selection to all possible samples of size n. In order to attain the "lofty goal" of equal opportunity, we need some device for doing this. In particular, a device is required which gives all numbers $1,2,\cdots,N$, where N = the number of possible samples, an equal chance of being selected. One way of doing this is to obtain N ping-pong balls and number them consecutively $1,2,\cdots,N$. The ping-pong balls are then put into a hypothetical "hat", for example a wastebasket. The balls in the "hat" are thoroughly mixed, for example by throwing all those in the "hat" against an uneven wall and then sweeing them up and putting them back into the "hat". The idea of thoroughly mixing the balls is crucial, as this produces the independence of selection between two balls. (In a proposed draft-lottery system, Jim, Bob, Bill and Sam would each have an equal chance of selection, but the selection of one should not affect the chances of any of the others being drafted.) Then we close our eyes, reach into the "hat", and blindly draw out a ball. The ball could have any number from 1 to N on it; any number from 1 to N has an equal chance or equal opportunity of being selected, that is one out of N or 1/N.

Now if we return the ball to the "hat", and after thoroughly mixing the balls, again draw one in the manner described previously, any number $1,2,\cdots,N$ has an equal chance of selection, even the one we returned to the "hat". Thus the same ball could be drawn the second time by mere chance alone. When such a scheme is utilized repeatedly, a series of random numbers can be generated. Also, when equal opportunity of selection by a chance process is possible for all elements of the universe, this is defined to be random or chance allotment and the process is called a random process. By returning the ball to the "hat" after every draw and repeating the sampling we have been sampling with replacement of the ball.

If the process described above is used to produce a series of random numbers but the ball is not returned to the "hat" after each draw, this is described as sampling without replacement. Thus no item is repeated, and we have n different numbers resulting from n drawings. In most surveys we do not wish to have any one person or item appearing or being interviewed in the sample more than once; thus, sampling without replacement would be utilized for many surveys. For the random number generator described above, we would thoroughly mix the ping-pong balls and blindly draw one; then the remaining N-1 balls would be thoroughly mixed and a second ball drawn; and so on, until n balls have been drawn. The numbers on the n balls would be the random sample of n numbers from N numbers. The elements in the population, which had been previously numbered from 1 to N, then correspond to the numbers on the N balls. The selected random sample would correspond to those numbers on the n randomly selected balls.

In the examples considered in this chapter, we have been talking mostly about sampling without replacement. Hence, in drawing a random sample of three different letters from the six letters a, b, c, d, e, and f, we would put six ping-pong balls numbered a, b, c, d, e, and f in the "hat". After thoroughly mixing the balls, we blindly draw one from the "hat"; the remaining five balls are again thoroughly mixed and again we blindly draw one ball from the "hat", and then the remaining four balls are thoroughly mixed again, and again a ball is blindly drawn from the "hat" This results in a random sample of three letters from the universe of six letters, this being one of the 20 possible samples listed in section V 8. Since the thorough mixing allows an equal and independent opportunity of any ball being selected, we have a random sample of size $n = 3$.

Alternatively, we could have numbered the 20 possible samples from 1 to 20, and we could have obtained 20 ping-pong balls numbered consecutively from 1 to 20. The 20 balls could have been put into a "hat" and thoroughly mixed; one of the balls blindly drawn would have resulted in the random sample of $n = 3$ letters corresponding to the number on the ball. The process would result in a random sample of $n = 3$ letters. The randomization procedure here selects from the same 20 possible samples and with the same opportunities as the procedure described in the preceding paragraph.

Another device for obtaining a set of random numbers is through the use of the telephone directory. Suppose we number the digits starting from the right of a seven-digit telephone number as the 1st, 2nd, 3rd, 4th, etc. digits. If we drop the 1st digit and use only the 2nd, 3rd, and 4th numbers of a telephone directory for a large city, it has been demonstrated that these three digits may be used as random numbers of 000 to 999 or for $N = 1000$ numbers. Using any two of these three columns of digits would result in random numbers from 00 to 99 or for $N = 100$ numbers. Likewise, use of any one of the three columns may be used for random numbers from 0 to 9 or for $N = 10$ numbers. If N is smaller any unused numbers are ignored. Numbers in bold face type should be omitted, as they usually contain consecutive or repeating digits, and an arbitrary start (randomly selected if possible) in the directory should be made. One method of starting is to pick the next word you hear, find its alphabetical place in the directory, and start from there, either proceeding forward or backward. To illustrate, let us select a random number from the set 01, 02, $\cdots$, 20. From the Madison, Wisconsin directory, let us start after the name U. R Hoarse to obtain the following numbers (only last 4 columns given):

7646
7203
7422
2224
7688
4083
3560
2537
⋮

The first number between 01 and 20 that occurs in the second and third columns is 20 and the second number appearing is 08. If one wished to use all numbers between 00 and 99, each number could be divided by 20 and the remainder used to obtain the number, with 00 being equated to the number 20. In the above list of numbers, the remainders would be (columns numbered from the right):

3rd and 4th columns	2nd and 4th columns	2nd and 3rd columns
16	14	04
12	10	00
14	12	02
02	02	02
16	18	08
00	08	08
15	16	16
05	03	13
⋮	⋮	⋮

Any number that repeats is ignored, as is a number not in the set $1, 2, \cdots, N$. The use of remainders allows many more numbers to be used without so much skipping.

There are many so-called random tables published in books: many programs for generating random numbers on high-speed computers are available. One of the most extensive, if not the most extensive, publication contains a million random digits; it was put out by the Rand Corporation, and is entitled "A Million Random Digits with 100,000 Normal Deviates".

V.15. Problems

V.1. Figure V.1 has been drawn (by D. S. Robson, Cornell University) to illustrate a population (e.g., fish). The parallelograms are of various lengths ranging from 1/8" to 2". By visual or judgment selection alone, select one parallelogram to represent the mean of the population, and measure it in mm. As a second part of the problem, stratify the parallelograms into large, medium, and small, then select by judgment one parallelogram in each to represent the mean of the stratum; obtain the average length of the three sampling units selected. How does the mean of the three compare with the one selected previously? As a third part of the problem, number the 100 parallelograms from 00 to 99 and randomly select three sampling units. Use a local telephone directory to obtain a set of random numbers. Measure the three sampling units and compare the mean of the simple random sample with the judgment sample obtained previously. (Note: It is a good idea to summarize the sample results from all class members to illustrate variability among samples and possible biases in judgment samples. As a fourth part of the problem, the class members are to select a simple random sample of size ten to illustrate the decrease in variability in samples of size ten as compared to samples of size 3. Then compare all methods with the true mean of the population. All sampling should be without replacement.)

V.2. This is to be an independent survey to be conducted by the student. The results are to be checked against all other surveys of the class for independence. The survey is given the weight of one class examination and therefore is considered to be an important problem. It is to be submitted prior to the end of the course in order to allow time for grading. In the write-up of the survey, indicate how each of the ten steps has been followed. The survey may be on any topic of interest, but the number of interviews is limited to ten as ten is an easy number to use in division, it is a large enough number to obtain experience in survey work, and people are already interviewed too frequently.

V.3. Devise a procedure for generating random numbers for the set 0,1,2,3,4,5,6,7,8, and 9. Use identical circular tags, construct a ten-sided die, or develop any other procedure. From 100 trials on the random number generator, record the number of times each of the ten digits occur. Does the procedure appear to be unbiased in that each number occurs equally frequently? Why or why not?

Figure V.1. 100 Parallelograms

V.16. References and Suggested Reading

Huff, D. [1954]. How to Lie With Statistics. W. W. Norton and Company, Inc., New York, pp. 142.

(Chapter 1 is recommended reading in relation to sampling.)

McCarthy, P. J. [1970]. Sampling. Elementary principles. 3rd Printing, Bulletin No. 15, New York State School of Industrial and Labor Relations, Cornell University, Ithaca, New York, pp. iii + 31.

(This bulletin is more technical than Slonim. The presentation is elementary and is directed more toward the analysis aspects.)

McCarthy, P. J. [1957]. Introduction to Statistical Reasoning. McGraw-Hill Book Company, Inc., New York, Toronto, and London, pp. xiii + 402.

(Chapters 6 and 10 are recommended reading. The basic ideas of sample design are presented in these two chapters.)

Moe, E. O. [1952]. New York farmers' opinions on agricultural programs. Cornell University Extension Bulletin 864.

(An example of the planning, conduct, and analysis of a survey.)

Slonim, M. J. [1960]. Sampling (original title was Sampling in a Nutshell). Simon and Schuster, New York, pp. xii + 144.

(The entire book should be read. The presentation is elementary, and the ideas and concepts of sampling are easy to grasp.)

For those who wish more of the theory, applications, and analyses of sampling designs, the following books are recommended. The reader will need to have a relatively good background in statistical theory and methodology in order to comprehend the material presented.

Cochran, W. G. [1963]. Sampling Techniques, 2nd edition. John Wiley and Sons, Inc., New York and London, pp. xvii + 413.

Deming, W. E. [1950]. Some Theory of Sampling. John Wiley and Sons, Inc., New York, pp. xvii + 602.

Hansen, M. H., Hurwitz, W. N., and Madow, M. G. [1953]. Sample Survey Methods and Theory. Volumes I and II. John Wiley and Sons, Inc., New York.

Sukhatme, P. V. [1954]. Sampling Theory of Surveys with Applications. The Iowa State College Press, Ames, Iowa, pp. xix + 491.

Yates, F. [1960]. Sampling Methods for Censuses and Surveys. 3rd edition, rev. and enl. Charles Griffin and Company, Ltd., London, and Hafner Publishing Company, New York, pp. 440.

CHAPTER VI. EXPERIMENTAL DESIGNS

VI.1. Introduction

The experimental designs considered in this chapter are for comparative experiments involving two or more treatments, where the object of the investigation is to obtain information on the treatments relative to each other. In other words, the interest is on differences between treatment averages (or effects) rather than on the averages (or effects) *per se*. One set of characteristics of designing the experimental arrangements or procedures we shall consider is:

1. Arrangement of the procedure to increase the *efficiency* of the experimental investigation. One design is said to be more *efficient* than a second if the variation in average response for any treatment is smaller than in the second design.
2. Grouping of the experimental material in such a manner that the units within a group are more alike than are units in different groups. This kind of grouping is called *blocking* or *stratification*.
3. Fairness to each treatment by subjecting all treatments to as nearly equal conditions as possible and by utilizing chance allotments or *randomizations* thereafter.

First, these characteristics will be exemplified with three illustrative examples. Then we shall present a number of types of experimental designs which control various types of heterogeneity among the individual items or units in the investigation.

We have not said anything about the size and the shape of the smallest unit of observation, the *observational unit*, nor about the smallest unit to which one treatment is applied, the *experimental unit*. In some cases the observational and experimental units are the same; in others the experimental and/or the observational unit size is fixed and cannot be varied. When the size and shape can be varied according to certain criteria, one can talk about optimum size and shape, but this is a topic that will not be discussed in this text. We shall assume that the size and the shape of the observational and experimental units are given. Examples of investigations wherein the unit is fixed are the animal in physiological and nutritional studies, the plant in physiological studies, the individual in learning experiences, a cake or pie in baking studies (a whole cake or whole pie must be baked, even if size and shape can be varied), the classroom for teaching methods (the number of students can be varied, but the classroom is fixed), the automobile or the tire for road endurance tests, a piece of equipment used to produce or evaluate a product, fixed farms or pastures in certain management investigations, and so forth.

The first experimental design we shall consider deals with the weighing of very light objects on a scale, for instance a spring scale. Suppose that one has seven objects (a,b,c,d,e,f,g) to be weighed. One experimental design for weighing them would be (any order of weighing could be utilized, but they are ordered here for easy reading):

Weighing	Object weighed
1	determination for zero-correction
2	a
3	b
4	c
5	d
6	e
7	f
8	g

The weight of any object is the scale reading for its weight minus the scale reading for the first weighing. Such a design would require eight weighings, and each object would be weighed only once.

Alternatively, let us consider a different "weighing design" first suggested by Dr. Frank Yates [1935] of the Rothamsted Experimental Station in England. Suppose that the following experimental design is utilized to weigh the seven objects:

Weighing	Objects weighed
1	a,b,c,d,e,f,g
2	a,b,d
3	a,c,e
4	a,f,g
5	b,c,f
6	b,e,g
7	c,d,g
8	d,e,f

The same number of weighings is used here as for the previous weighing design, but each object has been weighed four times rather than only once as in the previous design. This means that the variation in weights from the above design is only 1/4 that of the first design and hence 4 times as efficient. The weight of each of the objects is obtained as follows:

Weight of object	Coefficients for weights from weighing 1	2	3	4	5	6	7	8
a	+	+	+	+	-	-	-	-
b	+	+	-	-	+	+	-	-
c	+	-	+	-	+	-	+	-
d	+	+	-	-	-	-	+	+
e	+	-	+	-	-	+	-	+
f	+	-	-	+	+	-	-	+
g	+	-	-	+	-	+	+	-

If the object is present in the weighing, it receives a plus sign and if not, a minus sign. The sum of the first 4 weighings minus the sum of the last four weighings gives the weight of object a; the sum of weighings 1,2,5, and 6 minus the sum of weighings 3,4,7, and 8 gives the weight of object b; and so on.

A number of scientific papers on weighing designs for a chemical balance, a spring balance, and other weighing devices have been published in statistical journals. Reference to these papers may be found in section XV-4 of Federer [1955]. Our purpose here is not to discuss weighing designs but to illustrate the pay-off in efficiency that is sometimes possible when an appropriate experimental design is used. Characteristic (1) above is exemplified by this example.

As a second example used to illustrate characteristics (2) and (3) above, let us suppose that the investigator is comparing four nutritional treatments (for example, standard ration - S, S + vitamin A, S + vitamin B, S + vitamin D), and is using the rat as the experimental animal or unit. Suppose that for design I he randomly selects 10 rats for each treatment without any regard to the rat's parentage. This allows all four treatments a fair or equal chance to be allotted any 10 of the 40 rats. Suppose that another investigator takes account of the rat's parentage and uses ten litters of four male rats each. (The word litter is used to designate the members born to a mother within a short period of time, say one day. Thus, twins in humans would be a litter of size two, triplets would be a litter of size three, etc. In certain types of animals such as rabbits, dogs, cats, swine, and so forth, the members of a litter are brothers and sisters, or half sibs, and therefore not identical in genetic composition.) The four treatments are then allotted by chance to the four male rats of each of the ten litters to form design II. This is "fair" to all four treatments as each has an equal chance of being allocated to any rat in the litter. In this design, the comparison among treatments is within a litter (on members of the same litter) and on rats of the same sex. The variation for many characteristics including nutritional response among members of the same litter or family is less than it is among members of different litters. Hence, design II would be expected to yield treatment means which are less variable than the corresponding means from design I. In fact, it was found from nutritional experiments on swine that the variation of treatment means compared on individuals of the same litter was about one-half of that obtained when the animals were not grouped or stratified into litters. Practically, this means that investigators using design II would require only one-half as many animals to obtain the same degree of variation among treatment means as those using design I. A simple change of design from I to II would cut the cost of experimentation by one-half; alternatively, for a fixed amount of experimentation it would decrease the variation among treatment means by one-half.

The use of blocking or stratifying experimental material into relatively homogeneous groups can greatly increase the efficiency of experimentation, since total variation is equal to that due to assignable causes plus bias plus random error. By blocking, a portion of the random error is placed into the assignable or controllable category, thereby reducing the amount of variation in the chance or random category.

As a third example, we shall utilize an illustration demonstrating characteristics (2) and (3); it is adapted from one described by W. J. Youden (formerly of the National Bureau of Standards and now deceased) in a lecture at Cornell University a number of years ago. The owner of a large fleet of cars wished to compare the effect of four brands of motor oil on the performance of cars in the fleet. His first experiment was conducted as follows. A new car was purchased; motor oil K was used in the car from 0 - 20,000 miles; motor oil C was used from 20,000 to 40,000 miles; motor oil P was used from 40,000 to 60,000 miles, and motor oil M was used from 60,000 to 80,000 miles. From the measures of performance used, it appeared that motor oil M was definitely inferior to the other three motor oils. The sales representative was called in, and the fleet owner told him that he would not be purchasing their product any longer. Upon inquiring about the reason for this decision, the salesman was informed about the experiment performed to compare the four brands of motor oils and about the poor performance of his company's product. After thinking about the experiment a moment, the salesman retorted, "Your experiment was unfair to our product. The car was all worn out and the poor performance you observed was not due to the oil used but to the dilapidated condition of the car!"

The fleet owner promised to take this into consideration before he made a final decision. It happened that four new cars for the fleet were purchased at this time. The owner decided to use these cars to perform an experiment which would be fair to all four motor oils. He decided to assign the numbers 1,2,3 and 4 in a random manner to the four oils by rolling a six-sided die and ignoring 5's and 6's. The number on the die obtained from the first roll would be assigned to oil K, the number obtained on the second roll, excluding the number used for oil K, would be assigned to oil P, and so on. He obtained the following code K=3, P=1, C=4, M=2. Since he knew nothing about the performance of each of the four new cars, he allotted the numbers 1,2,3 and 4 to the cars in an arbitrary manner. Also, the numbers 1,2,3 and 4 were allotted to the mileage groups in a random manner such that 3 = 0 to 20,000 mile group, 1 = 20,000 to 40,000 mile group, 4 = 40,000 to 60,000 mile group, and 2 = 60,000 to 80,000 mile group. He then drew up the following plan:

Mileage		Car number			
group number		1	2	3	4
(20,000 to 40,000 miles)	1	1 (P)	2 (M)	3 (K)	4 (C)
(60,000 to 80,000 miles)	2	2 (M)	1 (P)	4 (C)	3 (K)
(0 to 20,000 miles)	3	3 (K)	4 (C)	1 (P)	2 (M)
(40,000 to 60,000 miles)	4	4 (C)	3 (K)	2 (M)	1 (P)

When rearranged in the order of mileage, the grouping becomes:

Mileage group	Car number 1	2	3	4
0 to 20,000	K	C	P	M
20,000 to 40,000	P	M	K	C
40,000 to 60,000	C	K	M	P
60,000 to 80,000	M	P	C	K

The experiment was run according to the above experimental design, and oil M still ranked considerably lower in performance tests than the other three motor oils. The salesman for brand M motor oil was told that his oil would not be used any longer. Again he wanted to know how the experiment had been conducted. He was informed of the experimental procedures. He was quite disappointed, but after some time his face brightened, and he said, "Ah, yes, but how do I know that you didn't assign me a horrible driver, and this is what is causing the low performance for my company's product?" Well, the owner of the fleet had considered this variable as well, and he replied, "I took care of that, too. The drivers were assigned to the cars according to the following plan:

Mileage group	Car number 1	2	3	4
0 - 20,000	K S	C R	P D	M G
20,000 - 40,000	P R	M S	K G	C D
40,000 - 60,000	C G	K D	M R	P S
60,000 - 80,000	M D	P G	C S	K R

where drivers were assigned to the letters as follows: S = Smiley, R = Red, D = Demon, and G = Grumpy. The plan was fair in all respects, and extraneous variation was controlled. Your oil just doesn't stand up in comparison with the other oils."

The salesman went away very depressed about losing a customer; however, in a few days he was back again, because his company had developed a new oil called Super Special M. The new oil had been tested in the manner described above; it was found equal or superior in all respects to oils K, C, and P. Everyone lived happily ever after, until a better oil was developed and tested!

From problem III.2 the "Georgia Peach Squeezing Experiment" we noted that the experimenter had not been fair in all respects to the four treatments (the four quarters of the peach). He could have been fair to all treatments by picking four random samples of 250 peaches, assigning a treatment (quarter) to each sample of 250 peaches, and measuring the pressure necessary to puncture the skin on a given quarter of the peach. Thus, only one treatment would be performed on each peach. More efficient procedures of comparing all four treatments on each peach will be discussed later in this chapter.

VI.2. The Completely Randomized Design -- Zero-way Control or Elimination of Heterogeneity

If the experimental material available to an investigator contains only nonassignable variation, then it is impossible to block or to group the material into subgroups such that the variation among subgroups is larger than among individuals within subgroups with regard to the response being considered in the investigation. Any grouping would be no more effective than a random assignment of individuals to the subgroups and, hence, would be useless. In such cases we randomly assign the treatments to the experimental units. Usually the random assignment is restricted in such a manner as to have an equal number of experimental units assigned to each treatment.

To illustrate the above, suppose that our treatments consist of eight different kinds of cooking fats, one of them a standard. The characteristic to be observed is the quantity in grams of fat absorbed by doughnuts during cooking. The experimental material consists of one large batch of doughnut mix which is enough to make more than $8 \times 6 \times 5 = 240$ doughnuts. The experimental unit is 6 doughnuts, as these will all be baked at one time. The observational unit is also the set of six doughnuts, since no data are available on the grams of fat absorbed by an individual doughnut. In order to be fair to all treatments, let us assign consecutive batches of 6 doughnuts to treatments in a random fashion, until each treatment is observed on 5 sets, or replicates, of 6 doughnuts each. We can do this by putting numbers 1 to 8 on round tags which are as nearly alike as possible. These numbers are put in a hat or covered jar and thoroughly mixed. A number is drawn blindly, and this is the treatment number to receive the first set of 6 doughnuts. The number is returned to the hat, the tags are thoroughly mixed, and a tag is again drawn blindly. This second number represents the treatment to receive the second set of 6 doughnuts. This process is continued until the $8 \times 5 = 40$ sets of 6 doughnuts have been allotted to the $v = 8$ treatments and each treatment has received 5 experimental units of 6 doughnuts each. As soon as $r = 5$ sets of doughnuts have been allotted to a given treatment, the tag with that number may be removed from the hat.

Relative to the three characteristics to be considered in designing experiments, fairness is exhibited by allowing any treatment to receive any set of 6 doughnuts. No selection or "intelligent selection" by the experimenter is practiced. The material is relatively homogeneous in that a single batch is used and the order of cooking the doughnuts has no effect. Therefore, no blocking or stratification is required. Relative to the first characteristic, it does not appear that the procedure could be made more efficient by utilizing another procedure.

As a second example of the completely randomized design, suppose that 100 chicks from a single hatch of eggs of a single strain of dams and one sire are randomly divided into 4 groups of 25 chicks each. Suppose that 4 types of single-dose vitamins in capsule form represent the $v = 4$ treatments. The 4 treatments are randomly allocated to the $r = 4$ groups of 25 chicks, and a capsule is given to each of the 100 chicks. The response is weight at 8 weeks of age. The 100 chicks are treated alike in all other respects except for type of capsule, that is they are all in the same pen and have the same food and water sources. The chicks intermingle and so all are subjected to the elements of the environment in the enclosure. The treatments are compared in as nearly equitable manner as possible.

As a third example, suppose that a large oven is available for baking purposes. Suppose further that there are no gradients or heat pockets in the oven and the heat remains at the designated temperature once the oven has been heated to this point, and therefore temperature fluctuations are minor. Suppose that 5 different amounts of thickening in pies are to be used, and that these represent the $v = 5$ treatments, say A,B,C,D,E. Suppose further that four pies are to be baked for a given amount of thickening and that the 20 pies can be baked at one baking. The treatments are randomly allocated to the 20 places in the oven. One possible arrangement is shown in figure 6.1.

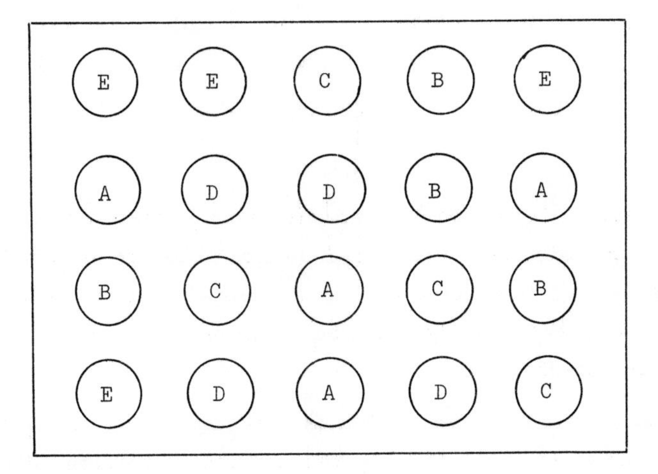

Figure 6.1. Arrangement of 20 pies in an oven for $v = 5$ treatments and $r = 4$ replicates of each treatment.

If the 20 pies cannot be baked at one time, we might use a single oven to bake the 20 pies in the following sequence:

Order	Treatment	Order	Treatment	Order	Treatment	Order	Treatment
1	E	6	A	11	B	16	E
2	E	7	D	12	C	17	D
3	C	8	D	13	A	18	A
4	B	9	B	14	C	19	D
5	E	10	A	15	B	20	C

For this situation there should be no gradients in the successive bakings, in order for the completely randomized design to be the appropriate one. The above layout for an experiment in a completely randomized design might be appropriate for 20 pots on a greenhouse bench or for a series of soil analyses involving 5 treatments.

The completely randomized design is the simplest of all experimental designs. It involves zero-way or no elimination of heterogeneity in the experimental material. The total variation in the experiment may be written as:

Total variation = variation among treatment means + error variation.

= assignable cause + nonassignable cause.

The yield of any experimental unit may be written as treatment mean + an error or discrepance term. It is permissible to use the above form when the different components of variation are additive in their effects. We may need to partition the assignable cause in the variation among the treatment means and a bias factor. An ordinary arithmetic mean is biased by the amount of the bias factor. Now let us consider an individual observation from an experiment and consider the various sources or contributing factors to the variation among observations. From the above, we may write the ij^{th} observation Y_{ij} as:

An observation = estimated treatment mean + random error, or $Y_{ij} = \bar{y}_{i.} + (Y_{ij} - \bar{y}_{i.} = e_{ij})$, where $\bar{y}_{i.}$ is the arithmetic mean of the r observations $Y_{i1}, Y_{i2}, \cdots, Y_{ir}$. Also, an observation = the over-all average of all treatment means + a deviation of an estimated treatment mean from the estimated over-all average + random error, or $Y_{ij} = \bar{y}_{..} + (\bar{y}_{i.} - \bar{y}_{..}) + e_{ij}$. The symbols are given above to indicate that we can and will use a shorthand notation in writing equations about variation. To illustrate these ideas, consider the following experiment:

Let treatment A = no crust on the pie top = "open-faced", let treatment B = crossed strips of pie dough on the top = "cross-hatched", and let treatment C = pie dough on the top = "kivered". Let $\bar{y}_{A.}$, $\bar{y}_{B.}$, and $\bar{y}_{C.}$ = arithmetic mean scores on consistency for treatments A, B, and C, respectively. Let Y_{ij} = j^{th} score on the ith treatment for j = 1,2,3,4 pies. Suppose that there is a bias factor of 10 because the scoring started at 10 instead of zero, that is all scores are read 10 too large. Let the 12 observations, or scores, be those given in table VI.1. The various estimated treatment effects $\bar{y}_{i.} - \bar{y}_{..}$ are computed in the top part of the table. The estimated random error deviations are given in the middle part of the table while the various components of each observation are given in the bottom part of the table.

In order to obtain some intuition about an equation with additive effects for the score or yield of an observation, consider that we would have a situation wherein the average of the three treatment means is 22, and to this we add a bias factor of 10. All 12 of our observations would have a score of 32. Then, suppose that we add -3 to the first four, zero to the next four, and +3 to the last four observations. Lastly, suppose that we assign random error components of -5, +1, 0, and +4 to the first 4 observations, +4, 0, -4, and 0 to the second four, and +3, -3, -1, and +1 to the last four to obtain the 12 observations. The addition of effects, or sources of variation, rather than their multiplication for example, produces an additive effects equation for an observation.

The above additive effects equation for the yield of an observation has been presented in terms of averages of observations, that is in terms of statistics from the sample. In terms of the parameters of an entire population we may write

$$Y_{ij} = \mu_i + \epsilon_{ij} \quad \text{for } i = 1,2,\cdots,v \; ; \; j = 1,2,\cdots,\infty \; .$$

$$= \mu + (\mu_i - \mu) + \epsilon_{ij}$$

$$= \mu + \tau_i + \epsilon_{ij} \; .$$

In the above $\epsilon_{ij} = Y_{ij} - \mu_i$ = a random error, $\tau_i = \mu_i - \mu$ = a treatment effect, and $\mu = (\mu_1 + \mu_2 + \cdots + \mu_v)/v$. μ_i is the true population mean

Scores for 12 pies; treatment means and effects

	Treatment A	Treatment B	Treatment C	All scores
	$Y_{A1} = 24$	$Y_{B1} = 36$	$Y_{C1} = 38$	
	$Y_{A2} = 30$	$Y_{B2} = 32$	$Y_{C2} = 32$	
	$Y_{A3} = 29$	$Y_{B3} = 28$	$Y_{C3} = 34$	
	$Y_{A4} = 33$	$Y_{B4} = 32$	$Y_{C4} = 36$	
Total	116	128	140	384
Treatment Mean $\bar{y}_{i\cdot}$	$\bar{y}_{A\cdot} = 29$	$\bar{y}_{B\cdot} = 32$	$\bar{y}_{C\cdot} = 35$	$\bar{y}_{\cdot\cdot} = 32$
$\bar{y}_{i\cdot} - \bar{y}_{\cdot\cdot}$	$\bar{y}_{A\cdot} - \bar{y}_{\cdot\cdot} = -3$	$\bar{y}_{B\cdot} - \bar{y}_{\cdot\cdot} = 0$	$\bar{y}_{C\cdot} - \bar{y}_{\cdot\cdot} = 3$	0

($\bar{y}_{\cdot\cdot} = 384/12 = 32$ = average of all observations = bias + estimated true mean.)

Estimated random error deviations = $e_{ij} = Y_{ij} - \bar{y}_{i\cdot}$

	Treatment A	Treatment B	Treatment C
	24 - 29 = -5	36 - 32 = 4	38 - 35 = 3
	30 - 29 = 1	32 - 32 = 0	32 - 35 = -3
	29 - 29 = 0	28 - 32 = -4	34 - 35 = -1
	33 - 29 = 4	32 - 32 = 0	36 - 35 = 1
Total	0	0	0

Observation = Y_{ij} = bias + estimate of true mean + $(\bar{y}_{i\cdot} - \bar{y}_{\cdot\cdot})$ + e_{ij} values

Treatment A	Treatment B	Treatment C
24 = 10 + 22 - 3 - 5	36 = 10 + 22 + 0 + 4	38 = 10 + 22 + 3 + 3
30 = 10 + 22 - 3 + 1	32 = 10 + 22 + 0 + 0	32 = 10 + 22 + 3 - 3
29 = 10 + 22 - 3 + 0	28 = 10 + 22 + 0 - 4	34 = 10 + 22 + 3 - 1
33 = 10 + 22 - 3 + 4	32 = 10 + 22 + 0 + 0	36 = 10 + 22 + 3 + 1

Table VI.1. Observations, means, and error deviations.

associated with the ith treatment. Now, $Y_{ij} = \mu + \tau_i + \epsilon_{ij} = \bar{y} + (\bar{y}_{i.}-\bar{y}) + e_{ij}$ does not mean that $\bar{y} = \mu$, $\tau_i = (\bar{y}_{i.}-\bar{y})$, $e_{ij} = \epsilon_{ij}$ but merely that the sum of three components equals the sum of three other components, for example $4 + 5 + 6 = 15 = 13 + 2 + 0$.

VI.3. The Randomized Complete Block Design -- One-way Elimination of Heterogeneity

If it is possible to group the experimental material or conditions in a manner such that the variation among experimental units within a group is less than the variation would have been without grouping, this should be done in order to compare treatments on the less variable material or under less variable conditions. The second illustrative example in the introduction illustrates this point. Suppose that the 4 nutritional treatments were labeled A,B,C, and D. The 4 rats in each litter would be randomly allocated to a treatment. One possible arrangement for design II would be:

Litter number	Rat number and treatment number			
1	1 - B	2 - A	3 - D	4 - C
2	5 - B	6 - C	7 - A	8 - D
3	9 - C	10 - A	11 - B	12 - D
4	13 - A	14 - B	15 - D	16 - C
5	17 - D	18 - C	19 - A	20 - B
6	21 - D	22 - C	23 - A	24 - B
7	25 - B	26 - A	27 - D	28 - C
8	29 - C	30 - B	31 - A	32 - D
9	33 - D	34 - C	35 - A	36 - B
10	37 - D	38 - A	39 - C	40 - B

The rats could be tagged in some manner in order to retain their identity, or they might be housed in individual cages with the cages exposed to as nearly equal environments as possible. If this is not possible, then the 4 caged rats of a litter should be put in one environment, those from a second litter in a second environment, and so forth; then the observed variation among the 10 groups is composed of variation among litters + variation among environments. However, as far as the treatments are concerned, they are compared within a group, and the variation among treatment means is less than it would have been without grouping by litter + environment.

If 10 of the above 40 rats are randomly allocated to the 4 treatments without any regard to parentage, it is possible that an arrangement of the following form could be obtained:

Litter + environment	Rat number and treatment number			
1	1 - A	2 - A	3 - A	4 - A
2	5 - A	6 - A	7 - A	8 - A
3	9 - B	10 - B	11 - B	12 - B
4	13 - B	14 - B	15 - B	16 - B
5	17 - C	18 - C	19 - C	20 - C
6	21 - C	22 - C	23 - C	24 - C
7	25 - D	26 - D	27 - D	28 - D
8	29 - D	30 - D	31 - D	32 - D
9	33 - A	34 - A	35 - B	36 - B
10	37 - C	38 - C	39 - D	40 - D

The difference between the arithmetic means of treatments A and C would be: $\bar{y}_{A.} - \bar{y}_{C.}$ = (effect of treatment A + effects of litters + environments 1,2, and one-half of 9) - (effect of treatment C + effects of litters + environments 5,6, and one-half of 10). In this case, it cannot be determined whether the difference is due to differences in treatments, in litters, or in environments. Such a complete mixing of effects is known as complete confounding. The difference between the means of treatments A and B is unconfounded or unmixed in only one, number 9, of the 10 litter + environment groups. Such an arrangement results in a partial mixing or partial confounding of responses of effects. The lack of confounding such as in the previous designs leads to informative and less variable treatment responses.

As a second example, suppose that the relative effectiveness of 9 different herbicides in eliminating dandelions from home lawns is to be tested, and that 12 different lawns have been selected for the investigation; the 12 lawns are relatively uniform in topography, grass cover, and dandelion infestation. Each lawn forms a relatively uniform block of land. (It was for situations like this that the randomized complete block design was first described, used, and named by Sir Ronald A. Fisher.) Each of the 12 blocks, or lawns, is divided into 9 experimental units which are as alike as possible. Then, the 9 treatments are randomly allocated to the 9 experimental units in each of the 12 blocks or lawns, with the numbers 1,2,···,9 representing the treatments. One possible arrangement is shown in figure VI.2. The characteristic to be measured is number of dandelions in the plot or experimental unit, at monthly intervals, up to one year after an application with the herbicide. In the above experiment there could be considerable variation in dandelion count among the 12 lawns, but this would not affect the differences between treatments, since all treatments are compared with each other on each of the 12 lawns.

If the treatments had been randomly allocated to the 12 × 9 = 108 experimental units or plots, and if the lawns differed in dandelion count, all 12 plots of some treatments could by chance have been allocated to lawns with a low dandelion count and other treatments to lawns with a high dandelion count. The comparison between treatments would then be mixed up with differences between lawns. In the randomized block design, each of the 9 treatments appears on each of the 12 lawns. Given that lawns differ in dandelion count, this arrangement makes the differences between means less variable than if there had been no stratification or blocking.

The count at a given time ($t = 0,1,2,\cdots,12$ months) may be expressed as the sum of the block and treatment estimated means and an error term minus the overall mean, thus: Count = block mean + treatment mean - overall mean + error, or symbolically as $Y_{ij} = (\bar{y}_{.j}$ = block mean) + ($\bar{y}_{i.}$ = treatment mean) - ($\bar{y}_{..}$ = overall mean) + $e_{ij} = \bar{y}_{..} + (\bar{y}_{.j} - \bar{y}_{..}) + (\bar{y}_{i.} - \bar{y}_{..}) + e_{ij}$.
In terms of the parameters of the population, we may write the above as:

$$\begin{aligned} Y_{ij} &= \mu_{i.} + \mu_{.j} - \mu + \epsilon_{ij} \\ &= \mu + (\mu_{i.} - \mu) + (\mu_{.j} - \mu) + \epsilon_{ij} \\ &= \mu + \tau_i + \beta_j + e_{ij}, \end{aligned}$$

where $\mu = \sum_{i=1}^{v} \mu_{i.}/v$, $\mu_{i.}$ = population mean of treatment i over all blocks or environments in the population, $\mu_{.j}$ = population block or environment mean over all individuals in a population and over the 9 treatments 1,2,3,···,9, $\epsilon_{ij} = Y_{ij} - \mu_{i.} - \mu_{.j} + \mu$ = random error, τ_i = the true effect for treatment i, and β_j = true effect for the jth block. Of course, if these parameters were known, there would be no need to conduct an experiment.

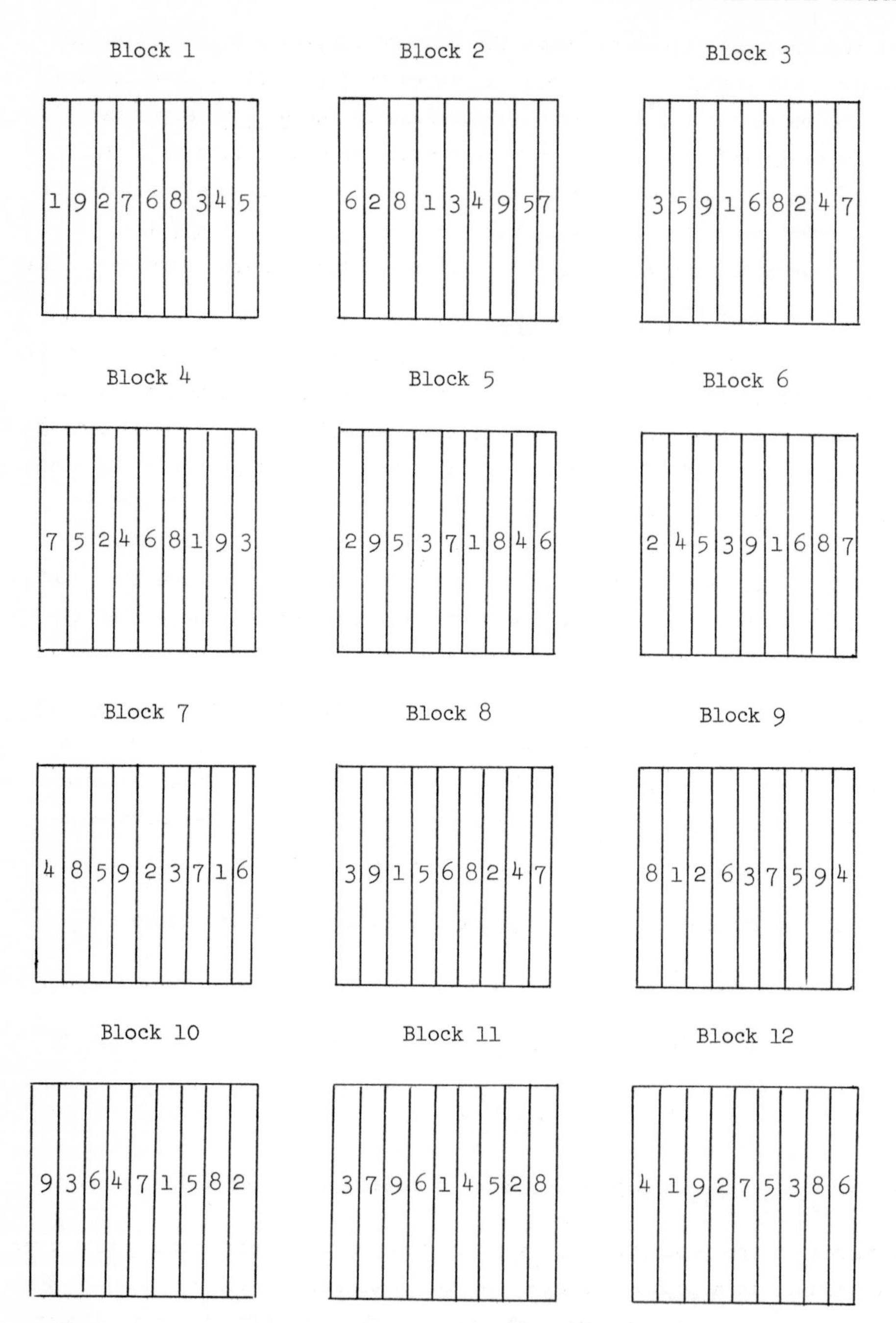

Figure VI.2. Randomized complete block design arrangement for $v = 9$ treatments with $r = 12$ replicates each.

The sum of all counts for two given treatments, say 1 and 2, in the experiment described above is:

treatment one sum + sum of 12 block means - 12 (overall mean)
+ 12 error terms

treatment two sum + sum of 12 block means - 12 (overall mean)
+ 12 other error terms.

The difference between two treatment means is:

$$\frac{1}{12}\{\text{treatment one sum - treatment two sum + 12 error terms - 12 other error terms}\}$$

$$= \bar{y}_{1.} - \bar{y}_{2.} + \frac{1}{12}\sum_{j=1}^{12}(e_{1j} - e_{2j}) ,$$

$$= (\bar{y}_{1.} - \bar{y}_{..}) - (\bar{y}_{2.} - \bar{y}_{..}) + \frac{1}{12}\sum_{j=1}^{12}(e_{1j} - e_{2j}) ,$$

$$= \tau_1 - \tau_2 + \frac{1}{12}\sum_{j=1}^{12}(\epsilon_{1j} - \epsilon_{2j}) ,$$

where $\sum_{j=1}^{12} e_{ij}$ is equal to $e_{i1} + e_{i2} + e_{i3} + e_{i4} + e_{i5} + e_{i6} + e_{i7} + e_{i8} + e_{i9} + e_{i10} + e_{i11} + e_{i12}$. Here we may note that the effects of the true overall mean and of the true block means do not appear in the difference between two treatment means. Since we are comparing herbicide treatments for effectiveness of dandelion control, the differences between means are the statistics of interest. All designs having the property that differences between two arithmetic means for any of the categories do not contain any effects other than the category effects and random errors are known as <u>orthogonal designs</u>. Also, if the differences between arithmetic means of treatments contain only differences due to true treatment effects plus differences of random errors, then the treatment effects are said to be <u>orthogonal</u> to the other sources of variation in the experiment. This is an important property of experimental designs.

The above definition of orthogonality relates to the parameters in an experiment design. An alternate definition of orthogonality is a combinatorial one which is: "If n_{ij} is the number of times that the ith treatment occurs in the jth block and if the ratio $n_{1j}:n_{2j}:n_{3j}:\cdots:n_{vj}$ stays constant for every value $j=1,2,\cdots,r$, then the treatment effects are said to be orthogonal to the block effects." To illustrate this, consider the first example in this section. Each treatment A, B, C, and D occured once in each of the 10 blocks or litters. Hence, $n_{Aj}:n_{Bj}:n_{Cj}:n_{Dj} = 1:1:1:1$ for every value of $j=1,2,\cdots,10=r$. As a second illustration, consider the following design consisting of r = 5 litters of six rats each with v = 3 treatments A, B, and C and with treatment A occurring on three rats in each litter, B occurring on two rats in each litter, and C occurring on one rat in each litter. With the rats randomly allotted to each letter in each litter, the design would be:

Litter Number	Rat number and treatment					
1	1-A	2-B	3-A	4-A	5-B	6-C
2	7-A	8-A	9-A	10-C	11-B	12-B
3	13-B	14-A	15-A	16-A	17-C	18-B
4	19-A	20-A	21-A	22-B	23-C	24-B
5	25-A	26-C	27-B	28-B	29-A	30-A

Thus, $n_{A1} = n_{A2} = n_{A3} = n_{A4} = n_{A5} = 3$; $n_{B1} = n_{B2} = n_{B3} = n_{B4} = n_{B5} = 2$; and $n_{C1} = n_{C2} = n_{C3} = n_{C4} = n_{C5} = 1$. The ratio $n_{Aj}:n_{Bj}:n_{Cj} = 3:2:1$ for every value of $j=1,2,3,4,5$, and hence treatment effects are orthogonal to block effects.

Note that the individual observation in the preceding experiments involving blocks and treatments is assumed to be the sum of four terms. This need not be the case, as some observations may be the product of these terms rather than their sum. The appropriateness of the assumption of additive effects must be questioned for every type of experiment. If the observation is the product of terms instead of the sum, one could use another function of the observations to obtain additive effects. In this case, one could transform the observation to log of observation. Thus, if Y = abcd, then log Y = log a + log B + log c + log d. One might wonder why additivity of effects is desirable. The answer is simply that computations and interpretations are simpler on the additive scale. Despite the desirability of simplicity, it may be necessary to work on a nonadditive scale; this greatly complicates the statistical procedures.

In field and laboratory experimentation on biological material the randomized complete block design is probably the most frequently used experimental design. Ease of construction, layout, and analysis of results contribute heavily to its frequent use. Also, it has been found to be considerably more efficient for the above type of experiment than the completely randomized design. Summarization of several hundred field experiments over a period of years indicates that 6 blocks or replicates of a randomized complete block design are approximately equivalent to 10 replicates of a completely randomized design in attaining the same degree of variability associated with a treatment mean. For field experimentation, the blocking or stratification into blocks reduces the variability among treatment means to six-tenths of the variation without blocking.

The value of blocking material is dependent upon the type of experimental material under consideration. Each type of experimentation requires individual evaluation. One can always block as a form of insurance against heterogeneity, but overstratification results in some disadvantages which will be discussed later. As a rule, one should use the minimum blocking to control the heterogeneity or the suspected heterogeneity present in the experimental material.

As an illustration of the above consideration, suppose that one were interested in only three herbicides instead of nine and that the size of the experimental unit were fixed, in that the lawns were divided into nine plots or experimental units instead of three. One could use blocks of size three and have three blocks per lawn. However, if the nine experimental units were relatively homogeneous, one could use a completely randomized design of three treatments and three replicates on each treatment for each lawn. This would result in minimum blocking which would control the lawn to lawn variability. It should be pointed out, however, that one would probably divide the lawn into thirds and have larger experimental units.

To illustrate another variation on the randomized complete block design, suppose that only five herbicides were of interest with four of these (1, 2, 3, and 4) being of more interest than the fifth one (no. 5), and suppose that nine experimental units were available on each lawn. Treatments 1, 2, 3 and 4 could be included twice on each lawn and treatment 5 could be put in once. If we let numbers 1 and 6 be the plots for treatment 1, numbers 2 and 7 be the plots for treatment 2, numbers 3 and 8 be the plots for treatment 3, numbers 4 and 9 be the plots for treatment 4, and number 5 be the plot for treatment 5 in the original dandelion design, then the arrangement in the first three blocks or lawns would appear as shown in figure VI.3.

Both of the above variations on the randomized complete block design are orthogonal designs. That is, differences between treatment means do not involve the block effects. As long as the orthogonality of block and treatment effects is a property of the design, the analysis remains simple.

As an example of checking orthogonality of effects suppose that we have a randomized complete blocks design of three treatments A, B, and C such that A is included three times, B twice, and C once in each of two blocks. We then express the yields symbolically and in a systemmatic manner as follows:

block 1	block 2
$Y_{A11} = \bar{y}_{...} + t_A + b_1 + e_{A11}$	$Y_{A21} = \bar{y}_{...} + t_A + b_2 + e_{A21}$
$Y_{A12} = \bar{y}_{...} + t_A + b_1 + e_{A12}$	$Y_{A22} = \bar{y}_{...} + t_A + b_2 + e_{A22}$
$Y_{A13} = \bar{y}_{...} + t_A + b_1 + e_{A13}$	$Y_{A23} = \bar{y}_{...} + t_A + b_2 + e_{A23}$
$Y_{B11} = \bar{y}_{...} + t_B + b_1 + e_{B11}$	$Y_{B21} = \bar{y}_{...} + t_B + b_2 + e_{B21}$
$Y_{B12} = \bar{y}_{...} + t_B + b_1 + e_{B12}$	$Y_{B22} = \bar{y}_{...} + t_B + b_2 + e_{B22}$
$Y_{C11} = \bar{y}_{...} + t_C + b_1 + e_{C11}$	$Y_{C21} = \bar{y}_{...} + t_C + b_2 + e_{C21}$

where $t_i = \bar{y}_{i..} - \bar{y}_{...}$, $b_j = \bar{y}_{.j.} - \bar{y}_{...}$, $e_{ijh} = Y_{ijh} - \bar{y}_{i..} - \bar{y}_{.j.} + \bar{y}_{...}$, $\bar{y}_{...}$ = overall mean, $\bar{y}_{i..}$ = mean of ith treatment for i = A, B, and C, and $\bar{y}_{.j.}$ = mean of jth block for j = 1,2. The treatment means are:

$$\bar{y}_{A..} = \bar{y}_{...} + t_A + \tfrac{1}{2}(b_1 + b_2) + \tfrac{1}{6}(e_{A11} + e_{A12} + e_{A13} + e_{A21} + e_{A22} + e_{A23})$$

$$\bar{y}_{B..} = \bar{y}_{...} + t_B + \tfrac{1}{2}(b_1 + b_2) + \tfrac{1}{4}(e_{B11} + e_{B12} + e_{B21} + e_{B22})$$

$$\bar{y}_{C..} = \bar{y}_{...} + t_C + \tfrac{1}{2}(b_1 + b_2) + \tfrac{1}{2}(e_{C11} + e_{C21})$$

where the symbols are as previously defined. Comparing all possible differences among arithmetic means of treatments we note that treatment effects t_i are orthogonal to the $\bar{y}_{...}$ and the b_j effects, and also that the population parameter values may be used in place of the sample values.

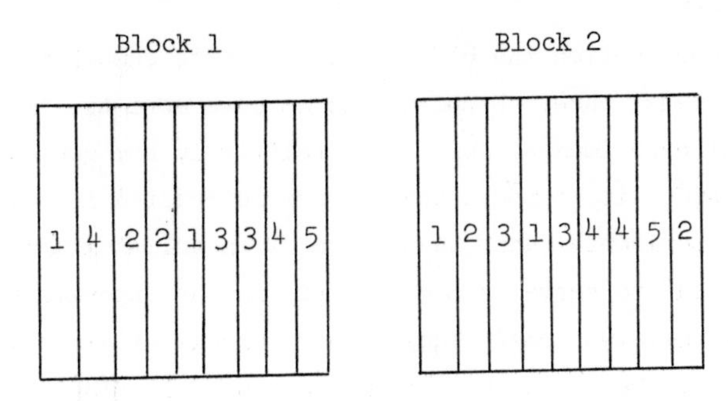

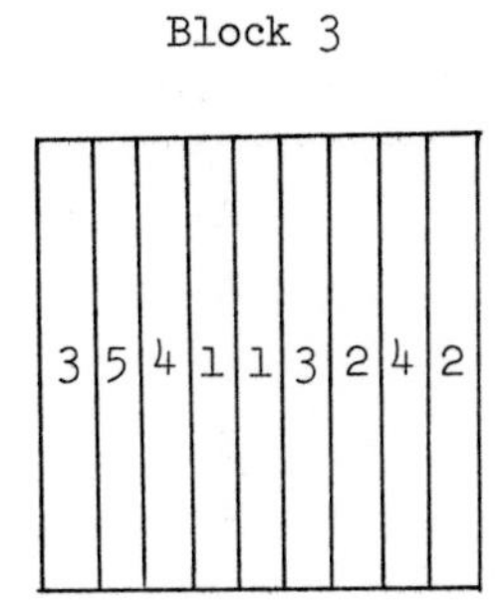

Figure VI.3. Arrangement of $v = 5$ treatments with 6 replicates on treatments 1 to 4 and 3 on number 5 in a randomized complete block design.

VI.4. The Balanced Incomplete Block Design -- One-way Elimination of Heterogeneity

In the previous section the block size was large enough to accommodate all treatments at least once. Suppose that we have more treatments to test than the number of experimental units in a relatively homogeneous block of material. To be specific, suppose that we are interested in five mosquito repellents; the experimental unit is the arm of an enlisted army private who has "volunteered" to serve as a guinea pig in the experiment. Obviously enlisted men have only two arms! Thus we have blocks of size two, but there are five treatments. We solve this dilemma by the following scheme, as "volunteers are plentiful":

Block number or "volunteer"	Left arm	Right arm
1	A	B
2	A	C
3	D	A
4	E	A
5	B	C
6	B	D
7	E	B
8	C	D
9	C	E
10	D	E

where the five treatments are numbered A, B, C, D, and E. Let us suppose that the left arm is no more or less susceptible to mosquitoes than the right arm. (If not, then we could include the mirror images of each of the above pairs on another 10 "volunteers".) Then the designation of right or left arm is omitted. The experiment is conducted as follows: There are 10 pairs of cages each containing the same number of mosquitoes. Each of the 10 "volunteers" is randomly allocated a block number or pair number from 1 to 10; this amounts to randomly assigning a pair of treatments to the "volunteer". Then a coin is flipped to determine which member of the pair of treatments falls on the left arm and which on the right arm. The specified treatment is applied to the arm of the "volunteer" who puts the left arm into one cage and the right arm into a second cage of hungry mosquitoes!

The arms are left in the cage for a specified length of time, say one hour. The number of landings and bites by hungry mosquitoes are recorded.

It should be noted that we were fair in allocating a pair of treatments to each volunteer and the arms to the members of the pairs. This experimental design is fair in another respect. It should be noted that every treatment occurs an equal number of times, once in this case, with every other treatment in one of the 10 blocks. To check this, note that treatment D, for example, appears with A in block 3, with B in block 6, with C in block 8, and with E in block 10. This same type of relationship holds for any other treatment.

For designs such as the one above, statisticians have coined a special name. If there are v treatments in b blocks of size k, and if every pair of treatments occurs together the same number of times, λ, in the b blocks, the design is called a balanced incomplete block design = bibd. When v = k and the treatment pairs occur together once in each block, we have a randomized complete block design = rcbd. Thus we see that the rcbd is a special case of the bibd. The bib designs are balanced but are not in general orthogonal. The statistical computations are more complicated than those for the rcbd, but they still are not difficult.

For blocks of size two and for v > 2 treatments, this design is known in some literature citations as a paired comparison design. A listing of bib designs for v treatments in b blocks of size k may be found in several places (for example, Cochran and Cox [1957], Cox [1958], and Federer [1955]).

As a second example, suppose that we wish to compare v = four brands of shoes. In order to be fair to each brand of shoes, we select a two-footed subject, the block, and we have him wear a left shoe of Brand A, say, and a right shoe of Brand B for a specified period, say six weeks. Then the shoes are reversed in that the second member of the pair of Brand A is worn on the right foot, and Brand B on the left; these are also worn for six weeks. The wear measurements are recorded for both pairs of shoes. Six subjects are required to obtain a bib design. The design is:

	Treatment			
Subject	A	B	C	D
1	x	x		
2	x		x	
3	x			x
4		x	x	
5		x		x
6			x	x

If 12 subjects were available, the above design could be repeated or could be conducted in one half the time by letting a second individual wear the remaining left and right shoes of two pairs. The randomization scheme described for the first example should be followed here also.

The design with blocks of two units may be used in many types of investigations, since there are many instances where blocks of size two occur naturally. Besides the two arms and the two legs of an individual the following come to mind:

1. identical twins
2. opposite leaves on a plant
3. roasts and other cuts of meat from opposite sides of an animal
4. double-yolked eggs
5. opposite halves of a leaf
6. opposite halves of a fruit
7. two eyes, eyelids, ears, etc. of an individual.

The natural pairing of entities in the biological world is of importance in designing experiments so that treatments are compared on relatively homogeneous material.

An example of a bib design was obtained from J. C. Moyer, Geneva Experiment Station. He was interested in comparing the flavor of juice obtained from hand-harvested and from mechanically harvested grapes held for various periods of time. Five treatments were used. To evaluate the flavor of the juice, a panel of judges was available. Since grape juice is rather tart, each judge could taste no more than three, or at most four samples, at one sitting. If four samples could be tasted, a bibd for $v = 5$, $k = 4$, $b = 5$, $\lambda = 3$, and $r = 4$ replicates is:

Blocks = Judges				
1	2	3	4	5
1	1	1	1	2
2	2	2	3	3
3	3	4	4	4
4	5	5	5	5

However, if only three samples could be tasted at one sitting, then blocks of size three must be utilized. A bibd for this case is:

Blocks = Judges									
1	2	3	4	5	6	7	8	9	10
1	1	1	1	1	1	2	2	2	3
2	2	2	3	3	4	3	3	4	4
3	4	5	4	5	5	4	5	5	5

Here $v = 5$, $b = 10$, $\lambda = 3$, and $r = 6$ replicates on each treatment.

It was decided to use the above design. The five treatments for one series of experiments were:

1 = mechanical harvest; held for 5 hours
2 = hand harvest; held for 0 hours
3 = hand harvest; held for 2 hours
4 = hand harvest; held for 12 hours
5 = hand harvest; held for 21 hours.

Each judge scored the grape juice samples from 1,2,3,···,10, with 1 being at the top of the scale and 10 at the bottom of the scale relative to desirable taste. The results of one of these experiments are given below:

	Score of block = panelist or judge										
Treatment	1	2	3	4	5	6	7	8	9	10	Sum
1	5			1	8		8	7		4	33
2		2		4		3	3		5	7	24
3	9	6	9	9	9	8					50
4			3		6	1		3	4	8	25
5	8	9	8				10	10	10		55
sum	22	17	20	14	23	12	21	20	19	19	187

To determine the effect on flavor of the resulting grape juice, 34 such sets of data as the above were obtained for different times and methods of harvesting grapes. Many other examples may be found in published literature.

The estimated treatment means in a balanced incomplete block design are not simply the arithmetic averages. This is because there is nonorthogonality between the incomplete block and the treatment effects. The formula for computing the treatment mean adjusted for block effects is treatment mean adjusted = k (sum of all observations for the treatment - sum of the means of the blocks in which the treatment occurred)/(kr - r + λ) + $\bar{y}$. For treatment 1 above the adjusted mean is computed as:

$$\frac{3(5 + 1 + 8 + 8 + 7 + 4) - (22 + 14 + 23 + 21 + 20 + 19)}{3(6) - 6 + 3} + \frac{187}{30}$$

$$= \frac{99 - 119}{15} + \frac{187}{30} = \frac{-20}{15} + \frac{187}{30} = \frac{147}{30} = 4.9$$

whereas the unadjusted arithmetic mean is 33/6 = 5.5. The other adjusted treatment means are similarly computed. The more nonorthogonal a design becomes, the more complex the computation of effects.

VI.5. The Simple Change-over Design -- Two-way Elimination of Heterogeneity

Suppose that one were to compare two merchandising treatments, say two different displays, simultaneously, in stores where the two treatments are on the same counter and the line of traffic moves from left to right past the counter. If the two treatments were identical, then the one in the first position would be purchased more frequently than the one in the second position. In order to be fair to both treatments, say A and B, one could set up the following design for 10 stores (Conditions of this nature were found in an actual marketing experiment.):

	Store number									
Position	1	2	3	4	5	6	7	8	9	10
1	B	B	A	B	B	A	A	A	B	A
2	A	A	B	A	A	B	B	B	A	B

In the above design, both treatments appear in each store once, and each treatment appears five times or one-half of the time in each position. Thus, we have been fair to both treatments, and variation between positions and among stores has been balanced out or controlled. This removal of variation from two sources from the differences between treatment means decreases the variation between treatment means. To illustrate this, let a single observation or measurement be expressed in terms of estimated means or effects, as follows:

$$\begin{aligned}\text{Observation} &= \text{position mean} + \text{store mean} + \text{treatment mean} - 2(\text{overall mean}) + \text{error}\\ &= \bar{y}_{...} + (\bar{y}_{h..} - \bar{y}_{...}) + (\bar{y}_{.i.} - \bar{y}_{...}) + (\bar{y}_{..j} - \bar{y}_{...}) + e_{hij}\\ &= \text{overall mean} + \text{position effect} + \text{store effect} + \text{treatment effect} + \text{error} .\end{aligned}$$

Using this form, the first observation Y_{11B} may be written as

$$Y_{11B} = \bar{y}_{...} + (\bar{y}_{1..} - \bar{y}_{...}) + (\bar{y}_{.1.} - \bar{y}_{...}) + (\bar{y}_{..B} - \bar{y}_{...}) + e_{11B} .$$

The sum of the 10 observations for treatment A is:

$$\begin{aligned}&Y_{21A} + Y_{22A} + Y_{13A} + Y_{24A} + Y_{25A} + Y_{16A} + Y_{17A} + Y_{18A} + Y_{29A} + Y_{10A}\\ &= 10\bar{y}_{...} + 5(\bar{y}_{1..} - \bar{y}_{...}) + 5(\bar{y}_{2..} - \bar{y}_{...}) + \sum_{i=1}^{10} (\bar{y}_{.i.} - \bar{y}_{...})\\ &\quad + 10(\bar{y}_{..A} - \bar{y}_{...}) + \sum_{i=1}^{10} e_{hiA} ,\end{aligned}$$

and the sum of the 10 observations for treatment B is:

$$\begin{aligned}&Y_{11B} + Y_{12B} + Y_{23B} + Y_{14B} + Y_{15B} + Y_{26B} + Y_{27B} + Y_{28B} + Y_{19B} + Y_{20B}\\ &= 10\bar{y}_{...} + 5(\bar{y}_{1..} - \bar{y}_{...}) + 5(\bar{y}_{2..} - \bar{y}_{...}) + \sum_{i=1}^{10} (\bar{y}_{.i.} - \bar{y}_{...})\\ &\quad + 10(\bar{y}_{..B} - \bar{y}_{...}) + \sum_{i=1}^{10} e_{hiB} .\end{aligned}$$

The means are obtained by dividing by 10. Then the difference between the mean of treatment A and the mean of treatment B is:

$$\text{treatment A effect} - \text{treatment B effect} + \frac{1}{10} \sum_{i=1}^{10} e_{hiA} - \frac{1}{10} \sum_{i=1}^{10} e_{hiB} .$$

By the definition of orthogonality given previously, we see that the above design is orthogonal. This design is known as a simple change-over design for two treatments. The schematic plan of the simple change-over design for three treatments (A,B,C) in three rows and in 3s = 12 columns is:

	Columns											
Row	1	2	3	4	5	6	7	8	9	10	11	12
1	A	A	A	A	B	B	B	B	C	C	C	C
2	C	C	C	C	A	A	A	A	B	B	B	B
3	B	B	B	B	C	C	C	C	A	A	A	A

The design for v treatments in v rows and vs columns may be constructed in a manner similar to that described for two and for three treatments. Simple change-over designs may be used in many situations. For example, suppose that we wish to compare v foods in a cafeteria line in v different positions on vs different days. As another example, consider the comparison of v programs or subjects at v different hours of the day in vs schools (teachers, years, etc.).

In the simple change-over design, all treatments appear once in each column and s times in each row. To randomize a given plan, allot the letters (treatments) to the first row in the same manner as for a completely randomized design. Since each letter must appear once in each column, this completes the randomization for two treatments, and we simply write in the remaining letter in the second row. For three treatments, randomly allot the letters to the 3s different positions such that each letter appears s times in each of the first two rows and no letter appears more than once in a column; treatments in the third row are inserted so that all treatments appear once in each column. For more than v = 3 treatments, simply extend the above process.

VI.6. The Latin Square Design -- Two-way Elimination of Heterogeneity

The latin square is a plan of k rows and k columns of a square with k symbols arranged such that each symbol appears once in each row and once in each column. If the symbols are Latin letters we could, as Sir Ronald A. Fisher did, call this a Latin square plan. If the symbols used were Greek letters we could call the plan a Greek square. If the symbols used were Arabic symbols we could call the plan an Arabic square, and so on. By common usage, this plan is used with Latin letters and when properly randomized is called a latin square design. Furthermore, this design for the removal of row and column variation from treatment differences controls variation from two sources and not necessarily from rows and columns. The row and column designation merely refers to the two sources. To illustrate, suppose that three different pie recipes represent the treatments, that a large oven is available for baking the nine pies simultaneously, and that the treatments are arranged in the oven as follows:

C	B	A
B	A	C
A	C	B

The above design controls variation in two directions. Now suppose that only one pie can be baked at one time, that three pies can be baked on a given day, and that the order of baking has an effect. Then the following plan of baking would be useful:

	Day								
	1 order			2 order			3 order		
	1	2	3	1	2	3	1	2	3
Treatment	C	B	A	B	A	C	A	C	B

which when rearranged looks "like a square":

Order of baking	Day 1	2	3
1	C	B	A
2	B	A	C
3	A	C	B

Each treatment appears once on each day and once in each order of baking.

Considerable use has been made of the latin square design for studying merchandising innovations as they affect the sale of grocery store products. We shall consider an experiment (Dominick [1952]) involving the following four treatments on McIntosh apples:

A = regular apples

B = apples $2\frac{1}{4}$ inch in diameter

C = apples $2\frac{1}{2}$ inch in diameter, carefully selected

D = apples $2\frac{1}{2}$ inch in diameter, highly colored and uniform.

Four stores from the same chain of grocery stores were to be used. This is often a necessity for experiments of this type in order to disentangle management factors and building arrangement factors from their differential effects on the treatments. Stores from different chains may introduce difficulties in assessing treatment responses; it is therefore preferable to run the experiment on stores of a single chain. The stores are the columns and the first four days of the week are the rows in the following latin square design:

	Store			
Day of week	1	2	3	4
Monday	A	B	C	D
Tuesday	B	A	D	C
Wednesday	D	C	B	A
Thursday	C	D	A	B

It was felt that this experiment was too small to estimate properly the differences in sales of apples per 100 customers for the four treatments. Therefore, four latin square designs for these four treatments were used as follows:

	Week 1				Week 2			
Day or part of day	Store				Store			
First part of week	1	2	3	4	1	2	3	4
Monday	A	B	C	D	B	D	C	A
Tuesday	B	A	D	C	D	A	B	C
Wednesday	D	C	B	A	A	C	D	B
Thursday	C	D	A	B	C	B	A	D
Second part of week								
Friday a.m.	B	A	D	C	D	C	B	A
Friday p.m.	C	D	B	A	B	A	C	D
Saturday a.m.	D	C	A	B	C	D	A	B
Saturday p.m.	A	B	C	D	A	B	D	C

The week was divided into two parts as described above, because the sale of apples in the first four days of the week was approximately equal to the sale in the last two days of the week. Friday and Saturday were split to equalize sales in the two parts of a day as nearly as possible. Also, since the purchase of apples is generally not a daily but rather a weekly event, the purchase of apples from a given treatment on Monday, for instance, would not affect sales of apples during the rest of the week; that is, the purchase of apples from any given treatment would eliminate the purchase of apples by that customer for another week. If the purchase of apples one week were to affect the purchase of apples the following week, all treatments would be affected equally. Also, the shopping habits of the customers from the four experimental stores were similar with respect to such variables as frequency of shopping, volume of purchases, and proportion and number of customers per day.

A possible randomization procedure for the above design is to

1. construct a latin square,
2. randomly allot the letters to the treatments,
3. randomly allot the column numbers to stores, and
4. randomly allot the row number to the days.

To illustrate, suppose that we have three treatments. We number three circular tags of the same size and shape as 1, 2, and 3 and place these into a hat. Let our square be

1	2	3
3	1	2
2	3	1

. Shake the hat with the tags, draw out one tag blindly, and assign that number to the first treatment; draw a second number from the hat and let that be the second treatment number; the remaining number in the hat is assigned to the third treatment. Put the tags back into the hat, shake, and again draw out the three tags, for example, 2,1,3 which is the allotment of the stores to the columns; do likewise for the rows, such as 3,1,2. Following the last two steps, we have:

	Store		
Days	2	1	3
3	1	2	3
1	3	1	2
2	2	3	1

to produce the plan

	Store		
Days	1	2	3
1	1	3	2
2	3	2	1
3	2	1	3

The observation or measurement in the orthogonal latin square design is of the same form as the simple change-over design; that is

$$\begin{aligned}\text{Observation} &= \text{overall mean} + \text{row effect} + \text{column effect} + \text{treatment effect} \\ &\quad + \text{error} \\ &= \bar{y}_{...} + (\bar{y}_{h..} - \bar{y}_{...}) + (\bar{y}_{.i.} - \bar{y}_{...}) + (\bar{y}_{..j} - \bar{y}_{...}) + e_{hij} \\ &= \bar{y}_{...} + r_h + c_i + t_j + e_{hij}\end{aligned}$$

where $\bar{y}_{...}$ is the mean of all observations, $\bar{y}_{h..}$ is the mean of the hth row, $\bar{y}_{.i.}$ is the mean of the ith column, $\bar{y}_{..j}$ is the mean of the jth treatment, and where each term in one equation is replaced by its alternative form in the other equations. Special analyses are necessary when the above form of additivity of effects does not hold.

As another illustrative example, the Georgia peach squeezing experiment in problem III.2 could have been set up as follows:

Group of peaches	Order of measurement			
	1	2	3	4
1 (250 peaches)	left front	left back	right back	right front
2 (250 peaches)	right front	left front	left back	right back
3 (250 peaches)	right back	right front	left front	left back
4 (250 peaches)	left back	right back	right front	left front

The order of performing the measurements would be orthogonal to the treatments (quarters of a peach) in this design, whereas in the design used the treatment effects and the order of performing the measurement effects were completely confounded or mixed-up.

W. J. Youden, formerly of the National Bureau of Standards, presented a lecture entitled "How statistics improves physical, chemical, and engineering measurement" to United States Department of Agriculture personnel on 12/14/49. In his lecture he gave many examples in various fields of the use of the latin square design. The following is an excerpt from a mimeographed copy of his lecture:[1]

> I am going to stop here in my discussion of how to estimate errors. Everyone is much more interested in how you reduce them. This is much

[1] With the permission of W. J. Youden.

more challenging; and really, we are more useful as statisticians, I think, at this phase of the work.

I would not dare to claim that statisticians will help reduce the error of measurements if I was not fortified by my own personal experience and by the experience of scientists I have worked with on their projects. Let me enter this phase of it with a momentary digression and tell you about a farmer who had four sons. He offered a prize to that son who got the best yield with some crop. The boys entered this contest with enthusiasm; but when the farmer had set aside a field for this contest, a question came up immediately. How should they divide the field to make sure the various portions allotted were as closely equal as possible in their fertility?

Suppose we divide the field into a checkerboard by marking off 4 horizontal strips with 4 vertical strips. That gives us 16 plots or 4 plots for each boy. Something like this would do:

A	B	C	D
C	D	A	B
D	C	B	A
B	A	D	C

This is an attempt to make sure that each of the 4 boys has a fair sample of the field. Son A gets 4 plots and samples every vertical and every horizontal strip. This is also true for every other boy. After the harvest is in, each boy takes the average yield of the 4 plots assigned to him. A scientist would immediately ask if the difference between the averages for the boys are great enough to indicate a real difference in farming ability for the 4 boys. Suppose I defer the answer to that for a moment. I will only tell you that this particular arrangement is very widely used in experimental agriculture. It is very successful in reducing the error of the comparisons in spite of the fact that there is something artificial and arbitrary, something almost hopeful, in the idea that the fertility can be considered to go by strips.

For the rest of the talk I am going to show you that this same arrangement is even more successful in the physical, chemical, and engineering laboratories. Indeed, I think I was one of the first to take it out of the field and bring it in as far as the greenhouse. I found pathologists were studying tobacco mosaic virus, and, in order to compare the toxicity of different solutions, they would smear the virus solutions over the leaves of tobacco plants. In 3 or 4 days little spots came out on the leaves; the stronger the virus, the more spots. To compare the solutions then, they would smear them on the leaves and count the spots.

For some reason or other most of these tobacco plants were grown to the point where they had about five leaves. By smearing the same solution on all leaves, for several plants, it was revealed immediately that there were certain natural groupings. The leaves from the same plant, as might well be expected, had a common quality of susceptibility to the production of spots. Another plant would be resistant; all the leaves on that would give smaller counts. The total count on the five leaves of one plant might be one-fifth or one-third what it was on another plant. But even more striking was the fact that there was a positional effect. The top leaves tended to be alike (as did the second, the third, and the fourth, and the bottom leaves) in the sense that all the top leaves might give about half the count of their corresponding bottom leaves from the same plant.

Here, then, you see the familiar rows and columns made to order. Nothing hopeful about this regularity, it is there. And to compare five virus solutions, we will simply make sure, if we label them A, B, C, D, E, that they are allotted to the leaves in the same kind of pattern I had a moment ago for the farmer's sons.

Leaf Position	Plant Number				
	1	2	3	4	5
Top	A	B	C	D	E
2nd	B	E	D	C	A
3rd	C	A	E	B	D
4th	D	C	A	E	B
5th	E	D	B	A	C

Note the arrangement of the 5 letters: all 5 on every plant, all 5 in each leaf position. The net result of this was to so improve the accuracy of the comparisons, that it is quite conservative to say it was like presenting the pathologist with an extra greenhouse. He did not need to test as many plants with each solution. This was a case where these strips really paid off.

. . .

Now I am going to just briefly run over some other cases where this same design -- which is called the latin square because Latin letters are used in it -- has been used.

It is being used in the measurement of standards of radioactivity and in rating the samples that are sent in to be compared with those standards. These measurements are made by first putting the known standard in front of a Geiger Counter and getting a count, and then in turn placing the unknowns and getting counts, and comparing these counts. It takes a certain time to make these measurements. During this time the voltage changes, and conditions change. That's one of the troubles in doing experiments. If a standard and three unknowns are each measured four times the familiar arrangement in a latin square makes it possible to consider each column as a period of time. The rows, which correspond to the 1st, 2nd, 3rd, and 4th measurement in each time period correspond to positional effect on the enamel panels. This is a precision type of measurement.

Another case that is rather interesting comes from physical chemistry where they were comparing sources of temperature and have some standard cells which will set up a temperature with tremendous faithfulness, probably even to four decimal places. It stretches the best thermometers in the land to the uttermost to try to detect differences among these cells. They want to know whether they can make a series of cells that are really all alike. One trouble is that the resistance thermometer has to be married to a cell for a whole day to come to equilibrium. So to compare two cells using the same thermometer you must make measurements on successive days. Or if you want to compare them on the same day, you must use two different thermometers. Another study revealed that there were day to day effects, for example, comparing two cells using the same thermometer on different days brought in an error from the different days. When they wanted to compare two cells and avoid this error by doing both measurements on the same day, they had to use two different thermometers. Then they had to take somebody's word for it that the thermometers were the same. They were right at that borderline where they were making such precise measurements that this assumption seemed to be questionable. Let cell I, cell II, cell III, and cell IV correspond to rows in the latin square and thermometer A, thermometer B, thermometer C, thermometer D refer to the columns. On the first day every cell gets a thermometer. The assignment of thermometers to cells for each day is shown in the latin square.

Cell No.	Thermometer A	B	C	D
I	1	2	3	4
II	3	4	1	2
III	2	3	4	1
IV	4	1	2	3

The numbers refer to days. We are not interested so much in the difference between days. They appear where treatments usually appear in an agricultural design. In agricultural work we are not interested in the differences in fertility between the horizontal strips and the vertical strips; but in this experiment it is the differences among the row averages and the differences among the column averages in which we are interested. Now we want to remove from the problem the difference between days. This design does it.

The latin square has also been used in studies of electroplating where the rows and columns correspond to positions on the metal plates and the letters to different laboratories (in an interlaboratory test). It has been used in comparing makes of tires, and this is another case where it is a natural. Let us take four makes of tires from each manufacturer, and take four automobiles. How are you going to compare them fairly? You should not assign all 4 tires of one make to one automobile because there are differences in drivers. It is so easy to set up this same latin square arrangement. Let each column equal an automobile, each row a wheel position, each letter a tire manufacturer.

The latin square was used in the worst possible way once in studying 8th Air Force Bombing. Rows became targets, and columns represented the order in which the bomb groups went over the target, and the entries in the square were the bomb groups themselves. We learned about bombing that way.

VI.7. Latin Rectangle Designs -- Two-way Elimination of Heterogeneity

In a marketing investigation by Dr. Max E. Brunk, Cornell University, seven different merchandising practices were studied to determine their effects on sales of sweet corn. When the experiment was started, it was not known how long the sweet corn season would last, although it was fairly certain to last at least four weeks. It was therefore necessary to set up a design such that as much balance as possible would be retained if some of the rows in a 7 × 7 latin square design were to be deleted or added. The following design was used:

	Store						
Week	1	2	3	4	5	6	7
1	E	F	G	A	B	C	D
2	B	C	D	E	F	G	A
3	D	E	F	G	A	B	C
4	F	G	A	B	C	D	E
5	G	A	B	C	D	E	F
6	A	B	C	D	E	F	G
7	C	D	E	F	G	A	B
8	A	B	C	D	E	F	G

Every treatment (A,B,C,D,E,F,G) appears once in each row. Hence treatments and rows are orthogonal. If seven weeks are used, the design is an ordinary latin square design, and rows, columns, and treatments are all orthogonal to each other. If the design ends at four weeks, columns and treatments are not orthogonal, but it should be noted that the design of $v = 7$ treatments in $b = 7$ blocks of size $k = 4$ forms a bib design with each treatment pair occurring together twice in the seven blocks. Thus, a latin rectangle design of 4 rows by 7 columns can be obtained from the first four rows of the above plan; the treatments and row effects are orthogonal, the row and column effects are orthogonal, and the treatment and column effects are associated in the same manner as in a bib design when the treatment pairs occur together in the columns the same number of times. Such a design is dubbed the <u>Youden square</u> after W. J. Youden who created a number of these designs while working at the Boyce Thompson Institute. This is one of two experimental desgins bearing the name of a man. The other is a special type of latin square called the Knut Vik square. It appears that the influence of Sir Ronald A. Fisher is responsible for the names in both cases.

Likewise, the first six rows also form a Youden square such that treatment pairs occur together in the columns five times. The deletion of any row of a k × k latin square design produces a Youden square design. Also, the addition of any row of the square to the latin square produces a design with the same balanced properties as the Youden square.

Several additional types of latin squares and latin rectangles have been constructed and are available in published literature. The grouping of treatments in rows and/or columns may not be balanced, resulting in nonorthogonality of effects. Consequently, the statistical analyses will be more complicated than those for orthogonal designs such as the latin square and the simple change-over. The requirements of the experiment, not the ease of analysis, determine the appropriate design.

The equation for yield of an observation and the randomization procedure follows that for the latin square design. It should be noted that the simple change-over design is also a latin rectangle design.

VI.8. Latin Cube Designs -- Three-way Elminiation of Heterogeneity

Although latin squares were first studied in the latter part of the eighteenth century by Professor L. Euler, the famous Swiss mathematician, the concept of latin cube arrangements is rather recent. K. Kishen, Indian statistician, and Sir Ronald A. Fisher independently presented latin cube designs in the early 1940's. A practical use for these designs has been found by M. E. Brunk and his co-workers at Cornell University in setting up a series of latin square designs. The following design for three merchandising treatments (A,B,C) was set up in three stores in Ithaca, New York:

x_3 = Day of week	Week = x_1								
	1			2			3		
	store = x_2			store = x_2			store = x_2		
	1	2	3	1	2	3	1	2	3
Monday	A	B	C	B	C	A	C	A	B
Tuesday	B	C	A	C	A	B	A	B	C
Wednesday	C	A	B	A	B	C	B	C	A

If we set up a design with the three axes representing a cube, we obtain the diagram shown in figure VI.4. It should be noted that a plane perpendicular to any axis results in a latin square design, and so in the above, the design in each store, the design on any day of the week, and the design in any week each forms a latin square. The three sources of variation removed from the differences between treatment means are days of the week, weeks, and stores. The statistical analyses for latin cube designs have been partially worked out; the analysis can take many forms depending upon the purposes of the experiment. An exhaustive evaluation of all analyses has not yet been made in published literature.

The latin cube design has many practical applications. A study on rotating practices in field experiments has been designed in a $5 \times 5 \times 5$ latin cube design for a soil conservation study. The 4^3 latin design has been used in a merchandising study on paper products for which several statistical analyses were developed.

The full randomization procedure has not been studied, but an approximate randomization procedure after the construction of a latin cube is the random allotment of the letters to the treatments, of the weeks to the planes on the x_1 axis, of the stores to the planes on the x_2 axis, and of the days to the planes on the x_3 axis.

VI.9. <u>Magic Latin Square Design -- Three-way Elimination of Heterogeneity</u>

In addition to latin square plans, Professor L. Euler discussed the magic latin square plan. The 4×4 and 6×6 magic latin squares are:

A	B	C	D
C	D	A	B
B	A	D	C
D	C	B	A

A	B	C	D	E	F
D	E	F	A	B	C
C	A	B	E	F	D
F	D	E	C	A	B
B	C	A	F	D	E
E	F	D	B	C	A

In the above 4×4 square all four treatments (A,B,C,D) appear once in each row, once in each column, and once in each 2×2 square. In the above 6×6 square all six treatments (A,B,C,D,E,F) appear once in each row, column, and 2×3 rectangle. The variation controlled is row, column, and the 2×2 and 2×3 rectangles in the above designs. The randomization procedure and

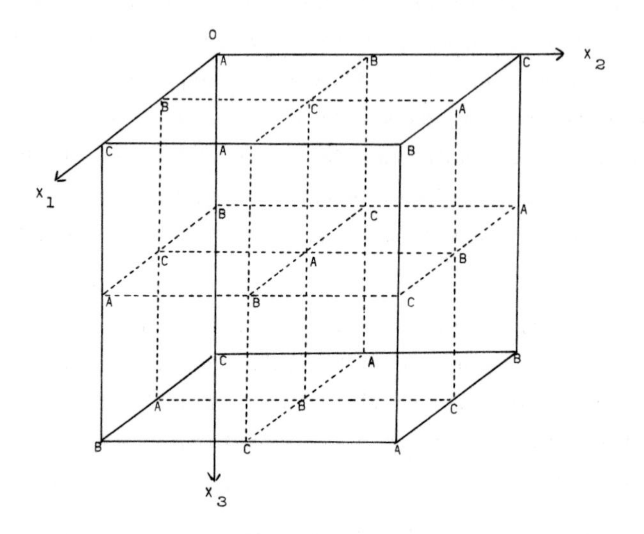

Diagrammatic representation

I			II			III		
A	B	C	B	C	A	C	A	B
B	C	A	C	A	B	A	B	C
C	A	B	A	B	C	B	C	A

Schematic representation

Figure VI.4. 3 × 3 × 3 Latin cube of first order.

analysis is described by Federerer [1955]. Although the magic latin square design has been utilized in field experiments, one wonders whether it is not a case of over-stratification.

VI.10. Design for Treatments Applied in Sequence When the Treatment Effect Continues into the Next Period

Suppose that three merchandising treatments,

A = a display of 4-pound Polythene bags,

B = a display of 6-pound Polythene bags, and

C = a display of 8-pound Polythene bags,

were used to determine the effect of size of bag on the sale of McIntosh apples in grocery stores. Six stores from a given chain in central New York are selected for the study. The period of observation on sales of apples is one week. If a person purchases two four-pound or one eight-pound bag of apples it is possible that this would affect their purchase of apples during the following week, that is, the effect of a treatment might last for more than the treatment period; this is called a residual effect of the treatment. The sale of apples for a given treatment above or below the mean during the treatment period (the week the treatment is in the store) is the direct effect of the treatment. The following design, known as a double change-over design, is used (see Federer [1955], example XIV-2):

	Store					
Week	1	2	3	4	5	6
1	A	B	C	A	B	C
2	B	C	A	C	A	B
3	C	A	B	B	C	A

In the above it will be noted that treatment B follows A twice and treatment A follows B twice. The same balance is attained for pairs A and C and B and C.

The randomization procedure is to randomly allot the stores to the columns and the letters to the treatments. The rows are not randomized, because the sequence of treatments in each column must be maintained. The statistical analysis is somewhat complicated, since not all effects are orthogonal. The design is useful in many types of experiments; the dairy cow, the patient, the worker, the rat, the hospital, and so on, replace the store category, and the period of treatment replaces the week.

There are several variants of designs for measuring direct and residual effects of treatments. One such design is:

Period	Store 1	2	3
1	A	B	C
2	B	C	A
3	C	A	B
4	A	B	C
5	C	A	B
6	B	C	A
7	A	B	C

The randomization is the same as that for the previous design. The yield equation is of the same form as the latin square design except that is contains an additional term for a residual effect of a treatment.

VI.11. Augmented Designs -- n-way Elimination of Heterogeneity

In certain areas of investigation, an experimenter may desire to use an unequal number of experimental units for each of the treatments. This may be desired because of lack of material or because of the desire to screen out undesirable treatments with minimum effort. For example, the Hawaiian Pineapple Research Institute was interested in soil fumigants as they affected the growth of the pineapple plant. Chemists can produce many, say 400 to 500 new possible soil fumigants each year. Since a satisfactory soil fumigant was already available, and since any fumigant superior to the present one would have to be quite exceptional, the researcher decided that he would allocate only one experimental unit for the trial of any new possible soil fumigant; this was necessary since an ineffective soil fumigant would produce no pineapples and a fumigant that was too strong would kill the plants. The economic loss resulting from plots with no pineapple fruit could not be tolerated on a large scale. Also, the effective dose of a chemical could be determined fairly closely by reference to known levels of a similar chemical compound. Any new soil fumigant that was not as good as the standard or better would be rejected on the basis of its performance from one experimental unit. If it was not rejected, the new soil fumigant was a candidate for further testing and would go into the group that would be entered in the second stage of screening or testing.

Each year, in addition to the 400 to 500 new possible soil fumigants available, the investigator has 2, 3, 4, or more promising new soil fumigants, the standard commercially used soil fumigant (say A), and an extremely effective soil fumigant that was not practically useful (say B). Suppose that the three promising new soil fumigants are labelled C, D, and E, and that the 400 new soil fumigants to be tested are numbered 1,2,···,400. The investigator decides that he needs 20 replicates of A,B,C,D and E, and only one of treatments 1 to 400. He sets up the following schematic arrangement:

Block							
1	2	3	4	5	···	19	20
A	A	A	A	A		A	A
B	B	B	B	B		B	B
C	C	C	C	C		C	C
D	D	D	D	D	···	D	D
E	E	E	E	E		E	E
1	21	41	61	81		361	381
⋮	⋮	⋮	⋮	⋮		⋮	⋮
20	40	60	80	100		380	400

In the blocks of 25 experimental units (plots) and 25 treatments, he randomly allots the letters to the 25 experimental units in each of the 20 blocks. Then he randomly assigns the numbers 1,2,···,400 to the 400 possible new soil fumigants and fills in the remaining 20 plots in blocks of 25 as indicated above. The treatments A,B,C,D, and E, which may be called <u>standard treatments</u>, are arranged as in a randomized complete block design. The name <u>new treatments</u> or the name <u>augmented treatments</u> could be applied to the 400 possible new soil fumigants. Such a design has been dubbed an <u>augmented randomized complete block design</u>. Each block of a standard rcb design has been augmented with the new treatments. (In Hawaiian the word for augmented is "hoonuiaku". Hence, we could, as the author has done (Federer [1955]), call this design a hoonuiaku rcb design.)

If the investigator had wished to control variation in two directions, he could have set up four 5 × 5 augmented latin square designs as follows:

square I ... square IV

A 1-4	B 37-40	C 41-44	D 77-80	E 81-84
B 5-8	C 33-36	D 45-48	E 73-76	A 85-88
C 9-12	D 29-32	E 49-52	A 69-72	B 89-92
D 13-16	E 25-28	A 53-56	B 65-68	C 93-96
E 17-20	A 21-24	B 57-60	C 61-64	D 97-100

A 301-304	B 337-340	C 341-344	D 377-380	E 381-384
D 305-308	E 333-336	A 345-348	B 373-376	C 385-388
B 309-312	C 329-332	D 349-352	E 369-372	A 389-392
E 313-316	A 325-328	B 353-356	C 365-368	D 393-396
C 317-320	D 321-324	E 357-560	A 361-364	B 396-400

Instead of there being only one experimental unit in the row-column intersection, there are five experimental units or plots. The standard treatment is randomly allocated to one of these 5 plots. The latin square design for standard treatments follows that given in section VI.7, and the new treatments are numbered in the same manner as for the augmented rcb design and allocated to the remaining plots.

An alternative method of a schematic layout of an augmented latin square design for 20 replicates of the standard treatments and 300, say, of the new treatments each appearing once, would be as indicated on the following page. This design is not as flexible as the previous one, but it is more efficient in controlling variation among treatment effects. A randomization procedure has been devised for the following design, but the analysis, although not too difficult, has not been published as of this date. All five standards appear once in every row and once in every column, and all other treatments occur only once; such an arrangement leads to a relatively simple analysis.

Any experimental design with a standard set of treatments can be augmented to form an augmented design, for instance the augmented bib design, the augmented latin rectangle design, etc. Furthermore, if some treatments are included once, some are included twice, and some r times, certain augmented designs with fairly simple analyses can be devised. An analysis, though complex, can be developed for any augmented design consisting of v_1 treatments included once, v_2 treatments included twice, v_3 treatments included three

Row	Column 1	2	3	4	5	6	7	8	9	10	11	12	13	14	15	16	17	18	19	20
1	A	B	C	D	E	1	2	3	4	5	6	7	8	9	10	11	12	13	14	15
2	16	A	B	C	D	E	17	18	19	20	21	22	23	24	25	26	27	28	29	30
3	31	32	A	B	C	D	E	33	34	35	36	37	38	39	40	41	42	43	44	45
4	46	47	48	A	B	C	D	E	49	50	51	52	53	54	55	56	57	58	59	60
5	61	62	63	64	A	B	C	D	E	65	66	67	68	69	70	71	72	73	74	75
6	76	77	78	79	80	A	B	C	D	E	81	82	83	84	85	86	87	88	89	90
7	91	92	93	94	95	96	A	B	C	D	E	97	98	99	100	101	102	103	104	105
8	106	107	108	109	110	111	112	A	B	C	D	E	113	114	115	116	117	118	119	120
9	121	122	123	124	125	126	127	128	A	B	C	D	E	129	130	131	132	133	134	135
10	136	137	138	139	140	141	142	143	144	A	B	C	D	E	145	146	147	148	149	150
11	151	152	153	154	155	156	157	158	159	160	A	B	C	D	E	161	162	163	164	165
12	166	167	168	169	170	171	172	173	174	175	176	A	B	C	D	E	177	178	179	180
13	181	182	183	184	185	186	187	188	189	190	191	192	A	B	C	D	E	193	194	195
14	196	197	198	199	200	201	202	203	204	205	206	207	208	A	B	C	D	E	209	210
15	211	212	213	214	215	216	217	218	219	220	221	222	223	224	A	B	C	D	E	225
16	226	227	228	229	230	231	232	233	234	235	236	237	238	239	240	A	B	C	D	E
17	E	241	242	243	244	245	246	247	248	249	250	251	252	253	254	255	A	B	C	D
18	D	E	256	257	258	259	260	261	262	263	264	265	266	267	268	269	270	A	B	C
19	C	D	E	271	272	273	274	275	276	277	278	279	280	281	282	283	284	285	A	B
20	B	C	D	E	286	287	288	289	290	291	292	293	294	295	296	297	298	299	300	A

times, $\cdots$, v_r treatments included r times. Such designs allow considerable freedom in the use of all experimental material in the desired proportions.

Although there are numerous other types of experimental designs created by statisticians, the above should be sufficient to convey to the reader some of the principles for construction and some properties of experimental designs.

VI.12. Summary of Principles of Experimental Design

At the beginning of this chapter we listed three desirable characteristics for designing experiments; we discussed the concepts of randomization, blocking, efficiency, orthogonality, and balancing. Sir Ronald A. Fisher defined and developed these concepts of experimental design. Figure VI.5 is a small adapted replica of one that is said to have hung on the wall of his office at the Rothamsted Experiment Station in England; it illustrates three

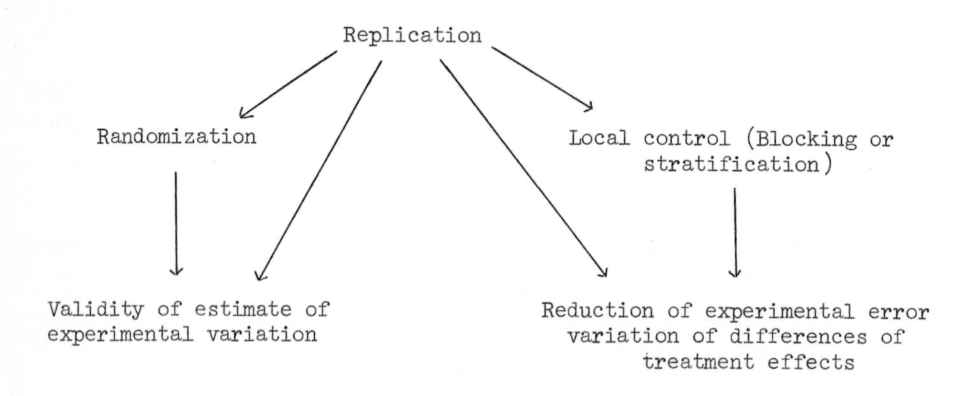

Figure VI.5. Relationships of three basic principles of experimental design.

of the basic principles of design; viz. replication, randomization, and local controls. Randomization and replication are necessary to obtain a valid estimate or measure of the experimental variation. Replication and "local control" (= blocking or grouping) are necessary to achieve a reduction in the random variation among treatment effects in the experiment. Use of "local control" has been made throughout this chapter in blocking or grouping to eliminate or to control the various sources of variation.

We stated that randomization was used in order to be fair to the treatments. If the variation among the experimental units within a group is due entirely to random error with no known method of grouping, then the random allotment of treatments to the experimental units is entirely fair. Any treatment then has an equal chance of receiving the highest or the lowest observation; any other method of assignment would lead to unfairness to some treatments in that there would be discrimination. Also, any device of allotting treatments that tends to make them more alike or more unlike than they would have been by a random allocation leads respectively to larger or smaller estimates of error variation than for the random allotment of treatments. We may summarize the purpose of randomization as follows:

1. to obtain unbiased estimates of differences among treatment responses (means or effects), and
2. to obtain an unbiased estimate of the random error variation in the experiment.

With respect to the latter, a valid estimate of the experimental variation may be obtained if there is sufficient replication and additivity of effects. (See example 5.6, pages 77-8, in Cox [1958].)

There are several procedures for achieving a random allocation of treatments to the experimental units within the block. We have discussed the use of identical and numbered balls or tags in a jar or hat, the use of coins, and the use of a die or a pair of dice. There are also so-called random number tables avaialble from many sources, and some of these were discussed in section V.14.

Replication is the repetition of treatments in different blocks or sources of variation in the experiment. For example, in the randomized complete block design, we need at least two blocks in order to distinguish between the estimated overall mean $\bar{y}_{..}$ and the estimated block effect, say $\bar{y}_{.j} - \bar{y}_{..}$. If there is only one block then $\bar{y}_{.1} = \bar{y}_{..}$. This is true also for treatments in a comparative experiment, since two or more treatments are required in order to compare items. If there is only one treatment, then the estimated treatment mean, say $\bar{y}_{1.}$, and the overall mean $\bar{y}_{..}$ are indistinguishable and the estimated treatment effect $\bar{y}_{1.} - \bar{y}_{..}$ cannot be obtained.

Also, an increase in the number of replicates of a treatment tends to decrease the variation in the estimate of a difference between two treatment means from orthogonal designs. This is the manner in which replication leads to a reduction in the experimental error of differences of treatment effects.

In addition to the three basic principles of experimental design (randomization, replication, and blocking), the fourth principle of orthogonality is important in order to ensure that the estimate of the random variation between treatment means is the same for all pairs of treatments having equal replication and having the same degree or magnitude of random error variation. Statistical analyses are simpler than those for nonorthogonal designs, and orthogonal designs are the most efficient of all designs.

If orthogonal designs are not possible, then we strive for balanced designs which still ensures that differences between pairs of treatment effects all have the same variance. In the balanced designs discussed, all treatment pairs occur with each other equally frequently in the b blocks of size k for the v treatments. Since bk = total number of experimental units and since there are v treatments each repeated r times then bk = vr in balanced designs. Orthogonal designs may be balanced designs, but the reverse may not be true. The randomized complete block design is usually a balanced design, but the balanced incomplete block design is generally not an orthogonal design.

Figure VI.6 illustrates the relationships of six (the above five plus confounding) Fisherian principles of experimental design.

Partial confounding in incomplete block designs may lead to more efficient designs than complete block designs, depending upon the experimental variation and the blocking. The double-headed arrow indicates that blocking and orthogonality are related but that one does not necessarily lead to the other. If we have blocks with an appropriate design, for instance, the rcb

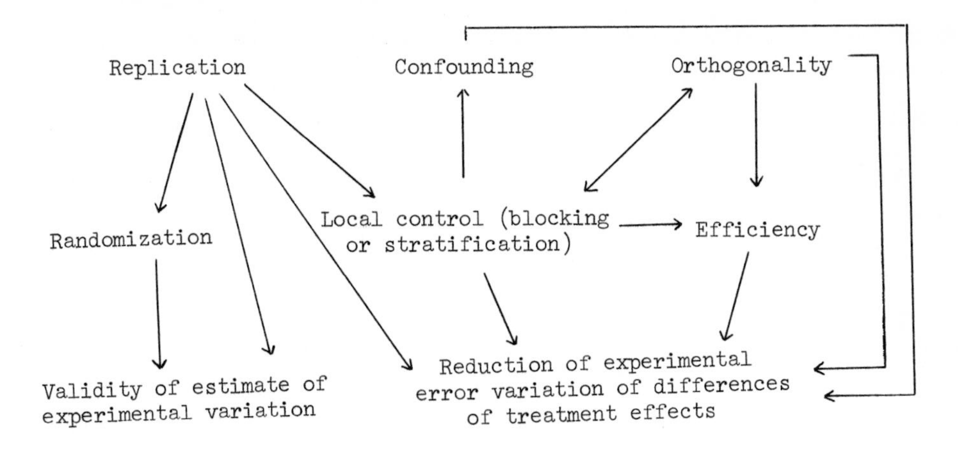

Figure VI.6. Interrelations among six principles of design.

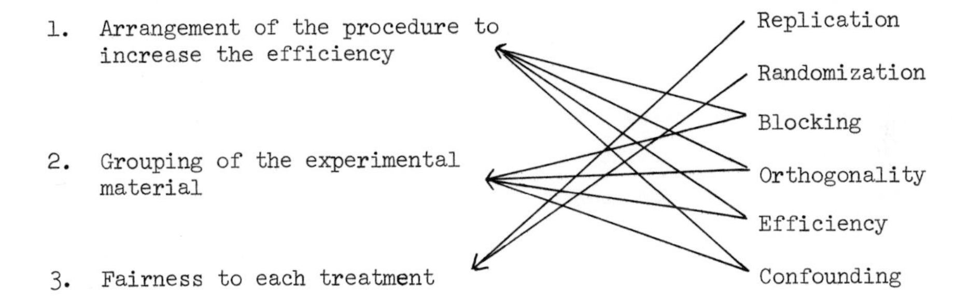

Figure VI.7. Interrelationships of principles of design and three criteria.

design, we can have orthogonality between treatment and block effects.

The interrelationships between the first three criteria listed in the chapter and the six in figure VI.6 may be represented as shown in figure VI.7.

The above discussion is an introduction to properties of experimental designs. Various designs may be compared in order to determine which are the better designs for experimental work.

One of the items to be considered in the fourth principle of scientific inquiry, the design of the investigation, is the selection of the appropriate experimental design. A number of experimental designs have been presented in this chapter to illustrate the kinds of designs available for different types of variation among the experimental units.

VI.13. An Experiment

In problem VI.2 of this chapter you are asked to design an experiment for comparing the speed of two calculating machines in performing a series of calculations, specifically the computation and accumulation of the squares of a set of numbers. The calculation is known as the computation of the "sum of squares" of a set of numbers. In a smaller class or with laboratory sections in the class this experiment could actually be performed with pairs of students. One student can do the computing, and the other student can perform the timing and record the time for calculation. The conduct of this experiment can be demonstrated with only one pair of students to determine which of two calculating machines is faster in performing the stated statistical computations (also, see problem XI.5).

This experiment has been performed routinely in the introductory statistics course at Iowa State University for many years. Such an experiment allows each pair of students a set of data on which to perform a variety of statistical computations. This experiment with a particular set of data is also discussed in chapter 1 of Cochran and Cox [1957] in connection with principles of scientific inquiry.

VI.14. Books on the Design and Analysis of Experiments

There are numerous books on statistical methods or analyses of the Snedecor and Cochran [1967] type, for instance, Steel and Torrie [1960], etc. These books, in general, stress analyses after the data are available rather than methods of procuring the data; examples of the rcbd and latin square designs are given, but the reasons for using these designs receive little or no discussion. The books which stress the purposes, construction, and layout of experimental arrangements or designs and treatment designs and the corresponding analyses are listed below. Unfortunately, all are necessarily written assuming some knowledge of statistical analyses on the part of the reader, the elementary facts of which can be obtained from Bevan [1968], Campbell [1967], McCarthy [1957], Moroney [1956], and Pearce [1965], among others.

Cochran, W. G. and Cox, G. M. [1957]. Experimental Designs, 2nd edition (1st edition, 1950), John Wiley and Sons, Inc., New York, pp. xiv + 617.

(This book requires a fair knowledge of the statistical methods in Snedecor and Cochran [1967]. It is the most comprehensive book on plans for designs; analyses and numerical examples are presented for a number of these.)

Cox, D. R. [1958]. Planning of Experiments, John Wiley and Sons, Inc., New York, London, and Sydney, pp. vii + 308.

(Professor Cox has attempted to avoid statistical and mathematical technicalities. The book is an elementary discussion of experiment and treatment design, with some discussion of the reasons and objectives behind each.)

Davies, O. L. (editor) [1954]. Design and Analysis of Industrial Experiments, Oliver and Boyd, Edinburgh.

(Basically this book is similar to Cochran and Cox in statistical content; it utilizes examples from industry and engineering.)

Federer, W. T. [1955]. Experimental Design -- Theory and Application, The Macmillan Company, New York, pp. xix + 593.

(A somewhat advanced text on the construction, layout, and analyses of types of designs. Emphasis is placed on the reasons for using a design as well as on the statistical analysis. A considerable amount of generality is achieved. Many numerical examples covering a variety of fields are presented to illustrate the various designs and analyses.)

Finney, D. J. [1955]. Experimental Design and Its Statistical Basis, The University of Chicago Press, Chicago and London, pp. xi + 169.

(Professor Finney has written a book at about the same level as Professor Cox, but it is slanted more toward biological topics. The design topics are treated with different emphasis and in a different manner than are those in Cox's book. Both books could be read to advantage.)

Finney, D. J. [1960]. An Introduction to the Theory of Experimental Design, The University of Chicago Press, Chicago and London, pp. xii + 223.

(This book covers a wide range of topics in the design and analysis of treatment and experimental designs. It is more mathematical than the book by D. R. Cox.)

Fisher, R. A. [1926]. The arrangement of field experiments, Journal of the Ministry of Agriculture, volume 33, page 503.

(This is a paper which is listed because it is the first publication on designing experiments using randomization in designs; it forms an excellent introduction to the subject of experimental designs.)

Fisher, R. A. [1935]. The Design of Experiments, 1st edition in 1935 with several subsequent editions. Oliver and Boyd, Edinburgh, pp. xi + 236.

(Nearly all the general ideas in experimental design stem from the pioneer work of Sir Ronald A. Fisher as expressed in his writings, lectures, consultations, informal conversations, and reflections. Fisher's works could profitably be read prior to other accounts of experimental and treatment designs, and again after each book read; this applies even to the most advanced and the most mathematical books. Do not expect to understand all of Fisher at first reading, at second, or even at the third reading. Some of the ideas are complex enough that people are still struggling to understand them.)

Kempthorne, O. [1952]. The Design and Analysis of Experiments, John Wiley and Sons, Inc., New York, pp. ix + 631.

(An advanced and general text on experimental and treatment designs, stressing the theoretical approach to the construction, randomization procedure, and analysis; few numerical illustrations are used.)

Kirk, R. E. [1968]. Experimental Design: Procedures for the Behavioral Sciences, Brooks/Cole Publishing Company, Belmont, California, pp. xii + 577.

(The level of difficulty is about the same as that of Finney [1960]; more emphasis is given to statistical methods, and a smaller range of topics on experimental design is covered.)

Mann, H. B. [1949]. Analysis and Design of Experiments, Dover Publications, Inc., New York, pp. x + 198.

(A highly mathematical and elegant treatment of the subject of design relative to construction and analysis.)

Quenouille, M. H. [1953]. Design and Analysis of Experiments, Griffin, London.

(On about the same level as Cochran and Cox and Davies with a discussion of long term experiments as well as of the more general topics covered by these authors.)

Yates, F. [1937]. The design and analysis of factorial experiments, Imperial Bureau of Soil Science, Technical Communication No. 35, Harpenden, England, pp. 95.

(This pamphlet is a classical writing on the subject of the treatment design known as a factorial. Many writers have taken their material from this excellent publication, which discusses many aspects of experimental design, including as yet unsolved problems.)

If one were to understand the principles and theory contained in the books listed above, one would be considered a competent statistician in the field of experimental and treatment design. One would be able to do research in this field and to work on some of the unsolved problems.

VI.15. Research Work in Design

The fields of treatment design and experimental design are being actively investigated by numerous statisticians and mathematicians. A bibliography of papers published in these fields has been prepared for the period 1950 to 1967; nearly 7000 references are listed therein. Before 1950, nearly 2000 citations relate to work in these fields.

VI.16. Problems

VI.1. Set up a randomized complete blocks design for an experiment involving 5 dental cleaning preparations developed by Dr. Iva Payne. Use the 5 preparations on one patient as the block and use 10 different patients. The characteristic measured is "cleanliness" of teeth after cleaning. Describe the randomization procedure and a method of conducting the experiment.

VI.2. Design an experiment for evaluating fairly the speed of computing sums of squares on two calculating machines, make A and make B. One operator and 10 different sets of numbers will be used. Each of the 10 sets of sums of squares will be computed on each of the machines. We know from past experience that an operator computes a sum of squares for a given set of numbers faster the second time than he does the first time.

VI.3. In section 36 of Sir Ronald A. Fisher's book, entitled *The Design of Experiments*, the following puzzle is given:

Sixteen passengers on a liner discover that they are an exceptionally representative body. Four are Englishmen, four are Scots, four are Irish, and four are Welsh. There are also four each of four different ages, 35, 45, 55, and 65, and no two of the same age are of the same nationality. By profession also four are lawyers, four soldiers, four doctors, and four clergymen, and no two of the same profession are of the same age or of the same nationality.

It appears, also, that four are bachelors, four married, four widowed, and four divorced, and that no two of the same marital status are of the same profession, or the same age, or the same nationality. Finally, four are conservatives, four liberals, four socialists, and four fascists, and no two of the same political sympathies are of the same marital status, or the same profession, or the same age, or the same nationality.

Three of the fascists are known to be an unmarried English lawyer of 65, a married Scots soldier of 55, and a widowed Irish doctor of 45. It is then easy to specify the remaining fascist.

It is further given that the Irish socialist is 35, the conservative of 45 is a Scotsman, and the Englishman of 55 is a clergyman. What do you know of the Welsh lawyer?

For the problem in this class you may answer the above questions or you may answer the simpler problem given below; it is an adaptation of the above puzzle.

Nine passengers on an airliner discover that they are an exceptionally representative group. Three are Englishmen, three are Scots, and three are Irish. By profession, three are lawyers, three are doctors, and three are clergymen. No two of the same profession have the same nationality. Furthermore, three are married, three are widowed, and three are divorced. No two of the same marital status are of the same profession or the same nationality. Finally, three are laborites, three are conservatives, and three are socialists. No two of the same political sympathies are of the same marital status, of the same profession, or of the same nationality.

If the married English clergyman and the widowed Irishman are conservatives, specify the remaining seven individuals if you can. If not, what additional information do you need?

VI.4. An investigator knows that there are two sources of variation inherent in his experimental material. He wishes to compare two treatments and to use 8 experimental units for each treatment. Give three experimental designs for controlling two sources of variation among the 16 experimental units.

VI.5. Construct a schematic arrangement of a balanced incomplete block design for v = 6 treatments in blocks of size k = 4. How many times do the individual pairs of treatments occur together in the b blocks?

VI.6. Construct a schematic arrangement of a balanced incomplete block design for v = 7 treatments in blocks of size k = 4. How many times do the individual pairs of treatments occur together in the b blocks? Repeat but use incomplete blocks of size 3. How many times do the individual pairs of treatments occur together in the b blocks?

VI.7. Write down the linear and additive yield equations using estimated effects for the completely randomized design, the randomized complete block design, and the latin square design. Which elements in each of the yield equations represent variation due to assignable causes, bias, and random error?

VI.8. Devise a 7 X 7 latin square such that the first 3 rows form a Youden square and the last 4 rows also form a Youden square. (See your solution in problem VI.6.)

VI.9. Use the telephone directory to obtain a randomization plan for the dandelion experiment described in section VI.3. Describe your procedure, including the place started.

VI.10. Use the same telephone directory used for problem VI.9 and obtain a randomization procedure for a 5 X 5 latin square design. Describe the procedure you used, including the place in the directory where you started.

VI.11. Conduct a comparative experiment wherein you follow the principles of scientific investigation. Select an experimental and treatment design and take a set of measurements or observations. Some topics for investigation might be to compare v individuals as the treatments and to measure the distance from a target or line that a penny falls when tossed; compare v pairs of individuals in a series of bridge games; compare v different brands of ping-pong balls for height of bounce when dropped from a height of 5 feet; compare v different kinds of seed for time from planting until germination occurs; compare weighings by v individuals as described on page 96 of W. J. Youden's [1962] book; or compare any other v treatments of interest.

VI.17. References and Suggested Reading in Addition to Those Listed in Section VI.14.

Bevan, J. M. [1968]. *Introduction to Statistics*. Philosophical Library, Inc., New York, pp. vii + 220.

Campbell, R. C. [1967]. *Statistics for Biologists*. Cambridge University Press, London and New York, pp. xii + 242.

Dominick, B. A., Jr., [1952]. Merchandising McIntosh apples under controlled conditions -- customer reaction and effect on sales. Ph.D. Thesis, Cornell University.

Federer, W. T. [1956]. Augmented (or hoonuiaku) designs. The Hawaiian Planters' Record 50:191-208.

Jowett, G. H. and Davies, H. M. [1960]. Practical experimentation as a teaching method in statistics, Journal of the Royal Statistical Society, Series A, 123:10-35.

Kahan, B. C. [1961]. A practical demonstration of a needle experiment designed to give a number of concurrent estimates of π. Journal of the Royal Statistical Society, Series A, 124:227-239.

McCarthy, P. J. [1957]. *Introduction to Statistical Reasoning*. McGraw-Hill Book Company, Inc., New York, Toronto, and London, pp. xiii + 402.

Moroney, M. J. [1956]. *Facts From Figures*. 3rd and revised edition. Penguin Books, Ltd., Baltimore, Toronto, Mitcham (Australia), and Harmondsworth (England), pp. iv + 472.

Pearce, S. C. [1965]. *Biological Statistics: An Introduction*. McGraw-Hill Book Company, New York, Toronto and Lodnon, pp. xiii + 212.

Snedecor, G. W. and Cochran, W. G. [1967]. *Statistical Methods*. 6th edition. The Iowa State University Press, Ames, Iowa, pp. xiv + 593.

Steel, R. G. D. and Torrie, J. H. [1960]. *Principles and Procedures of Statistics*. McGraw-Hill Book Company, Inc., New York, Toronto, and London, pp. xvi + 481.

Yates, F. [1935]. Complex experiments. Journal of the Royal Statistical Society, Series B, 2:181-247.

Youden, W. J. [1962]. *Experimentation and Measurement*. National Science Teachers Association and National Bureau of Standards, Scholastic Book Services, New York, pp. 127.

CHAPTER VII. TREATMENT DESIGN AND SELECTION OF CONDITIONS FOR THE EXPERIMENT

VII.1. Introduction

As described under the fourth principle of scientific inquiry, the treatment design or the selection of the treatments in an experiment may be one of the more important aspects of an experimental investigation; in many experiments the success of the experiment is vitally connected to the selection of the treatments. In this chapter we shall discuss

1. the presence of conditions in an experiment as related to a treatment design,
2. the relation of treatment design, the conditions of the experiment, and the population in which inferences are to be made, and
3. a number of treatment designs.

First, we shall consider an article written by E. M. Jellinek, Laboratory of Applied Physiology, Yale University, entitled "Clinical tests on comparative effectiveness of analgesic drugs". The article appeared in Biometrics, volume 2, pages 87-91, 1946. Excerpts of the article are reproduced below for discussion purposes.*

> A headache remedy, designated here as drug A, is composed of ingredients a, b and c. Ingredient b was running short, and the manufacturers wished to know whether or not the efficacy of this drug would be lowered through the omission of this ingredient. In order to answer this question, 200[1] subjects suffering from frequent headaches were to be treated for two weeks on each occurrence of headaches with drug A, two weeks with drug B, which was composed of ingredients a and c, two weeks with drug C containing ingredients a and b, and two weeks with drug D, a placebo consisting of ordinary lactate which is pharmacologically inactive.
>
> The four drugs were made to appear identical in color, shape, size, and taste. Neither the subjects nor the physicians administering the drugs were aware of the differences in the composition of the four drugs. Because of possible progressive sensitization or desensitization to the drugs, they were administered in different sequences as follows:

[1] Actually 199 subjects completed the tests.

* Reprinted with the permission of H. A. David, Editor of Biometrics.

Group I	50 subjects
Group II	49 subjects
Group III	50 subjects
Group IV	50 subjects

A full account of the selection of subjects, type of records kept, instructions and mode of administration as well as psychological implications of the experience with the placebo will be given elsewhere.[2]

The subjects took the tablets whenever a headache occurred. At the end of each two week period they reported to the physician the number of headaches they had in the course of that period and how many of these were relieved satisfactorily by the drug. They also reported the dosage taken on each occasion and the time elapsing between administration of the drug and the onset of relief from pain. Observations on psychological, gastric and heart reactions were noted, too. For each subject his "success rate" for each of the four drugs was computed as follows:

$$\frac{\text{Number of headaches relieved}}{\text{Number of headaches treated in the two week period}}$$

The potency of the drugs is expressed in terms of the arithmetic means of these individual "success rates". The analyses ··· were carried out on the individual rates.

Some subjects had only three headaches in the course of a two week period while others had up to ten attacks in the same period. Thus the individual rates are based on a varying number of headaches. This introduces an undesirable element into the analysis ···, but the great consistency of the data shows that the results may have been affected only to a small degree by this aspect of the tests. In other surveys, however, it may be desirable to stipulate the testing of each drug on four or five occasions rather than during a fixed period of time.

---

[2]To be published in The Journal of Psychology.

First 2 Weeks	Second 2 Weeks	Third 2 Weeks	Fourth 2 Weeks
A	B	C	D
B	A	D	C
C	D	A	B
D	C	B	A

The mean success rates of the three analgesics, A, B, and C, and the placebo, D, compared as follows:

	A	B	C	D
Mean Success Rate	.84	.80	.80	.52

. . .

Inasmuch as placebos have been used at all in clinical tests of drugs, the procedure was to express the efficacy or non-efficacy of a drug in terms of "how much better" the drug was than the placebo. Thus in the present instance it would have been said that drugs, A, B, and C were only "53 to 62 per cent better" than placebo. That such statements are meaningless and misleading will be seen from the further analysis of the data.

The success rate of .52 on placebo was due to 120 out of the 199 subjects. No relief whatever was reported by 79 subjects although they had three to ten headaches treated with placebo. On the other hand, these same 79 subjects when treated with one of the three analgesics reported from one third to all of their headaches relieved. The 120 subjects who reported relief at all through placebo did not do so only on one or two occasions, but rather consistently. The nature of response to placebo is seen best from the distribution of the number of headaches reported as relieved by placebo in subjects who had a constant number of attacks. In table I the distribution of the number of relieved headaches is given for 59 subjects of this study who were exposed on five headache attacks to placebo. The distribution is given also for 121 subjects including these 59 subjects and another 62 subjects from later studies who also had five exposures to placebo on the occasion of headache attacks.

Examples of the rare U shaped distribution are seen here. Thus there are individuals who definitely tend to respond to placebo. This difference in response to placebo must reflect a difference in the nature of headaches. The sample is drawn from at least two broad populations of sufferers from headaches. If subjects never report relief through a pharmacologically inactive substance but always report at least some attacks relieved through bona fide analgesics, it must be assumed that they represent a "pure culture" of physiological headaches not accessible to suggestion, while the 120 subjects who either always or most of the time responded to placebo represent, perhaps predominantly, psychogenic headaches and to some extent also milder physiological headaches coupled with a tendency toward suggestibility.

Table I. Distribution of the Number of Headaches Reported as Relieved by Subjects who had been Treated with Placebo on 5 Attacks of Headaches

Number of Relieved Headaches		Present Study Number of Subjects	Present and Later Studies Number of Subjects
0		22	49
1		1	1
2		5	6
3		7	12
4		8	18
5		16	35
	Total	59	121

Evidently persons suffering from psychological headaches lack the prerequisite condition for discrimination of potency among drugs, as any substance of the appearance of a drug and prescribed or administered by a physician will serve the purpose.

This finding suggested a separate analysis of the "success rates" of the three analgesics on subjects who did not and on subjects who did react to placebo. The mean "success rates" of the analgesics are shown in table II for these two classes of subjects in each of the four groups of different sequences of drug administration.

In spite of the small number of individuals in any of the four groups, the order of mean "success rates" in the class of subjects not reacting to placebo shows great consistency. In each of the four groups drug A occupies the first, drug C the second, and drug B the third place. The mean "success rates" of the entire class of subjects not reacting to placebo suggest definitely the importance of ingredient b which was lacking in drug B and a minor importance of ingredient c which was lacking in drug C as the mean "success rate" of the full formula, A, was much superior to that of B and somewhat superior to that of C.

Table II. Mean "Success Rates" on 3 Analgesic Drugs. 79 Subjects Not Reacting to Placebo and 120 Subjects Reacting to Placebo.

	Subjects not Reacting to Placebo				Subjects Reacting to Placebo			
	Number of Subjects	Drugs			Number of Subjects	Drugs		
		A	B	C		A	B	C
Group No.		"Success Rates"				"Success Rates"		
1	14	.90	.65	.86	36	.76	.87	.83
2	26	.88	.66	.70	23	.76	.84	.87
3	20	.85	.60	.71	30	.89	.85	.76
4	19	.91	.82	.86	31	.86	.90	.83
All groups	79	.88	.67	.77	120	.82	.87	.82

No consistency of mean "success rates" is seen in the class of subjects reacting to placebo; each of the three analgesics occupies first, second, and third places in one or the other of the four groups. As a matter of fact, the placebo, which is not shown in table II, occupied the first place in Group 1 with a mean "success rate" of .89, and for the entire class of 120 subjects the mean "success rate" of placebo was .86.

The sequences in which the drugs were administered had apparently no effect on their "success rates". The highest rates of drugs A occurred when it was the first and the fourth in order of administration. The highest rate of B was seen when it was third in sequence and of C when it was the third and second of the drugs administered in the respective periods.

In the class of subjects reacting to placebo the "drugs" variance was not significant, and it was only a fraction of the corresponding variance in the other class of subjects. The variation of the overall response to drugs was, however, significant even among those subjects who did react to placebo. This probably does not reflect a true response to the drugs but rather a difference between two types of sufferers from psychological headaches, namely, an erratic type who wish to impress the physician through the great variations in their condition, and those whose psychological headaches are not complicated by hypochondrasis. These types can be distinguished from the presence or absence of reports by the subjects on extremely minute detail. The subjects reporting such minute detail reported only one third to one half of their headache attacks relieved by placebo as well as by the bona fide analgesics, while the subjects not reporting minute observation as a rule reported complete success with placebo and the analgesics. In addition, there may have been some subjects with true physiological headaches but accessible to suggestion. The net result of these factors is a significant variation of overall susceptibility to drugs among the reactors to placebo.

Banal as it may sound, discrimination among remedies for pain can be made only by subjects who have a pain on which the analgesic action can be tested. The imagined pain, the psychological headache, may be a source of great discomfort to the subject, but it does not form the prerequisite condition for drug discrimination.

Through the use of placebo subjects who lack the basis of drug discrimination can be screened out, and the relative potency of drugs can be determined on the subjects in whom the essential condition for discrimination of analgesic action is given.

VII.2. Conditions Under Which the Experiment is Conducted

In order to assure success in an investigation, the conditions must be such as to allow treatment differences to be expressed if, indeed, they are present. The most elaborate and elegant treatment design can be rendered useless if the conditions are such that it is impossible to assess treatment differences. This fact was very much in evidence in the Jellinek experiment described in the preceding section. In order to compare the effectiveness of headache remedies, it is necessary for subjects to have headaches. Discrimination among pain remedies can be made only by subjects having a pain on which analgesic action can be tested. The screening of subjects by use of a placebo leads to the desired group. Also, if an investigator had obtained only the number of headaches not relieved over a two-week period and if the subjects had had no headaches, all headache remedies would have been equally effective.

In the dandelion experiment in the randomized complete block design discussed in chapter VI, all herbicides would receive a "perfect score" if there were no dandelions present in the lawns at the beginning of the experiment. Likewise, in fumigation and insect spraying experiments, organisms of the specified type must be present before it can be determined whether or not the fumigant and the sprays are effective.

As an example of the above, it was desired to learn whether soil fumigation with soil fumigants used on pineapples would be effective in increasing the yields of sugar cane. The use of soil fumigants for pineapples resulted in greatly increased yields of pineapple fruit. The pineapple fields were adjacent to fields on which sugar cane was grown. If the soil fumigant was killing a detrimental organism or was creating soil conditions for better plant growth in the pineapple fields, it was possible that the same result could be obtained on sugar cane. Consequently, it was decided to set up experiments to investigate this possibility; the sugar cane plantation managers offered fields which were high producing ones for use in the experiment. Their contention was that they wanted higher yields from these fields. There is no doubt that this was a desirable motive, but there is considerable doubt that the effectiveness of soil fumigation could be assessed on such fields. Presumably, high producing fields have nothing wrong with them and hence would be useless in assessing the effectiveness of a soil fumigant. Instead, the investigator asked for land which, because of unknown causes, produced the lowest yields. He wanted to conduct his experiment on so-called "sick" fields. Since it is not known what soil fumigation does for the pineapple plant, it was postulated that the highest chance of success for soil fumigation on sugar cane would be on "sick" fields.

In an experiment involving the determination of the nature of the ability of the tomato plant to set fruit shortly after fertilization, two tomato varieties, Porter and Ponderosa, were selected to be grown in Oklahoma. When grown side by side in fields there, Porter set fruit immediately, but Ponderosa did not do so until six weeks later. Grown in another environment, Porter and Ponderosa would have set fruit in the same length of time after flowering; it would have been impossible to study the reason for the difference under these conditions. It had been stated that the reason for the differential reaction of the two varieties was physiological, but the investigator found the difference to be genetic. He determined the nature of inheritance of this characteristic and showed that a Ponderosa type variety could be produced which would set fruit immediately under the weather conditions of Oklahoma.

VII.3. Experimental Conditions as Related to Application

The conditions of the experiment must be the same, or at least similar, to those to which the treatments are to be applied in practice. To illustrate this point we shall use an example in marketing research. Prior to 1949, when Professor M. E. Brunk and his colleagues started their marketing research investigations at Cornell University, it was common practice for market researchers to set up experiments in which all treatments appeared in the store at the same time. In practice, however, store managers would use only one of the treatments. Under the conditions of the experiment the customer was given a choice of all treatments; the commodity most often preferred by customers was reported to the store managers with the suggestion that they adopt this treatment. The fact that one treatment is preferred over another when the customer is given a choice is no assurance that this treatment will increase sales over the other treatment when the customer has no choice. This illustrates a cardinal point in experimentation involved in selecting a treatment for general use by farmers, teachers, businessmen, etc., and that is to *conduct the experiment under the same conditions that are to be used in practice*.

As a second illustration, in biological investigations an experiment conducted in a greenhouse or in a laboratory may not be comparable to an experiment conducted in a farmer's field and under the usual cultural conditions used by a farmer. The ease of obtaining results in a greenhouse, laboratory, clinical, small animal, or other experiment should not be construed to mean that these results are applicable under practical conditions. The results obtained by a trained set of panelists or judges may not be useful for the general public. Professor G. W. Snedecor, formerly of Iowa State University, was working with a trained set of butter tasters, some of whom had won international fame. He wondered whether the general public agreed with their tastes. To obtain some information on the question, he took a number of butter samples that had been rated by the "experts" to a meeting of a ladies club. The ladies did not agree with the experts and rated their top choice at the bottom. The experts may have picked the best butter using specified criteria, but this is no guarantee that the general public will agree with them.

Many other examples can be given. Some of our Cornell food experts, for example, had certain criteria for what constituted good applesauce. Their panelists rated the samples according to these criteria. The samples were tried on a sample of the general public that, it appeared, had become used to the strained applesauce fed to babies and ulcer sufferers; of course, they disagreed with the panelists. In addition, when a panel is set up from the general population, the panel itself may become unrepresentative of the population. This is especially true for such items as farm business records, nutritional studies (e.g. on milk), social habits, reading habits, and so forth. When people know that they are under observation, they frequently act quite differently than they do when they do not know it. For example, a person who works with the College Extension Service, keeping farm records, may make different decisions than when he keeps no records. When a person keeps a record of material read, he will automatically and consciously be more selective in his reading material than when no record is kept. As a member of the sample, he may read the *Times*, whereas he would read the local tabloid if he were not a member of the panel.

As another example of how people may change or may answer questions, a survey was conducted in a town in Western New York where the researchers had access to records of the total amount of milk sold, where it was sold, and the person to whom it was sold. There were a number of dairies in the town, and they had lists of all their customers, as well as of the amounts sold to each customer. The surveyors also knew the amount of milk bought and sold by all the stores in the town. A promotional campaign was launched to determine whether milk usage could be increased. A set of large tumblers with the word milk on the glass was given to each cooperating member. A questionnaire was taken before, during, and after the campaign, to ascertain the amount of milk reportedly consumed by families. It became evident that the mother, then the father, then the grandparents, and then an unrelated person living in the house were increasingly more reliable in reporting milk consumption by the children of the family. Evidently, the mother reported the amount she thought her children should use rather than what they actually used. The father was almost as biased as the mother. The grandparents and an unrelated occupant of the house were much more reliable than the parents in reporting the actual milk consumption of the child. Some mothers reported that they bought from dairy A, which reportedly sold high quality milk, when dairy B has just delivered milk to their door! Also, even though the people interviewed indicated that the milk promotional campaign had been a great success in increasing the consumption of milk, the records indicated that the campaign did not change milk usage in this town. If people are at an optimum usage level as far as they are concerned, a campaign will be useless in changing their habits unless it changes their opinions about optimum quantity. If milk usage is at full capacity there can be no increase.

VII.4. Controls

The necessity of using a control was amply illustrated by the Jellinek experiment. There the control or placebo was necessary to divide the participants of the study into two groups, the group that responds to a placebo and the group that does not. The type of control is also illustrated very well in this experiment, in that the placebo was like the drugs in all external appearances, but was lacking the active ingredient for relieving headaches.

In preceding sections there were illustrations of the necessity of having a control in determining whether a specified condition existed. This is true in the "squirt and count" type of experiments involving sprays and counts of number of insects remaining. If there are no organisms present or if they do no damage, it is necessary that this should be known in order to assess the effectiveness of the sprays being used. The control could be no spray or it could be the commercial or standard spray that is used for this type of material. Unless it is known that organisms are always present in sufficient numbers to damage the crop under question, it is essential that one control involving no spray and one control involving the standard spray should be included. Two such controls allow information on the number of organisms, or the amount of damage present, and on the effectiveness of new sprays as compared to the standard spray.

As an example of the above, sprays were being compared for their effectiveness in controlling damage caused by a specified insect to the potato plant. Damage is measured by the reduction in the weight of potato tubers obtained from a plant. The weights of tubers from 100 sprayed plants and from 100 unsprayed plants were compared. If the weights were the same within sampling variation, then we could say that

1. no damage resulted from not spraying,
2. the act of spraying was detrimental enough to offset the beneficial effect of the spray, or
3. no insects were present.

It was found that the act of running the spraying equipment over the ground between potato plants caused a reduction in the weight of tubers because the sharp lugs in the sprayer wheels destroyed roots near the surface. Therefore it was necessary to include two control treatments, one representing no spray and no machinery run over the ground, and the second representing no spray but in this case the sprayer would be run over the ground in the same manner as for the sprayed plots. With two such controls the investigator can assess the damage caused by running the sprayer over the ground and the effectiveness of the spraying treatment. Note that a spraying treatment could be both detrimental by damaging the plant due to the chemical compounds in the spray and beneficial by killing insects; these two effects could offset each other.

For organoleptic tests (e.g. taste tests, odor tests, etc.) it is necessary that a standard or control treatment should be included, in order to compare new treatments with a standard. Likewise, for fertilizer trials a treatment with no fertilizer and perhaps a treatment involving the standard fertilizer treatment are usually necessary. In variety trials one or more standard varieties are included as controls for the new varieties in the experiment. A standard laboratory technique needs to be included in an experiment involving a comparison of new laboratory techniques.

The many diverse examples listed above indicate the need for and beneficial effects from the inclusion of a control, _a point of reference_, in the experiment. Without this point of reference, the experiment may turn out to be only an experience rather than a procedure for obtaining factual and useful information. In some experiments, the problem of establishing standards can become a statistical one, as was the one described by Youden, Connor, and Severo [1959]; their study resulted from an inquiry from the United States Geological Survey concerning the choice of standards for estimating the amount of uranium in soil samples.

VII.5. Single Factor With Several Levels

The first item to be considered in single factor investigations is the factor or variable to be investigated. For example, let us suppose that we are interested in investigating the effect of inorganic chemicals on plant growth over the life period of the plant. Since many inorganic chemicals are involved, we will select one of them, say nitrogen, in the nitrate form, NO_3. Thus we need select not only the factor, say nitrogen, and the amounts or levels to be used, but also the form NO_3, of the factor to be used; quite different results may be obtained with a different form, say NH_3. The same amount of nitrogen could be used for the two forms, but the effect on plant response or growth could be quite different for the two forms of nitrogen.

After selecting the inorganic chemical and its form, we need to determine the range over which nitrogen will be studied. Suppose that we decide to use a sand culture experiment. Should we start at zero nitrate nitrogen in the sand culture? What should be the upper limit of amount of nitrate nitrogen used? The answers to these questions depend upon the objectives of the experiment, the type of plant used, and the stage of growth of the plant. Let us suppose that we are using the celery plant, that cuttings are to be used, and that growth measured as increase in weight will be studied over a 12-week period. We may decide to use a zero addition of nitrate nitrogen as the lowest level and to use 500 gms of nitrate nitrogen per ten kilograms of other material as the highest level. The other material will contain identical amounts of sand, water and all other elements considered necessary for plant growth. Sufficient amounts of all other elements will be included. The only item that varies will be amount of nitrate nitrogen applied. Do we want the complete response picture or curve over the entire range? If we do, then we utilize this range; if not then we either shorten the range or lengthen it to obtain the desired coverage.

After the range has been established, we have to determine the number of levels to be used. If we know the form of the response curve and the range, then we can say something definite about the choice of levels. For example, let us suppose that the response is linear, that is,

$$\text{response} = Y_i = a + bN_i + \text{error}$$

where N_i is the amount of nitrate nitrogen applied in grams per 10 kilograms of other material. This may be pictured as shown in figure VII.1. If the response is a straight line, we need only the zero application and the upper level, say 500 grams. These two different levels of the amount applied would be sufficient to estimate the linear plant response to amount of nitrate nitrogen. Likewise, if we know that $\log Y_i = a + bN_i + \text{error}$, or $Y_i = a + b \log (N_i+1) + \text{error}$, is the form of the response, we need only two levels, the lowest and the highest, in order to estimate the two constants a, the intercept, and b, the slope, of the response function.

An example of the functional form W_i = dry weight of a chick embryo at ages 6 to 16 days $= Ae^{bX_i}$, where X_i represents age in days, A and b are constants computed from data, and e is the base of the Naperian system of logarithms, is given by Snedecor [1946], page 376, and it is reproduced in table VII.2 and figure VII.2.[1]

The plot of the data and of the response function $W_i = 0.002046(1.57)^{W_i}$ indicates quite close agreement of the data with the functional form. As another way of looking at the data the form $\log_{10} W_i = \log_{10} .002046 + (\log_{10} 1.57)X_i = -2.689 + 0.1959X_i$.

[1] Reprinted by permission from Statistical Methods, 4th edition, by G. W. Snedecor, 1950 by The Iowa State University Press.

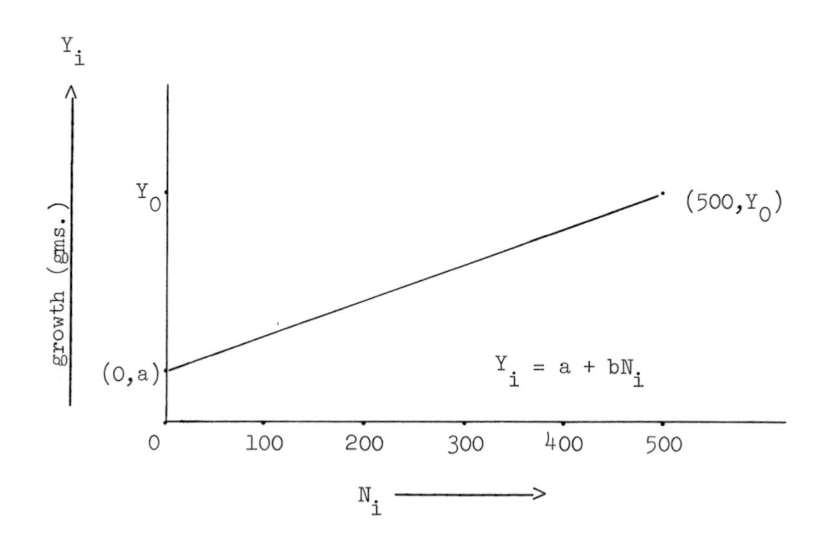

Figure VII.1. Grams of nitrate nitrogen (N_i) applied versus growth.

Table VII.1. Dry weights of chick embryos from ages 6 to 16 days, together with common logarithms

Ages in Days X	Dry Weight, W (grams)	Common Logarithm of Weight Y
6	0.029	-1.538*
7	0.052	-1.284
8	0.079	-1.102
9	0.125	-0.903
10	0.181	-0.742
11	0.261	-0.583
12	0.425	-0.372
13	0.738	-0.132
14	1.130	0.053
15	1.882	0.275
16	2.812	0.449

* From the table of logarithms, one reads $\log 0.029 = \log 2.9(10^{-2}) = \log 2.9 + \log 10^{-2} = 0.462 - 2 = -1.538$.

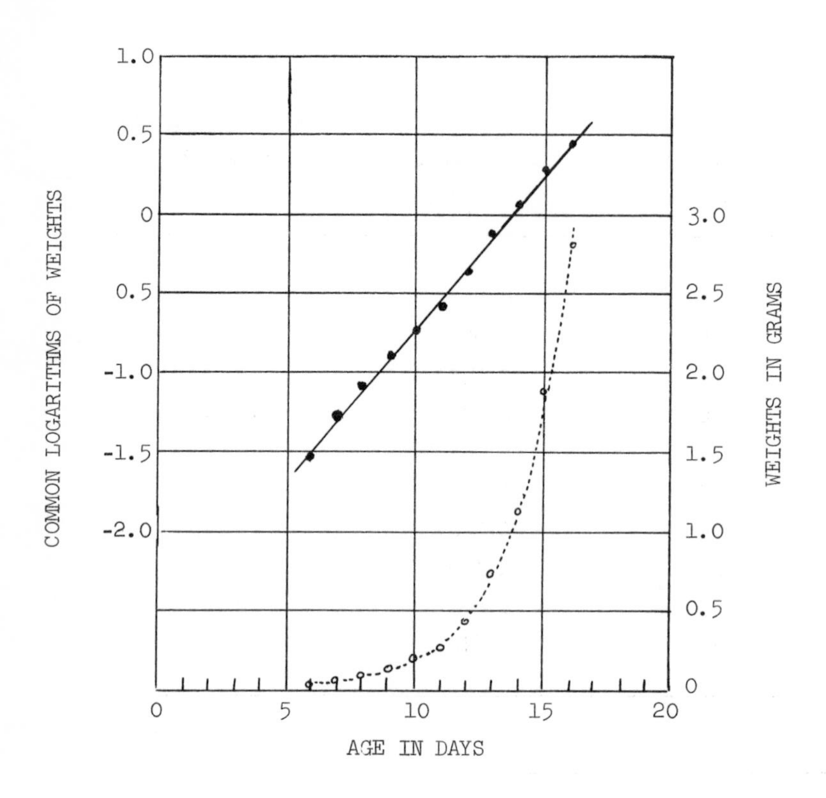

Figure VII.2. Dry weights of chick embryos at ages 6-16 days with fitted curves.

Uniform scale: $W = 0.002046(1.57)^X$

Logarithmic scale: $Y = 0.1959X - 2.689$

If the response is curved, then more than two levels of nitrate nitrogen may be required. Suppose that the response is of the form $Y_i = a + bN_i + cN_i^2 + \text{error}$. In this case the response may be similar to one of the forms shown in figure VII.3. In order to estimate the constants a, b, and c, we must use three or more levels of nitrate nitrogen. One procedure would be to put 4/9 of our observations at each end of the range and 1/9 in the middle. Given that the curve is second degree, or quadratic, the only need for observations in the middle is to determine whether c is negative or positive. The reason for a higher proportion of observations at the ends of the range is to tie down the ends of the curve.

The design of the number or proportion of observations at each point is the subject of past and current statistical research. The problem of efficiently allocating proportions of observations to the various levels of the variable becomes increasingly difficult as the degree, k, of the polynomial,

$$Y = a + \sum_{j=1}^{k} b_j N^j + \text{error}$$

$$= a + b_1 N + b_2 N^2 + b_3 N^3 + \cdots + b_k N^k + \text{error}$$

increases. However, an equal allocation of proportions of observations at equally spaced levels should be followed if the degree, k, of the polynomial is unknown. Also, an equal allocation of proportions and spacings may not be as efficient as some other procedure, but information is still obtainable just as it was in the first weighing experiment discussed in section VI.1 of the previous chapter.

So far we have been concerned with selecting levels at any point that we wish. In certain cases, however, the levels will not be presented by the investigator but may appear at random or be obtained as a result of other experimental conditions. As long as the level of the factor can be measured or determined and the error of determining the level is small compared to the range of the levels, the procedures discussed above all apply. For example, it would be simple to preselect levels of nitrate nitrogen applied to a greenhouse potted plant or to a group of plants in a 4-meters-square plot of ground, <u>but</u> it would be difficult if not impossible to do this with groups of plants having different prespecified levels

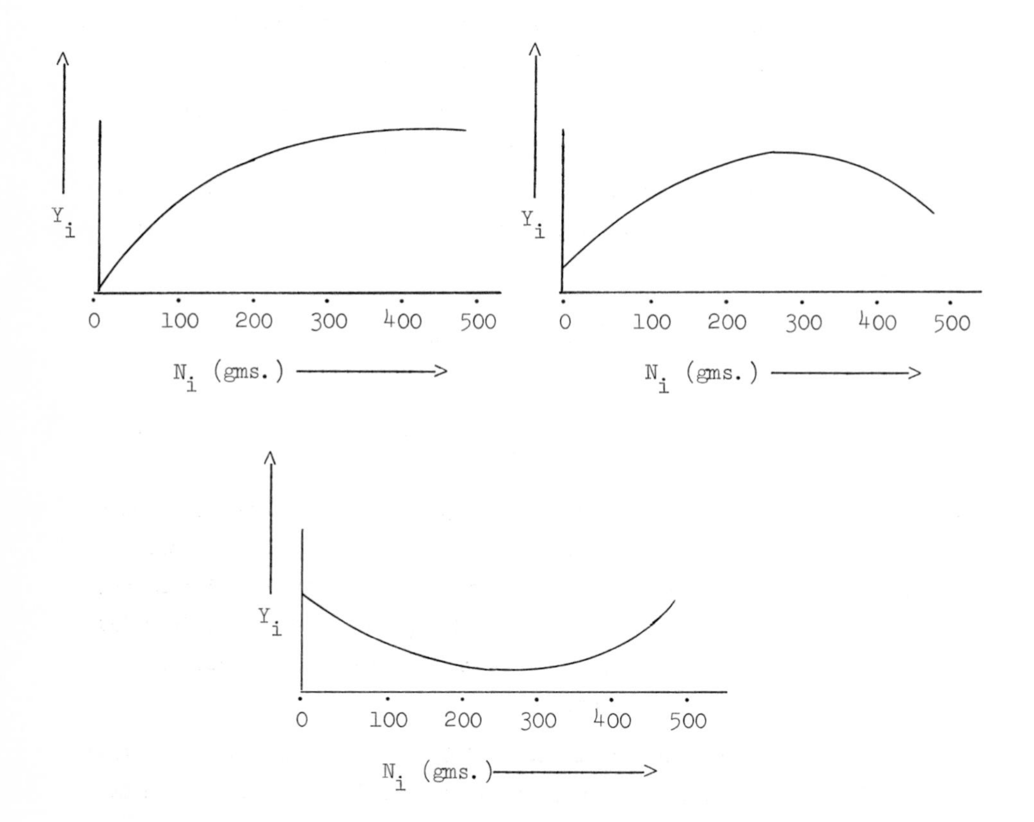

Figure VII.3. Forms of response.

of nitrate nitrogen. The mere fact that nitrate nitrogen is added to a plot of ground is no assurance that the plants will absorb this chemical in the proportion in which it is applied. Perhaps fertilizer studies should be conducted in terms of the amount of the chemical found in the plant rather than in terms of the amount that is applied to the ground.

There are many mathematical forms possible for an unknown response function. The investigator should try to obtain or to postulate the appropriate form prior to experimentation and then select an appropriate treatment design for the experiment. The emphasis in statistics textbooks is on the polynomial form, but this is not necessarily the most universal response function in real life. The polynomial form represents a rather easy and mathematically tractable form. Since this problem was solved before others, the results were the first to be incorporated into statistical writings, resulting, perhaps, in the emphasis placed on the polynomial response function.

In the main we have briefly illustrated that

1. the variable to be investigated must be selected,
2. the range of levels of the variable must be determined,
3. the levels must be selected, and
4. the proportion of observations at each level of the variable of interest must be determined.

In general one could use equal spacings of levels of the variable and equal proportions of observations at each level, but a more efficient treatment design usually results if a higher proportion of observations are placed at the end-points of the range and the remaining observations at points of maxima and minima which are peculiar to the particular form of the response function being studied.

VII.6. Treatment Designs Involving Two or More Factors and the Factorial Treatment Design

The experimenter may wish to study two or more variables jointly to observe the manner in which the response varies with the changing levels of the variables under study. Suppose that it is desired to study the effect on plant response of varying amounts of nitrogen and phosphorous. If the levels are selected, the same comments as discussed in the preceding section for a single variable apply here. Whether the levels are preselected or not, a level of each of the variables under study will be associated with each response observation. For example, N_1 and P_1 will be associated with Y_1, N_2 and P_2 with Y_2, N_3 and P_3 with Y_3, and so forth. The form of the response function will involve both variables and might be one of the following forms where N_i = ith level of nitrogen, P_i = ith level of phosphorous, and e_i is the error term associated with the ith observation:

$$Y_i = a + bN_i + cP_i + e_i$$

$$\log Y_i = a + bN_i + cP_i + e_i$$

$$\log Y_i = a + b \log N_i + c \log P_i + e_i$$

$$Y_i = a + bN_i + cP_i + dN_i^2 + fP_i^2 + gN_iP_i + e_i$$

$$Y_i = aN_i^bP_i^c \, e^{dN_i + fP_i + gN_iP_i + e_i}$$

$$Y_i = aN_i^b e^{cN_i} + dP_i^f e^{gP_i} + e_i \; .$$

In the last two equations e is the base for natural logarithms, whereas e_i is an error deviation. The response in the last equation would be the sum of two exponential forms.

From the above we note that the response in the experiment could be of various forms. From the laws and theories evolved for the subject matter in the area where the investigation is being conducted, it is often possible to eliminate one or more of the above forms and to choose a small subset of plausible mathematical forms. Sometimes it is possible to pick the exact form for the response function. Many examples of this occur in the physical and biological sciences. For example, the exponential growth law is written as $Y = ae^{bt}$ = weight at time t_i; the rate of increase of growth at a specified time, say t_0, is $dY/dt = abe^{bt}$ evaluated at $t = t_0$; Mendel's law of segregation has the form of the binomial distribution, and so on. When the form, or even an approximate form, is known, it should be utilized. When the mathematical function describing the phenomenon under observation becomes established beyond reasonable doubt then a law has been established; the goal of the empirical investigation has been achieved.

An example is given in table VII.2; the nature of the measurements obtained from an Hawaiian Sugar Planters' Association sugar cane experiment in Hawaii are illustrated.

As a second example, consider a set of hypothetical data where Y = grade-point average for four years at S-W University for the 20 most recent graduates from fraternity XYZ, X_1 = high school grade average, and X_2 = aptitude test scores as given in table VII.3.

Suppose that it has been decided to utilize the data for the 20 most recent graduates of the fraternity to determine how well one could predict a four-year grade-point average from high school standing and aptitude test scores. Suppose that the fraternity "research" team consulted with a statistician who obtained for them a least squares regression equation of the form:

$$\hat{Y}_i = a + b_{y1\cdot 2}X_{1i} + b_{y2\cdot 1}X_{2i} = -3.52 + .085\ X_{1i} - .0014\ X_{2i}\ .$$

(He actually used more significant figures to compute $\hat{Y}_i$ above.) For example, the first estimated grade-point average was computed as $-3.52 + .085(75) - .0014(480) = 2.2 = \hat{Y}_1$. From the equation we note that an increase of one percent point in high school grade increases the four-year grade-point average by .085 grade-point for constant aptitude test scores, whereas an increase of one point in the aptitude test score decreases the four-year grade-point average by .0014 point when X_1 is held constant. Both X_1 and X_2 exhibit a positive relationship with Y when considered individually, but X_2 exhibits a negative effect when the effect of X_1 is removed or held constant. For example, consider the two students with a high-school grade of 96; here, the one with the higher aptitude test score total had a lower estimated average than the one who had the lower score (3.9 vs. 4.0). If these data were to be used to predict the performance of "pledges", the pledgee would be well advised to obtain as high a high school grade and as low a test score as possible!

Observation Number	Sugar (tons/acre)[1]	Total Nitrogen in Base	Total Potash in 8-10th Node	Total Phosphorous in 8-10th Node
1	Y_{11} = 5.0	.0568	.554	.0723
2	Y_{12} = 4.3	.0621	.608	.0802
3	Y_{13} = 4.6	.0528	.653	.0650
4	Y_{21} = 10.0	.1143	.368	.0317
5	Y_{22} = 9.1	.1557	.463	.0322
6	Y_{23} = 9.8	.1359	.576	.0299
7	Y_{31} = 12.3	.2443	.190	.0184
8	Y_{32} = 11.6	.2738	.324	.0187
9	Y_{33} = 14.3	.2790	.473	.0258

Table VII.2. Measurements obtained from a sugar cane experiment.

[1] The first subscript of Y_{ij} refers to level of nitrogen application and the second refers to level of potash application.

Student number	High school grade = X_1 in %	Number of points for aptitude test = X_2	4-year grade-point average = Y	Estimated grade-point average = $\hat{Y}$	Error of estimate = $Y - \hat{Y}$
1	75	480	2.0	2.2	-0.2
2	99	700	4.1	4.0	0.1
3	86	680	3.5	2.9	0.6
4	71	500	1.8	1.8	0.0
5	86	550	3.5	3.0	0.5
6	97	750	4.1	3.7	0.4
7	80	450	2.8	2.7	0.1
8	89	650	3.5	3.2	0.3
9	91	700	3.5	3.3	0.2
10	97	760	3.0	3.7	-0.7
11	94	700	2.8	3.5	-0.7
12	96	750	4.0	3.6	0.4
13	96	720	3.9	3.7	0.2
14	89	700	3.3	3.1	0.2
15	94	700	3.3	3.5	-0.2
16	93	720	3.1	3.4	-0.3
17	78	400	2.4	2.6	-0.2
18	60	350	1.0	1.1	-0.1
19	91	700	2.8	3.3	-0.5
20	97	780	3.5	3.7	-0.2

Table VII.3. Hypothetical data for 20 students of fraternity XYZ.

Although it would have been possible to study preselected levels for the above set of data, it would have been difficult or impossible to obtain the prescribed levels of the variables of interest. For example, suppose that we wish to study levels of X_1 equal to 60, 70, 80, 90, and 100, and levels of X_2 equal 500, 550, 600, 650, 700, and 750. We may note that there is no combination of X_1 and X_2 for any of these values in the preceding set of data. For our study and for most similar situations it is not necessary to have preselected levels of X_1 and X_2. However, it is necessary that the range of levels for X_1 and X_2 cover the range desired by the experimenter. For the above combinations a very large population is required and the desired combinations must be selected, a time-consuming and data-wasting procedure.

Although the difficulties of obtaining selected levels for the above examples are apparent, we should emphasize that there are many types of investigation that allow selection of the levels of the variables, or <u>factors</u>, of interest to the investigator. For example, suppose that the experimenter is investigating growth of celery plants in sand culture experiments where all items except nitrate nitrogen, NO_3, and potash in the form of K_2O are held constant. Various amounts by weight of NO_3 and K_2O will be added to the experimental units. Suppose that it is decided to use the following levels of NO_3 and K_2O:

levels of NO_3 : 10, 20, 40, 80 grams

levels of K_2O : 13, 39, 65, 91 grams

Now the question arises as to what combinations of the various levels of the two factors should be used in the investigation. These combinations could be used as follows:

Levels of K_2O (grams)	Levels of NO_3 (grams)			
	10 = 0	20 = 1	40 = 2	80 = 3
13 = 0	n_0k_0 = 00	n_1k_0 = 10	n_2k_0 = 20	n_3k_0 = 30
39 = 1	n_0k_1 = 01	n_1k_1 = 11	n_2k_1 = 21	n_3k_1 = 31
65 = 2	n_0k_2 = 02	n_1k_2 = 12	n_2k_2 = 22	n_3k_2 = 32
91 = 3	n_0k_3 = 03	n_1k_3 = 13	n_2k_3 = 23	n_3k_3 = 33

(Note: Instead of using 0,1,2,3 to designate levels one could use 1,2,3,4. This is an arbitrary convention.)

The combination of the ith level of NO_3 and of the jth level of K_2O is denoted as n_ik_j = ij. Once the order of the subscript has been defined, the subscript ij is sufficient to define the treatment corresponding to the combination n_ik_j of the two factors NO_3 and K_2O. The observation (yields or weights) for the ij^{th} combination of the two factors would be Y_{ij} as denoted in the following table:

Level of K_2O (grams)	Level of NO_3 (grams)				Sum	Mean
	0	1	2	3		
	(weight or yield in grams)					
0	Y_{00}	Y_{10}	Y_{20}	Y_{30}	$Y_{\cdot 0}$	$\bar{y}_{\cdot 0}$
1	Y_{01}	Y_{11}	Y_{21}	Y_{31}	$Y_{\cdot 1}$	$\bar{y}_{\cdot 1}$
2	Y_{02}	Y_{12}	Y_{22}	Y_{32}	$Y_{\cdot 2}$	$\bar{y}_{\cdot 2}$
3	Y_{03}	Y_{13}	Y_{23}	Y_{33}	$Y_{\cdot 3}$	$\bar{y}_{\cdot 3}$
Sum	$Y_{0\cdot}$	$Y_{1\cdot}$	$Y_{2\cdot}$	$Y_{3\cdot}$	$Y_{\cdot\cdot}$	
Mean	$\bar{y}_{0\cdot}$	$\bar{y}_{1\cdot}$	$\bar{y}_{2\cdot}$	$\bar{y}_{3\cdot}$		$\bar{y}_{\cdot\cdot}$

The treatment design for the two factors described above is known as a factorial treatment design; it is a design that contains all combinations of two or more levels of the various factors. From such a treatment design, with each combination repeated an equal number of times, we find that the arithmetic means may be used to estimate the various effects. For example, the estimated effects for each of the levels of NO_3 and of K_2O are:

Effect of 10 grams of NO_3	= 0 level of NO_3	:	$\bar{y}_{0\cdot} - \bar{y}_{\cdot\cdot}$
" " 20 " " "	= 1 " " "	:	$\bar{y}_{1\cdot} - \bar{y}_{\cdot\cdot}$
" " 40 " " "	= 2 " " "	:	$\bar{y}_{2\cdot} - \bar{y}_{\cdot\cdot}$
" " 80 " " "	= 3 " " "	:	$\bar{y}_{3\cdot} - \bar{y}_{\cdot\cdot}$
Effect of 13 grams of K_2O	= 0 level of K_2O	:	$\bar{y}_{\cdot 0} - \bar{y}_{\cdot\cdot}$
" " 39 " " "	= 1 " " "	:	$\bar{y}_{\cdot 1} - \bar{y}_{\cdot\cdot}$
" " 65 " " "	= 2 " " "	:	$\bar{y}_{\cdot 2} - \bar{y}_{\cdot\cdot}$
" " 91 " " "	= 3 " " "	:	$\bar{y}_{\cdot 3} - \bar{y}_{\cdot\cdot}$

This type of effect, or of linear combinations of these, has been denoted as a main effect or one-factor effect in a factorial experiment. In biological literature these are often called interactions but should more properly be called one-factor interactions.

In addition to main effects, this two-factor factorial treatment design contains another type of effect, viz. two-factor interaction effects. There are effects peculiar to a mutually beneficial or mutually detrimental effect of a particular combination. The interaction effect for the ith level of NO_3 with the jth level of K_2O is computed as

$$Y_{ij} - \bar{y}_{i\cdot} - \bar{y}_{\cdot j} + \bar{y}_{\cdot\cdot}$$

If the above quantity is zero, the two-factor interaction effects would be zero, and $Y_{ij} = \bar{y}_{i\cdot} + \bar{y}_{\cdot j} - \bar{y}_{\cdot\cdot}$. Graphically we may represent the zero two-factor interaction case as shown in figure VII.4.

Alternatively levels of K_2O may be plotted on the abscissa as shown in figure VII.5.

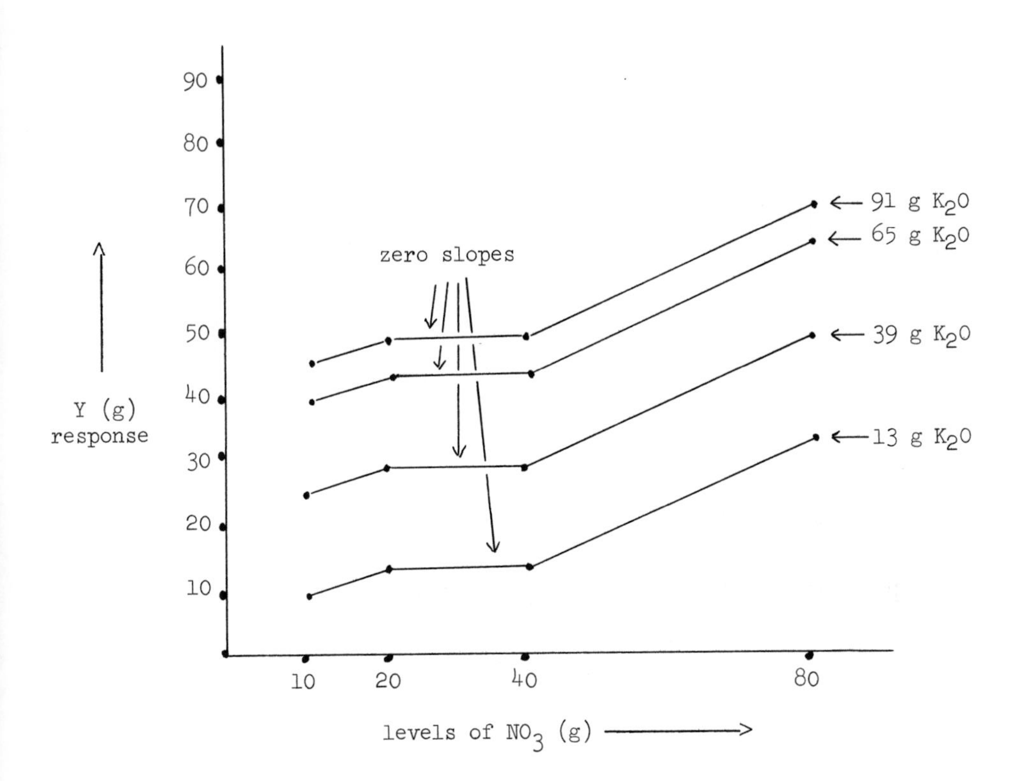

Figure VII.4. Representation of a zero two-factor interaction.

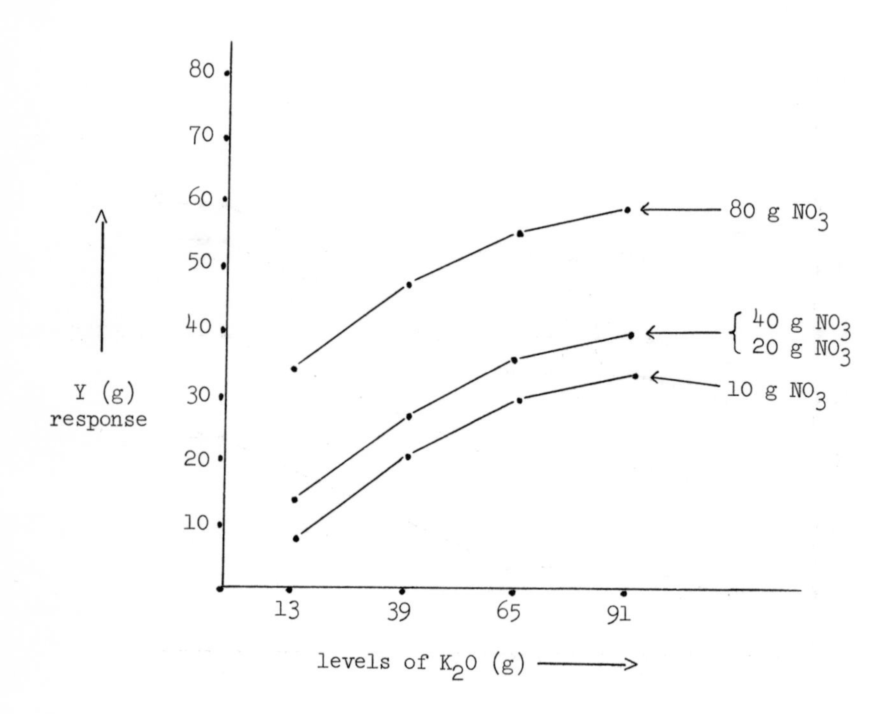

Figure VII.5. Alternative representation of zero two-factor interaction.

(Note that 20 grams and 40 grams of NO_3 gave identical values for all levels of K_2O in the former graph. Hence the points would be identical on the latter graph. This is an artificial example.) In the case of zero two-factor interaction the distance between any two curves remains constant throughout all levels of the factor represented on the abscissa.

In an actual experiment there will be sampling variation, and a zero two-factor interaction will not usually be observed even if the population interaction effects are truly zero. There are statistical procedures for setting bounds on sampling variations and to determine whether or not the effects may reasonably be expected to fall between a designated set of values. Some of these procedures will be considered later under methods of summarizing data from experiments and surveys.

If the purpose of an experiment utilizing a factorial treatment design is to estimate main effects and interactions, then it has been shown (Kempthorne [1952] Page 425) that the factorial treatment design is optimum, that is no other selection of treatments does this more effectively than a factorial with equal numbers of observations on each treatment. If the experimenter is given the particular form of the response function, then it is often possible to select a treatment design which is more efficient than the factorial.

The factorial treatment design listed above is called 4 levels of NO_3 × 4 levels of K_2O factorial, or alternatively, a $4 \times 4 = 4^2$ factorial of amounts of NO_3 and K_2O. In the term, 4^2, the four refers to the number of levels, and the superscript 2 refers to the number of factors. For example, a 4^3 factorial would represent 4 levels of three different factors in all combinations. In general we use symbols such as s^n to denote that there are s levels each of the n different factors. In the form 4 × 4 the first four represents the number of levels of the first factor and the second four those in the second factor. This form is convenient when the number of levels of the factors is not the same. For example, suppose that only the first three levels of NO_3 had been used above. Then the factorial would have been a 3×4 = 3 levels of NO_3 × 4 levels of K_2O factorial treatment design. Likewise, a $p \times q \times r$ factorial would represent a factorial with p levels of factor one × q levels of factor two × r levels of factor three.

The 4×4 factorial design was selected to introduce the idea of factorials. There are smaller factorials, such as 2×2, 2×3, 2×4, 3×3, and 3×4 but the 4×4 is large enough to illustrate several ideas which could not be illustrated as well with some of the smaller factorials. In practice, however, these smaller factorials are extensively used. In fact, the 2^n (n factors each at 2 levels) is a very popular factorial series. Statisticians also like the 2^n series because of certain mathematical group theoretic properties.

As an example of a 2×2 factorial treatment design, let us suppose that we are interested in comparing two types of rubber, that obtained from the hevea rubber tree = r_0 and that produced synthetically = r_1, manufactured by two different processes p_0 and p_1. The yield characteristic is number of miles a truck may be driven until only $\frac{1}{2}$ inch of the tread remains on the tire. Suppose that a latin square design is used on each of several trucks possessing eight sets of dual wheels with the rows of four tires corresponding to the rows of the latin square and the position of the tires from left to right corresponding to the columns. Suppose that the data in thousands of miles for the four treatments are of the form:

	Manufacturing process		Sum	Mean = $\bar{y}_{i\cdots}$
Type of rubber	p_0	p_1		
hevea = r_0	$\bar{y}_{00\cdot\cdot} = 15.1$	$\bar{y}_{01\cdot\cdot} = 19.5$	34.6	17.3
synthetic = r_1	$\bar{y}_{10\cdot\cdot} = 16.7$	$\bar{y}_{11\cdot\cdot} = 14.3$	31.0	15.5
sum	31.8	33.8	65.6	-
mean = $\bar{y}_{\cdot j\cdot\cdot}$	15.9	16.9	-	16.4 = $\bar{y}_{\cdots\cdot}$

$$\text{rubber main effects}: \hat{r}_0 = \bar{y}_{0\cdots} - y_{\cdots\cdot} = 17.3 - 16.4 = 0.9$$

$$\hat{r}_1 = \bar{y}_{1\cdots} - y_{\cdots\cdot} = 15.5 - 16.4 = -0.9$$

$$\text{process main effects}: \hat{p}_0 = \bar{y}_{\cdot 0\cdot\cdot} - \bar{y}_{\cdots\cdot} = 15.9 - 16.4 = -0.5$$

$$\hat{p}_1 = \bar{y}_{\cdot 1\cdot\cdot} - \bar{y}_{\cdots\cdot} = 16.9 - 16.4 = 0.5$$

interaction effects: $\widehat{rp}_{00} = \bar{y}_{00\cdot\cdot} - \bar{y}_{0\cdot\cdot\cdot} - \bar{y}_{\cdot 0\cdot\cdot} + \bar{y}_{\cdot\cdot\cdot\cdot} = 15.1 - 17.3 - 15.9 + 16.4 = -1.7$

$\widehat{rp}_{01} = \bar{y}_{01\cdot\cdot} - \bar{y}_{0\cdot\cdot\cdot} - \bar{y}_{\cdot 1\cdot\cdot} + \bar{y}_{\cdot\cdot\cdot\cdot} = 19.5 - 17.3 - 16.9 + 16.4 = 1.7$

$\widehat{rp}_{10} = \bar{y}_{10\cdot\cdot} - \bar{y}_{1\cdot\cdot\cdot} - \bar{y}_{\cdot 0\cdot\cdot} + \bar{y}_{\cdot\cdot\cdot\cdot} = 16.7 - 15.5 - 15.9 + 16.4 = 1.7$

$\widehat{rp}_{11} = \bar{y}_{11\cdot\cdot} - \bar{y}_{1\cdot\cdot\cdot} - \bar{y}_{\cdot 1\cdot\cdot} + \bar{y}_{\cdot\cdot\cdot\cdot} = 14.3 - 15.5 - 16.9 + 16.4 = -1.7$

Here we note that the sum of any set of main effects is zero (within rounding errors) and that the sum of any set of interaction effects for any given level of a specified factor is also zero (within rounding errors). The form of the calculations is the same as for the randomized complete block design since the blocks and the treatments correspond to the two factors and the interaction corresponds to what was called errors. Thus, the interaction may be thought of as discrepance from the additive main effects situations, that is, $\bar{y}_{ij\cdot\cdot} - \bar{y}_{\cdot\cdot\cdot\cdot} - \hat{r}_i - \hat{p}_j$ = ij^{th} mean minus the overall mean minus the ith rubber effect minus the jth process effect equals the ij^{th} interaction effect.

As an example of a 2 × 3 factorial treatment design, suppose that we are studying time spent watching television as compared to time spent in reading plus time spent on civic activities (for example, cub scouts, girl scouts, Red Cross, church visitations committee for shut-ins, Gray-Ladies at hospital, etc.) for three broad salary categories of married women with a family of four in the 20 to 40 age bracket. Suppose that our population has been stratified into six strata as described above and that a completely random sample of women is selected from each stratum. Suppose that the following data are obtained:

Activity	Total income for a family of size four t_0 less than \$6,000	t_1 \$6,000-\$12,000	t_2 \$12,000-\$18,000	Sum	Mean = $\bar{y}_{i..}$
	(hours spent per day)				
Watching TV = a_1	$\bar{y}_{10.} = 8.25$	$\bar{y}_{11.} = 3.50$	$\bar{y}_{12.} = 1.75$	13.50	4.50
Reading and civic organizations = a_2	$\bar{y}_{20.} = 0.25$	$\bar{y}_{21.} = 4.50$	$\bar{y}_{22.} = 6.25$	11.00	3.67
Sum	8.50	8.00	8.00	24.50	-
Mean = $\bar{y}_{.j.}$	4.25	4.00	4.00	-	$4.08 = \bar{y}_{...}$

activity main effects: $\hat{a}_1 = \bar{y}_{1..} - \bar{y}_{...} = 4.50 - 4.08 = 0.42$ hrs.

$\hat{a}_2 = \bar{y}_{2..} - \bar{y}_{...} = 3.67 - 4.08 = -0.41$ hrs.

income main effects: $\hat{t}_0 = \bar{y}_{.0.} - \bar{y}_{...} = 4.25 - 4.08 = 0.17$ hrs.

$\hat{t}_1 = \bar{y}_{.1.} - \bar{y}_{...} = 4.00 - 4.08 = -0.08$ hrs.

$\hat{t}_2 = \bar{y}_{.2.} - \bar{y}_{...} = 4.00 - 4.08 = -0.08$ hrs.

interaction effects: $\widehat{at}_{10} = \bar{y}_{10.} - \bar{y}_{1..} - \bar{y}_{.0.} + \bar{y}_{...} = 8.25 - 4.50 - 4.25 + 4.08 = 3.58$ hrs.

$\widehat{at}_{11} = \bar{y}_{11.} - \bar{y}_{1..} - \bar{y}_{.1.} + \bar{y}_{...} = 3.50 - 4.50 - 4.00 + 4.08 = -0.92$ hrs.

$\widehat{at}_{12} = \bar{y}_{12.} - \bar{y}_{1..} - \bar{y}_{.2.} + \bar{y}_{...} = 1.75 - 4.50 - 4.00 + 4.08 = -2.67$ hrs.

$\widehat{at}_{20} = \bar{y}_{20.} - \bar{y}_{2..} - \bar{y}_{.0.} + \bar{y}_{...} = 0.25 - 3.67 - 4.25 + 4.08 = -3.59$ hrs.

$\widehat{at}_{21} = \bar{y}_{21.} - \bar{y}_{2..} - \bar{y}_{.1.} + \bar{y}_{...} = 4.50 - 3.67 - 4.00 + 4.08 = 0.91$ hrs.

$\widehat{at}_{22} = \bar{y}_{22.} - \bar{y}_{2..} - \bar{y}_{.2.} + \bar{y}_{...} = 6.25 - 3.67 - 4.00 + 4.08 = 2.66$ hrs.

The sums of the various effects add to zero within rounding errors, that is, if the means had been carried out to more decimals the sum would be much closer to zero. The above means may be presented graphically as shown in figure VII.6. In this hypothetical example we note that the main effects for income category are very small; the activity main effects are not much larger, and the interaction effects are relatively large.

As a third example, consider the following data in tons per acre representing the total weight of sugar cane harvested from an experimental unit. Three sugar cane varieties (v_0, v_1, and v_2) were used in all combinations with three levels of nitrogen (150 lbs./A = n_0, 210 lbs./A = n_1, and 270 lbs./A = n_2) resulting in a 3 × 3 factorial treatment design. The treatment means in tons per acre for a sugar cane experiment are presented below:

Variety	Nitrogen			Mean
	n_0	n_1	n_2	
v_0	66.5	69.0	76.0	70.5
v_1	61.4	62.6	70.4	64.8
v_2	68.6	64.5	57.9	63.7
Mean	65.5	65.4	68.1	66.3

Graphically these results are represented in figure VII.7.

A pronounced variety by nitrogen level interaction is present in these data. Varieties v_0 and v_1 respond as might be expected from the effect of additional amounts of nitrogen fertilizer. However, variety v_2 does not behave as one would expect. Does additional nitrogen actually decrease yield? Or is the relationship due to other causes? The investigator found that variety v_2 was susceptible to a disease known as "red hot", so named because of the color of the internodal material of

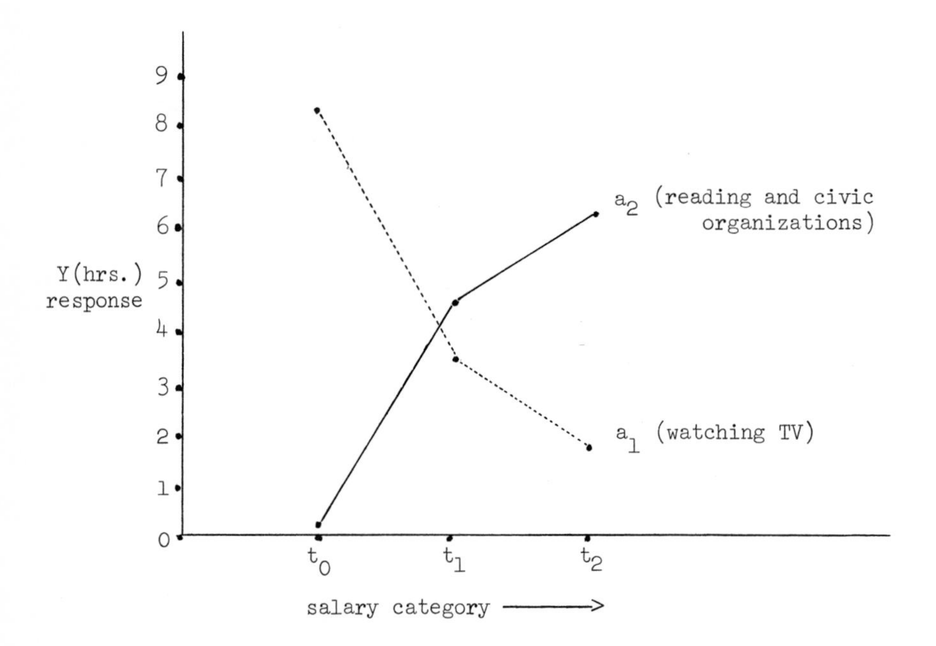

Figure VII.6. Graphical representation of salary versus hours of activity.

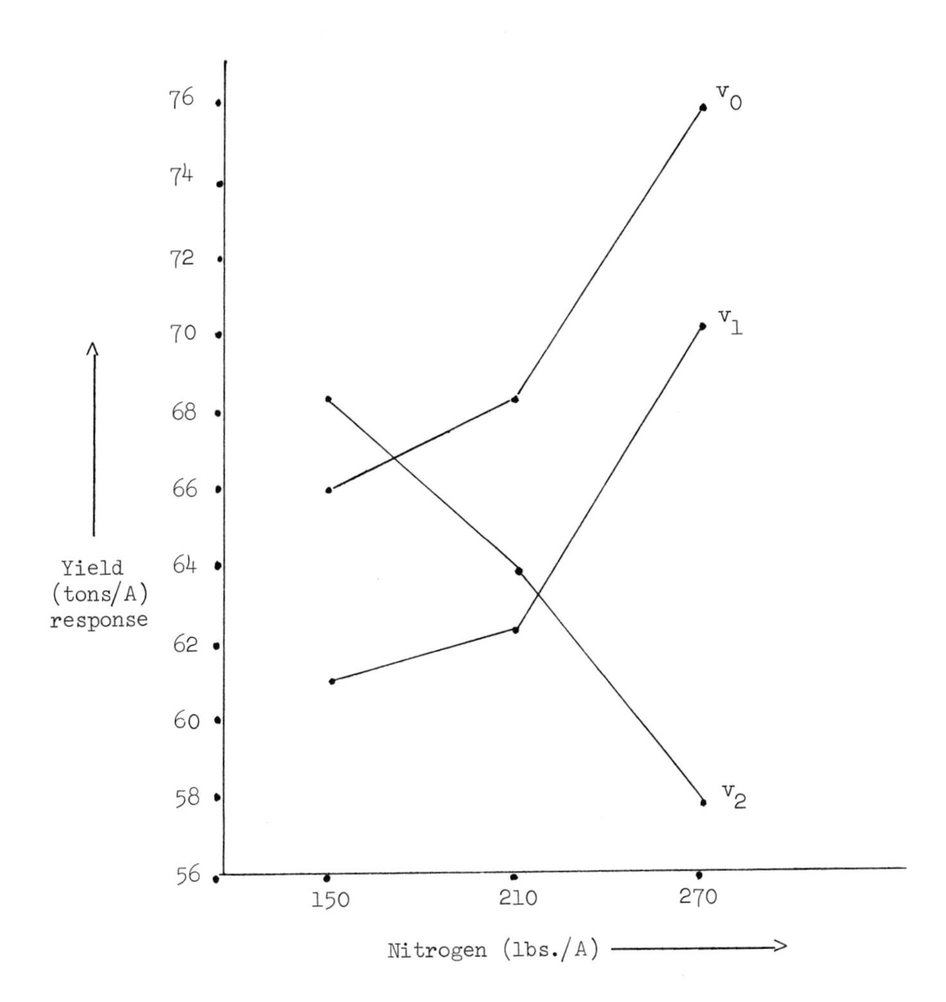

Figure VII.7. Graphical representation of relationship of nitrogen to yield of sugar cane.

plants affected with the disease. The incidence of the disease also increases with increased vigor and lushness of the plants. The addition of nitrogen caused the young plants to grow fast, and the incidence of the disease increased, resulting in more dead plants at harvest time for the plots receiving the higher amounts of nitrogen. The increased number of dead plants decreased yields. For this variety, life is just a vicious circle! If it tries to produce a lot of sugar, it is highly infected with the "red hot" disease!

As the last example, let us consider the 2^3 factorial treatment design. The illustration used here is taken from example VII-2 of Federer [1955]. The following three factors each at two levels in all combinations were utilized in an engineering education study on the use, preparation and effectiveness of stereophotographic aids. The data and treatments here are for the nonstereophotographs. The treatments are:

Photo and question set = factor b	Class = factor c			
	Freshman = c_0		Senior = c_1	
	Exposure = factor a		Exposure = factor a	
	Left = a_0	Right = a_1	Left = a_0	Right = a_1
b_0	$a_0b_0c_0$	$a_1b_0c_0$	$a_0b_0c_1$	$a_1b_0c_1$
b_1	$a_0b_1c_0$	$a_1b_1c_0$	$a_0b_1c_1$	$a_1b_1c_1$

In order to obtain a valid comparison of nonstereo with stereo pictures, an exposure was made a short distance to the left = a_0 and a short distance to the right = a_1 of the spot where the stereo picture was made. There were two sets of pictures with a different question set associated with each group of pictures; b_0 represented set I photos and the associated set of 27 questions and b_1 represented set II photos and the associated set of 18 questions. The freshman group of boys is represented by c_0 and the senior group of boys by c_1. Six freshmen boys out of 24 were randomly allocated to each of the a_ib_j combinations and six senior students out of 24 seniors were randomly allotted to each of the a_ib_j treatments. Each of the 48 boys was given a question and photo set which was one of the combinations $a_ib_jc_h$. The number of correct (R) and incorrect (W) answers is given below:

Treatment															
000		100		010		110		001		101		011		111	
R	W	R	W	R	W	R	W	R	W	R	W	R	W	R	W
10	17	10	17	2	16	10	8	6	21	4	23	6	12	9	9
10	17	6	21	1	17	0	18	5	22	8	19	11	7	5	13
6	21	10	17	5	13	4	14	7	20	9	18	5	13	11	7
12	15	5	22	10	8	4	14	12	15	11	16	7	11	8	10
8	19	2	25	8	10	6	12	11	16	6	21	2	16	8	10
5	22	7	20	3	15	6	12	8	19	5	22	6	12	9	9
51/162		40/162		29/108		30/108		49/162		43/162		37/108		50/108	
.315		.247		.269		.278		.302		.265		.343		.463	

The proportions of correct answers for the various treatments varied from about one-quarter to one-half. For the individual student, the proportion of correct answers varied from zero to 11/18 = 61%.

In the investigation more correct answers were obtained by <u>every</u> student on the stereophotos than on the nonstereophotos. The differences among category means for all other factors were relatively small.

VII.7. Fractional Replication in Factorials

A *fractional replicate* of a factorial treatment design is merely a subset of the complete factorial. For example, if the four treatments 000, 110, 101, and 011 were the only treatments included in an experiment, this would constitute a one-half replicate of a 2^3 factorial. The treatments not included form another one-half replicate and are 100, 010, 001, and 111. The two one-half replicates constitute the complete 2^3 factorial.

As a second example, suppose that the following numbers of treatments are selected from a 3^2 factorial

	a_0	a_1	a_2
b_0	1	0	1
b_1	0	n_1	0
b_2	1	0	1

The above fractional replicate of a 3^2 factorial is known as a <u>simple response surface treatment design</u> where the 1 represents one response for that combination, the zero represents no observations recorded, and n_1 means n_1 observations taken for a_1b_1. Likewise, the following fractional replicate of a 5^2 factorial is a simple response surface design (squares) plus 4 star points (circles):

	a_0	a_1	a_2	a_3	a_4
b_0	0	0	(1)	0	0
b_1	0	[1]	0	[1]	0
b_2	(1)	0	n_1	0	(1)
b_3	0	[1]	0	[1]	0
b_4	0	0	(1)	0	0

Thus, there are n_1 center points, 4 response surface points, and 4 star points in the above fractional replicate.

There are numerous types of fractional replicates, and an even more numerous list of published papers on the subject. The methods of constructing fractional replicates, the actual construction, the analyses for fractional replicates, and properties of various fractional replicates have been and are being actively investigated by statisticians.

Fractional replicate treatment designs are useful in specific instances. The fractional replicate is sometimes useful when the investigator wishes to study a large number of factors with each factor at several levels. For example, large factorials such as the 2^{10} with 1024 treatments and 3^7 with 2187 treatments may involve too many treatments for the experimenter to study. He may decide that the number of factors and/or the number of levels cannot be reduced. The alternative is to use a fractional replicate; for instance, a 1/8 replicate of a 2^{10} results in $2^7 = 128$ treatments and a 1/9 replicate of a 3^7 results in $3^5 = 243$ treatments.

In instances such as the above, fractional replication can be useful, but in cases where the treatments occur more than once in an experiment, it is inefficient to duplicate a fractional replicate. For example, a duplication of a star-point response surface design for $n_1 = 4$ would be:

	a_0	a_1	a_2	a_3	a_4
b_0	0	0	(2)	0	0
b_1	0	[2]	0	[2]	0
b_2	(2)	0	$2n_1=8$	0	(2)
b_3	0	[2]	0	[2]	0
b_4	0	0	(2)	0	0

If the form of the response were unknown, a much more efficient treatment design with the same number of observations would be the following:

	a_0	a_1	a_2	a_3	a_4
b_0	1	1	1	1	1
b_1	1	1	1	1	1
b_2	1	1	0	1	1
b_3	1	1	1	1	1
b_4	1	1	1	1	1

A general rule to follow in fractional replication is to take an additional fraction instead of repeating a fraction. In this manner the treatment designs nearer a complete factorial are achieved and a wider coverage of levels of treatment combinations is possible.

VII.8. Genetic Treatment Designs

In inheritance studies, the investigator must select the treatment (parents and crosses) included in the experiment. In studying simple Mendelian dominance in plants, the two parents, P_1 and P_2, and the cross $F_1 = P_1 \times P_2$ are sufficient. If, in addition, it is desired to know whether or not one, two, or three pairs of independent alleles are involved, then it would be necessary to include other treatments like, for example, the $F_2 = F_1 \times F_1$ progeny. The ratio of segregation in the F_2 individuals will indicate the number and nature of allelic pairs.

As the inheritance pattern becomes more complex, the greater will be the number of treatments required to ascertain the pattern. Also, as the number of generations of a cross increases, the number of possible crosses increases rapidly. For example, in the kth generation the following are possibilities for k = 0,1,2,3:

k^{th} generation	Crosses possible in generation k	No.=n_k
k=0	P_1, P_2	2
k=1	P_1, P_2, $F_1=P_1 \times P_2$	3
k=2	P_1, P_2, F_1, $F_2=F_1 \times F_1$, $B_1=F_1 \times P_1$, $B_2=F_1 \times P_2$	6
k=3	P_1, P_2, F_1, F_2, B_1, B_2, $P_1 \times F_2$, $P_1 \times B_1$, $P_1 \times B_2$, $P_2 \times F_2$, $P_2 \times B_1$, $P_2 \times B_2$, $F_1 \times F_2$, $F_1 \times B_1$, $F_1 \times B_2$, $F_2 \times F_2$, $F_2 \times B_1$, $F_2 \times B_2$, $B_1 \times B_1$, $B_1 \times B_2$, $B_2 \times B_2$	21
$k+1^{st}$		$n_k(n_k+1)/2$

Because the number of possible crosses quickly becomes too large to consider using all crosses, an investigator must necessarily select a subset of these; the correct subset depends upon the type of inheritance and genetics under study.

In breeding studies on variety evaluation, the genetic design includes selection of the proper controls. Several types of controls or standard varieties may be necessary to evaluate the new varieties being tested. Careful thought is required in obtaining adequate controls for screening new varieties.

An important type of genetic treatment design is the one known as the diallel crossing system for describing the crosses of k lines, for instance, strains of mice. Some possibilities for k=5 are:

Female lines	Design I male lines					Design II male lines					Design III male lines					Design IV male lines				
	1	2	3	4	5	1	2	3	4	5	1	2	3	4	5	1	2	3	4	5
1	x	x	x	x	x	x	x	x	x	x	-	x	x	x	x	-	x	x	x	x
2	x	x	x	x	x	-	x	x	x	x	x	-	x	x	x	-	-	x	x	x
3	x	x	x	x	x	-	-	x	x	x	x	x	-	x	x	-	-	-	x	x
4	x	x	x	x	x	-	-	-	x	x	x	x	x	-	x	-	-	-	-	x
5	x	x	x	x	x	-	-	-	-	x	x	x	x	x	-	-	-	-	-	-

In the above, x denotes the cross and - denotes no cross. All possible combinations are given in design I, which is the complete k^2 factorial. The other three designs are fractional replicates of the complete factorial. Design II represents "selfs" (a line crossed with itself) plus all possible crosses. Design III represents all possible crosses of the k lines plus all possible reciprocal crosses; for instance, if line 1 is the female and is crossed with line 2 as the male, then the reciprocal cross would be line 1 as the male crossed with line 2 as the female. Design IV gives the genetic treatment design of all possible crosses.

The genetic treatment design listed above as III has been used in measuring the amount of communication between individuals in psychological research. Obviously a person does not communicate with or talk to himself when he is a member of a group of individuals. This omits the "selfs". It is interesting that a number of genetic phenomena could be utilized to describe psychological phenomena associated with communication between individuals.

In classification and importance of job studies, it has been found that an individual thinks more highly of his own position or function than do people in other positions. This means that the "selfs" or the self-ratings should be omitted in an analysis of the data, resulting in design III. For example, suppose that we draw a sample from each of the following university groups and have them rate the relative importance of all groups other than their own:

Category	U	G	P	S	M	A
Undergraduate = U	-	x	x	x	x	x
Graduate student = G	x	-	x	x	x	x
Professor = P	x	x	-	x	x	x
Secretary = S	x	x	x	-	x	x
Maintenance personnel = M	x	x	x	x	-	x
Administrator = A	x	x	x	x	x	-

This would produce design III. This design has also been used to evaluate sequence of pairs of courses and educational experiences.

Design I has been used to study competition among varieties of wheat; the "selfs" represent a line in competition with plants of its own kind. Here again many of the genetic phenomena had direct counterparts in competition terms. If successive doses of drugs are given alone or in combination, the above design is directly useful. Design IV and genetic interpretations have been used in cockfighting experiments; it has also been used frequently in paired-comparisons experiments in the social sciences, and in round-robin tournaments. Design II has been utilized to compare mixtures of beans to determine whether higher yields can be obtained from mixtures than from a variety planted alone (the "selfs").

VII.9. Bioassay Treatment Designs

A _biological assay_, or _bioassay_, is an experimental procedure for identifying the constitution or for estimating the potency of materials by means of their reaction on living material. Examples of use of assays are:

1. identification of blood groups by serological tests,
2. estimation of potencies of vitamins from their effects on the growth of microorganisms,
3. comparison of insecticides by toxicity tests,
4. others.

An _analytical assay_ is a procedure of estimating the potency of a _test preparation_ (e.g., a natural source of a vitamin) relative to a _standard preparation_ containing the same active material (e.g., a pure chemical form). This type of assay is such that X units of the test preparation produce the same average response as RX units of the standard preparation. The value of R is then defined as the relative potency. One important type of response is the following:

$$\text{standard preparation} : Y = a + bX$$
$$\text{test preparation} : Y = a + bRX$$

Now the two response equations intersect when X=0 and the ratio of the slopes of the two lines, that is, $bR/b = R$, gives the relative potency. An experiment designed to estimate R in this way is denoted as a slope ratio assay. An example of this type of bioassay is reproduced below in figure VII.8 (from Finney [1955], page 125):[1]

An assay which probably has wider applicability than the slope ratio assay is the parallel line assay. In the latter type the relative potency R is measured as the horizontal difference between the following two parallel lines:

$$\text{standard preparation} : Y = a + b \log X$$
$$\text{test preparation} : Y = (a + b \log R) + b \log X$$

An example illustrating this type of assay is taken from page 127 of Finney [1955] and presented in figure VII.9.[2]

The problem of selecting the doses, X, to use, the subjects to be utilized in an experiment, and the number of doses represents the selection of the treatment design in bioassays. Many of the concepts and problems involved are discussed by Finney [1955]. In addition to the latter reference, Professor Finney has written an entire textbook on this important subject. Active statistical research is being pursued on bioassay at the present time.

VII.10. Sequential Selection of Treatments

In many types of experimentation it is possible to perform observations in sequence; each succeeding treatment is selected from the result for the present treatment. The object here is to reach some desired goal such as

---

[1] With the permission of the University of Chicago Press.

[2] With the permission of the University of Chicago Press.

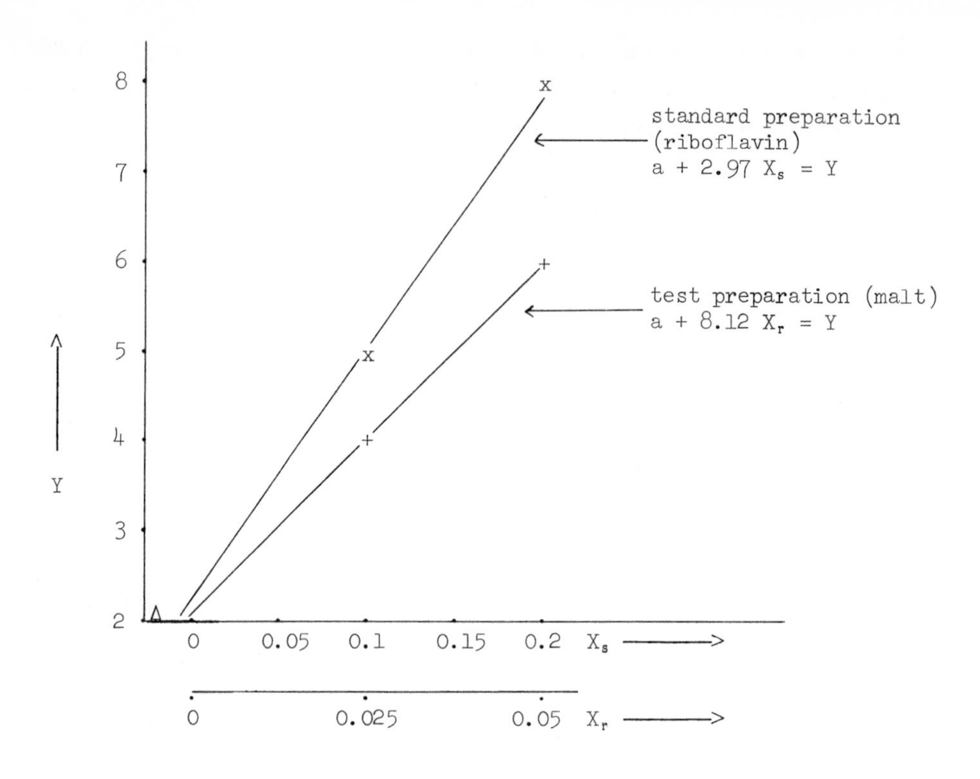

Figure VII.8. Assay of riboflavin in malt, using L. helveticus as subject. Upper horizontal scale (X_s): Dose of riboflavin per tube, in micrograms. Lower horizontal scale (X_r): Dose of malt per tube, in grams. Vertical scale (Y): Titer of N/10 sodium hydroxide in milliliters. Δ: mean response for 4 tubes without treatment; x: mean responses for 4 tubes on standard preparation; +: mean responses for 4 tubes on test preparation. Two lines intersecting at X=0 have been fitted by standard statistical techniques. The standard line rises by 2.97 ml per 0.1 mg riboflavin, the test line by 8.12 ml per 0.1 gm malt. Hence, the malt is estimated to contain 8.12/2.97, or 2.73 mg riboflavin per gram of malt.

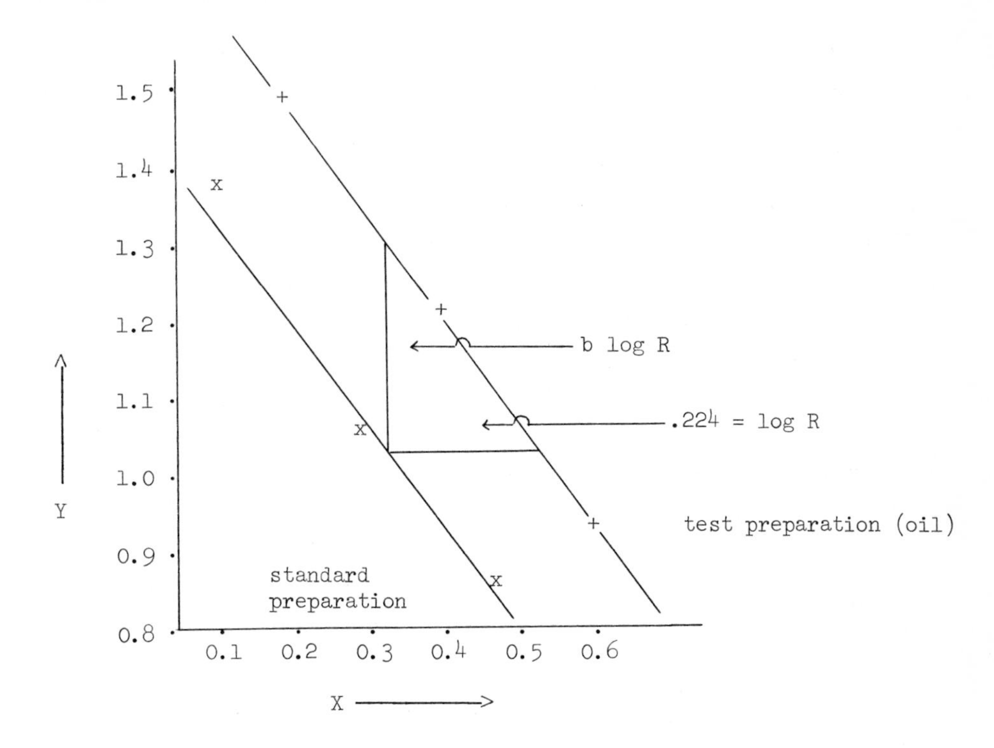

Figure VII.9. Assay of vitamin D in an oil by chick method. Horizontal scale (X): log daily dose per chick, in units vitamin D or milligrams oil. Vertical scale (Y): log tarsal-metatarsal distance, in 0.01 mm; x: mean responses for 28 chicks on standard preparation; +: mean responses for 28 chicks on test preparation. Two parallel lines have been fitted by standard statistical techniques. Measurement shows that the X values of the test line would have to be reduced by 0.224 in order to superimpose it on the standard line. Hence the oil is estimated to contain 0.597 units vitamin D per milligram (since log R = -.224 = 0.776 - 1 = log 5.97 - log 10 = log .597). Therefore, one gram of oil is equivalent to .597 units of vitamin D.

the dosage giving 50% kill (LD_{50} = mean lethal dose giving 50% kill), the combination of ingredients giving maximum load volume, the combination of ingredients making the softest (or hardest) alloy of materials, the combination of fertilizers giving maximum plant response, and so forth. In such investigations it is assumed that there is one combination or dosage which produces the desired results. For example, in seeking an optimum we may liken this to the following example. We proceed as if there were one hill in an area with all land sloping away from the summit. It is so foggy that we can observe only the point at which we are located. We take a reading with a level and proceed in the direction indicated, hoping to reach the summit. If there is more than one hill in the area, or if there are plateaus, we are in trouble and may never find our way to the summit with this procedure. So it is with the above statistical procedures designed for a particular situation. If, for example, the 50% kill is attained with a variety of dosages, there will be no unique dosage giving this prescribed value and there will also be a variety of doses giving other percentage kills.

One of the earlier sequential procedures described is the Dixon-Mood "up-and-down" method, so called because it was first used to determine the heights from which explosives could be dropped before they exploded. If the sample explosive did not explode when dropped at from 5 feet, say, the explosive was dropped from 6 feet. If it exploded, the next sample was tried from 5 feet. If this exploded the next sample was dropped from 4 feet. This process was continued, and it tended to concentrate the observations around the height at which explosives would explode when dropped.

The same procedure could be utilized to estimate the dosage for any specified percentage kill, for example, 50% kill. Samples of N = 100, say, insects would be used. A dosage would be tried and the percentage of insects killed would be observed. If less than 50% of the insects were killed, the dosage would be increased. If more, the dosage would be decreased. The process is continued until the percentage kill oscillates around the desired level.

Another procedure which achieves much the same effect as the above but which goes further in selecting the next treatment level for examination is the Robbins-Munro method. The first observation or treatment is selected from previous knowledge to be as near the desired point as possible. If there is no previous experience with the treatment, then a start is made at any point. The result of the first observation is denoted as Y_1 and of the first treatment as X_1. The result of the second observation is Y_2 observed as X_2; the third observation is Y_3 observed at X_3; etc. The resulting levels of X after the first level of X are to be determined as follows:

$$X_2 = X_1 + (\alpha - Y_1)$$

$$X_3 = X_2 + \tfrac{1}{2}(\alpha - Y_2)$$

$$X_4 = X_3 + \tfrac{1}{3}(\alpha - Y_3)$$

$$\vdots$$

$$X_{n+1} = X_n + \frac{1}{n}(\alpha - Y_n)$$

where α is the desired value, say 50% kill. The last value of X, that is X_{n+1}, is the estimated value of the dose giving the desired percentage kill.

Professors Kiefer and Wolfowitz of Cornell University have devised a procedure for obtaining the value of the treatment giving maximum (or minimum) response for situations involving mixtures of compounds such as proportions of two metals in an alloy and proportions of baking powder and soda in a sour milk cake. The procedure is almost as simple to apply as the Robbins-Munro procedure. The steps follow:

1. Select two sequences of numbers such as
$$a_i = \frac{1}{i} \text{ and } c_i = \left(\frac{1}{i}\right)^{\frac{1}{3}}.$$
2. Select a starting point of X, say Z_1.
3. Let $X_1 = Z_1 - c_1$ and $X_2 = Z_2 + c_1$.
4. Obtain responses Y_1 and Y_2 at X_1 and X_2 respectively.
5. Compute $Z_2 = Z_1 + (Y_2 - Y_1)$.
6. Let $X_3 = Z_2 - c_2$ and $X_4 = Z_2 + c_2$ and obtain the corresponding responses Y_3 and Y_4.
7. Let
$$Z_3 = Z_2 + \frac{\frac{1}{2}}{(\frac{1}{2})^{\frac{1}{3}}}(Y_4 - Y_3)$$

$$= Z_2 + (\tfrac{1}{2})^{\frac{2}{3}} (Y_4 - Y_3) .$$

8. The above process continues until n pairs of observations or responses have been obtained. The value of X (say X_{max}) giving the maximum response is then computed as

$$Z_{n+1} = Z_n + \left(\frac{1}{n}\right)^{\frac{2}{3}} (Y_{2n} - Y_{2n-1}) = X_{max}.$$

VII.11. Experimental Designs for the Various Treatment Designs

The nature of the treatment design has little or nothing to do with the experimental design used. For example, suppose that the experimenter uses a 2^3 factorial treatment design. Depending only upon the nature of the experiment and the nature of the heterogeneity present, he may use any one of the following experimental designs:

1. a completely randomized design with r replicates,
2. a randomized complete blocks design with r replicates,
3. an 8×8 latin square design with r=8 replicates,
4. a 4×6 latin rectangle design with r=3 replicates,
5. a 4×8 latin rectangle design with r=4 replicates,
6. an 8×7 Youden square design with r=7 replicates,
7. an 8×9 Youden square design with r=9 replicates,
8. a bib design with k=4, b=14, r=7, and λ=3,
9. a bib design with k=2, b=28, r=7, and λ=1, etc.

Thus, we see that there are many experimental designs available for any given treatment design.

VII.12. Problems

VII.1. (This problem is taken from examples 14.4 and 14.5 of Snedecor's book [1946], 4th edition.) [1] In a study of the baking quality of straight grade flour, Q, after being heated to 170°F. for various lengths of time T, in hours, the following data were obtained:

[1] Reprinted by permission from Statistical Methods, 4th edition, by G. W. Snedecor, © by The Iowa State University Press.

T (hrs.)	Q	log Q	log (Q - 14)	log T
0.25	93			
0.50	71			
0.75	63			
1.00	54			
1.50	43			
2.00	38			
3.00	29			
4.00	26			
6.00	22			
8.00	20			

Obtain log T, log Q, and log(Q-14) from logarithmic tables. Then prepare the following graphs:

1. Plot Q as the ordinate against T as the abscissa.
2. Plot log Q against log T and plot the line through the points for $\log Q = 1.7116 - 0.4678 \log T$.
3. Plot 1/Q against T.
4. Plot $Y = 1/(Q - 14)$ against T.
5. (Optional) If you have access to a calculating machine, plot Q against T using the following equation:

$$Q = \frac{1}{.02T + .0055} + 14 \quad .$$

VII.2. In an investigation of the relationship of the number of hours of preparation (T) for an examination and the percentage passing (Y) the following data were obtained:

Number of hours of preparation = T	0	10	20	30	40	60
Percentage passing = Y	66	85	91	95	97	97
log (100-Y)						

1. Plot Y, ordinate, against T, abscissa.
2. Plot log (100-Y) against T. (Note: This is a form of the Mitscherlich equation.)

Note the change in relationship obtained in the two plots.

VII.3. In a study of failure rates (F) of motorboat motors after T hours of continuous operation, the following data are available:

Number of hours of operation = T	2.5	5	10	40	120	240	480	720
Percentage of failures = F	55	60	63	66	69	73	74	75
log T								

1. Plot F, ordinate, against, T, abscissa.
2. Plot F against log T, abscissa.

Note the change in the relationship on the log scale.

Note: In the above three problems note that the type of the investigation is an absolute experiment in the sense that it is a single phenomenon experiment. The interest is on the response to a single factor rather than in comparing the individual levels *per se*.

VII.4. Suppose that Miss I. M. Fashion, "beauty" researcher, wishes to observe the effect of 3 kinds of oil (olive oil, lanolin, and saffron oil) with two types of pigments (plant pigment and laboratory pigment) on "presentability" of eye shadow after 12 hours, as measured by her procedure, for four different colorings (magnolia pink, blue-bonnet blue, avocado green, and ebony black). She wishes to use the various levels of these three factors in all combinations. Write out the combinations for her.

VII.5. Given the following effects from a 2×3 factorial experiment reconstruct the yield observations Y_{ij}.

	Two-factor Interaction Effects
$\bar{y}_{..} = 10$	a_0b_0 : -3
$\bar{y}_{0.} - \bar{y}_{..} = -7 = a_0$	a_0b_1 : +1
$\bar{y}_{1.} - \bar{y}_{..} = +7 = a_1$	a_0b_2 : +2
$\bar{y}_{.0} - \bar{y}_{..} = +13 = b_0$	a_1b_0 : 3
$\bar{y}_{.1} - \bar{y}_{..} = -12 = b_1$	a_1b_1 : -1
$\bar{y}_{.2} - \bar{y}_{..} = -1 = b_2$	a_1b_2 : -2

Graph the responses, Y_{ij}, using the three levels of the second factor as the abscissa and Y as the ordinate.

VII.6. The following represents the actual layout of an experiment and actual yields of wheat. The letters represent the treatments with n referring to nitrogen, p to phosphorous and k to potassium fertilizers. If a letter is present this means that element in the fertilizer was present. What appears to be the experimental design and the randomization procedure? What is the treatment design?

p 19	n 12	np 18	k 16	nk 11	(1) 12	npk 19	pk 19
n 13	nk 7	pk 17	npk 17	p 20	k 12	np 19	(1) 16
nk 11	np 18	n 10	p 18	(1) 10	npk 17	pk 18	k 14
pk 18	k 13	npk 14	(1) 12	n 11	np 15	p 17	nk 16
np 18	(1) 13	nk 13	n 11	pk 17	p 16	k 11	npk 17
k 15	pk 18	(1) 13	np 17	npk 16	n 10	nk 9	p 21
npk 19	p 19	k 11	pk 17	np 18	nk 9	(1) 10	n 15
(1) 18	npk 20	p 21	nk 16	k 17	pk 18	n 14	np 23

VII.7. Sometimes researchers unwittingly obtain an apparent relation which turns out to be spurious. For example, suppose that one has a measure of variation, say s, and a mean $\bar{y}$ for a series of values of $\bar{y}$ and s. Suppose that one plots relative variation = $s/\bar{y}$ against $\bar{y}$; if s stays relatively

constant compared to $\bar{y}$, then one is essentially plotting $1/\bar{y}$ against $\bar{y}$ which gives a beautiful curve. Another example of this appears when one plots new recruits relative to parent stock in the population against parent stock in the population. If the number of new recruits is relatively constant, then one is essentially plotting the reciprocal of numbers of parent stock against numbers of parent stock. Construct a set of 10 numbers, d_i, whose sum is zero, make the range in these 10 numbers small, say $-2 \geq d_i \geq 2$, select a set of 10 values of y_i whose range is 20, say 1 to 20, add the d_i randomly to the y_i, add a constant, say 4, to each of the $y_i + d_i$ values, and then plot $1/(y_i+d_i+4)$ against $y_i + d_i + 4$. Observe the relatively regular curve which, of course, merely indicates that one is plotting a set of values against a function of these same values.

VII.8. Obtain thickness measurements for pages 1-10, 11-30, 31-60, 61-100, 101-150, and 151-210 of this book. Plot the thickness against number of pages. Describe your measuring instrument. What functional relation exists between number of pages and thickness of pages? How would you estimate the thickness of a single page? Repeat the experiment but measure the thickness of page 1, pages 2-3, 4-6, 7-10, 11-15, and 16-21 each 10 times around the edge of the page. Plot the 10 measurements against number of pages. In both cases you measured 210 pages; which procedure do you believe to be the more accurate for determining the thickness of a page and why? Does the functional relation change? Why or why not? Which procedure gives greater variation in measurements?

VII.13. References and Suggested Reading

Cox, D. R. [1958]. Planning of Experiments. John Wiley and Sons, Inc., New York, pp. vii + 308

(Chapters 6 and 7, are, for the most part, at the level of this book. Dr. Cox has covered some of the points discussed in the above; these two chapters will serve to deepen the student's insight into problems of treatment design.)

Federer, W. T. [1955]. Experimental Design - Theory and Application. The Macmillan Company, New York, pp. xix + 544 + 47.

(Chapters VII and VIII treat the various aspects of factorial experiments and are more advanced and more detailed than the preceding reference. Additional material on factorial experiments may be found in Chapters IX and X.)

Finney, D. J. [1955]. Experimental Design and its Statistical Basis. The University of Chicago Press, Chicago and London, pp. xi + 169.

Kempthorne, O. [1952]. The Design and Analysis of Experiments. John Wiley and Sons, Inc., New York, pp. xix + 631.

(A theoretical and mathematical treatment of treatment design is given in chapters 13-21.)

Snedecor, G. W. [1946]. Statistical Methods. 4th edition. Iowa State College Press, Ames, Iowa, pp. xvi + 485.

(Although the numerous computational techniques are not to be considered in this text, a perusal of the many diverse examples in chapters 6, 7, 13 and 14 may be of interest to the reader.)

Yates, F. [1937]. Design and analysis of factorial experiments. Imperial Bureau of Soil Science, Technical Communication No. 35.

(A classical treatise on the subject of factorial experiments.)

Youden, W. J., Connor, W. S., and Severo, N. C. [1959]. Measurements made by matching with known standards. Technometrics 1:101-1-9.

VIII. SUMMARIZATION OF DATA: GRAPHS, CHARTS, AND FIGURES

VIII.1. Introduction

The previous chapters have been concerned with procedures for obtaining meaningful and reliable data. This chapter and chapter 8 describe some methods for summarizing data from a survey or an experiment. In the present chapter we describe different kinds of graphs, charts, diagrams, and figures useful in summarizing some relevant facts from a mass of data. As has been promised, the data obtained for the 1967 class survey are to be utilized to illustrate some statistical procedures. Since data collection often results in large quantities of numbers, some form of summarization is necessary.

VIII.2. A Partial Summarization of the Class Survey

One often hears the statement that "A picture is worth a thousand words" (attributed originally to an unknown Chinese of bygone days, or so Bob Hope said at the Oscar awarding ceremony on April 18, 1966). This statement applies so aptly to presenting the numerical results from an experiment or survey. The individual data may be too numerous to present, and/or the meaning may not be clear in the individual data whereas it is in summary form such as in a picture, graph, table, or chart. Summarization of individual observations is essential in order to glean the information from the data and to interpret the results from the survey or the experiment.

In the class survey on heights, ages, weights, eye and hair color, and class standing we obtained the data given in table VIII.1. The first items to note are that this represents a lot of numbers and that data on some individuals are incomplete. Of the 101 individuals registered for credit, 85 individuals were in class on the day of the survey, and an additional 13 who were not in class reported height, weight, age, hair and eye color, and class standing data to total 98 individuals. Of the 93 who submitted measurements on themselves, four failed to report a weight measurement and one failed to report his version of his eye and hair color. Also, in scanning the data of table VIII.1, student number 5 was measured as 1800 mm tall by team A and as 1706 mm tall by team B. It would appear from "own" height that team A made an error in measuring or in recording and that the height should have read 1700 mm instead of 1800 mm. Also, it is possible that this was a typographical error.

One method of summarizing some of the information obtained is given in figures VIII.1 and VIII.2, and tables VIII.2, VIII.3, and VIII.4. In figures VIII.1 and VIII.2 we have plotted the frequency of heights that occurred at any given measurement. Such frequency distributions are called <u>histograms</u>. For example, when measuring in feet, there were 9 individuals in the 5-foot class and 76 in the 6-foot class. Obviously, we had mostly 6-footers in the class! This most frequent class is called the <u>modal class</u> and the class center, 6 feet, is called the mode. The class interval is from 5'6" to 6'6". When heights are recorded in inches as reported on the class questionnaire, the range is from 62 inches to 75 inches. In 1966, the range was from 61 to 78 inches. One or more students were represented in all 14 classes, but girls appeared only in the first six classes. The modal classes here are the 68 and 71 inch classes and the modes are 68 and 71 inches. When the frequency distribution of heights in 2.5 cm class intervals for recording instrument and recorder A is constructed, we note that the range in class centers is from 160.0 to 192.5 cm whereas the range in mm is from 1600 to 1920 mm. The modal class is 175.0 cm with the mode equal to 175.0 cm. For recording instrument and recorder B the range is from 1613 mm to 1919 mm with a modal class of 172.5 cm and with a mode of 172.5 cm.

It should be noted here that measuring instrument A was biased. The metersticks were screwed to the pine board 3 mm above the bottom of the stick. Thus, before any comparisons are made between the heights obtained by the two recorders using the two instruments, 3 mm must be subtracted from each of the heights obtained by recorder A. Perhaps the measuring instrument should be biased by 3 cm to see if anyone observes this fact! If we take the differences between recorders, which are listed as A-B in table VIII.1, we note that the differences range from -62 mm to +94 mm, or from -6 to +9 cm. There are 41 differences with a minus sign, three ties, and 41 with a plus sign. After subtracting 3 mm from the A measurements, there are 51 differences with a minus sign, three ties, and 31 with a plus sign. Since we expect 85/2 = 42.5 pluses on the basis of a hypothesis of no difference in the measurements A and B, we note that this is a fairly poor fit with expectation. (Note for those who know about the chi-square statistic, χ^2 (one degree of freedom) = $[(52.5 - 42.5)^2 + (32.5 - 42.5)^2]/42.5 = \frac{200}{42.5}$ = 4.7, which is a rather large value of chi-square. One and a half of the ties were put into the minus class and one and a half into the plus class.

Table VIII.1. Measurements on 1967 class.

Student[a]	Own in.	mm (A)	mm (B)	ft. (A)	ft. (B)	yd. (A)	yd. (B)	Weight Own	Weight (B)	Age own (yr., mo.)	Eye color Own	Eye color (A)	Hair color Own	Hair color (A)	Student[b] Own	Hts. (mm) A-B	Hts. (mm) $\frac{A+B}{2}$ (in.)
1.	71	1810	1819	6	6	2	2	148	152	20, 1	Bl	Bl	Br	Br	So.	- 9	1814(71)
2.	65							120		32, 6	Br		Bk		Gr.		
3.	72	1800	1832	6	6	2	2	175	174	20, 7	Bl	Bl	Blo	Blo	Jr.	-32	1816(71)
4.	70	1800	1791	6	6	2	2	173	182	18, 5	Br	Br	Bk	Br	So.	9	1796(71)
5.	67	1800	1706	6	6	2	2	152	159	21, 11	Bl	Bl	Blo	Br	Sr.	94	1753(69)
6.	69	1750	1738	6	6	2	2	160	163	22, 6	Br	Br	Bk	Br	Jr.	12	1744(69)
7.	64	1660	1658	5	5	2	2	131	143	21, 2	Bl	Bl	Br	Br	Sr.	2	1659(65)
8.	72	1820	1820	6	6	2	2	180	184	20, 0	Bl	Bl	Blo	Blo	Jr.	0	1820(72)
9.*	66	1680	1680	6	6	2	2	136	136	18, 2	Bl	Bl	Blo	Blo	Fr.	0	1680(66)
10.	69	1760	1768	6	6	2	2	150	151	18, 10	Br	Br	Br	Br	Fr.	- 8	1764(69)
11.*	63	1640	1617	5	5	2	2	140	141	19, 6	Br	Br	Br	Br	So.	23	1628(64)
12.*	67	1730	1712	6	6	2	2	123	123	19, 11	Bl	Bl	Br	Br	Sr.	18	1721(68)
13.	70	1760	1770	6	6	2	2	150	154	19, 0	Gy	Bl	Br	Br	So.	-10	1765(69)
14.	71							171		18, 2	Br		Br		Fr.		
15.	75	1900	1890	6	6	2	2	170	169	22, 6	Bl	Bl	Bk	Br	Sr.	10	1895(75)
16.		1780	1784	6	6	2	2		144			Br		Br		- 4	1782(70)
17.	72	1840	1848	6	6	2	2	160	160	21, 10	Bl	Bl	Br	Br	Sr.	- 8	1844(73)
18.	68	1740	1737	6	6	2	2	160	169	32, 0	Br	Br	Bk	Bk	Gr.	3	1738(68)
19.*	65	1670	1681	5	5	2	2	145	150	20, 0	Bl	Bl	Br	Br	So.	-11	1676(66)
20.	71	1790	1779	6	6	2	2		181	21, 11	Gy	Br	Br	Br	Sr.	11	1784(70)
21.	72	1840	1839	6	6	2	2	153	158	19, 9	Br	Br	Br	Br	Fr.	1	1840(72)
22.	73	1890	1874	6	6	2	2	177	185	18, 6	Gr	Bl	Br	Br	Fr.	16	1882(74)
23.	72	1840	1833	6	6	2	2	224	220	32, 2	Bl	Bl	Br	Br	Other	7	1836(72)
24.		1700	1703	6	6	2	2		158			Gy		Br		- 3	1702(67)
25.	75	1900	1914	6	6	2	2	195	201	19, 5	Br	Br	Br	Br	So.	-14	1907(75)
26.	72	1840	1839	6	6	2	2	148	143	19, 8	Br	Br	Br	Br	So.	1	1840(72)
27.*	66	1690	1703	6	6	2	2	125	121	19, 5	Br	Br	Br	Br	So.	-13	1696(67)
28.	71	1800	1792	6	6	2	2	148	149	18, 2	Bl	Bl	Br	Br	Fr.	8	1796(71)
29.	70	1703	1765	6	6	2	2	130	129	19, 2	Gr	Br	Br	Br	So.	-62	1734(68)
30.	65	1603	1640	5	5	2	2	142	141	19, 4	Bl	Bl	Br	Br	So.	-37	1622(64)

[a] Asterisk indicates female student as determined by name.

[b] Fr. = Freshman, So. = Sophomore, Jr. = Junior, Sr. = Senior, Gr. = Graduate, Other = Other.

Table VIII.1. (2) Continued

Student[a]	Own in.	mm (A)	mm (B)	ft. (A)	ft. (B)	yd. (A)	yd. (B)	Weight Own	Weight (B)	Age own (yr., mo.)		Eye color Own	Eye color (A)	Hair color Own	Hair color (A)	Student[b] Own	Hts. (mm) A-B	Hts. (mm) $\frac{A+B}{2}$ (in.)
31.	71	1820	1800	6	6	2	2	168	170	19,	6	Gr	Br	Br	Br	So.	20	1810(71)
32.	69	1790	1765	6	6	2	2	180	186	21,	3	Bl	Bl	Br	Br	Sr.	25	1778(70)
33.	72							175		20,	0	Br		Br		So.		
34.	72	1840	1836	6	6	2	2	140	145	21,	7	Bl	Bl	Br	Br	Jr.	4	1838(72)
35.*	64	1660	1651	5	5	2	2	125	127	18,	9	Br	Br	Blo	Blo	Fr.	9	1656(65)
36.	71							145		21,	1	Br		Br		Sr.		
37.	69	1781	1783	6	6	2	2	145	149	19,	1	Bl	Br	Br	Br	So.	- 2	1782(70)
38.	71	1820	1828	6	6	2	2	146	148	21,	6	Br	Br	Br	Br	Sr.	- 8	1824(72)
39.	67	1680	1708	6	6	2	2	125	120	19,	5	Bl	Bl	Br	Br	So.	-28	1694(67)
40.*	66	1700	1706	6	6	2	2	128	129	19,	6	Br	Br	Br	Br	So.	- 6	1703(67)
41.	74	1910	1919	6	6	2	2		187	19,	7	Bl	Bl	Br	Br	So.	- 9	1914(75)
42.	74	1920	1895	6	6	2	2	175	184	19,	11	Bl	Bl	Br	Br	So.	25	1908(75)
43.	73	1850	1853	6	6	2	2	168	169	21,	0	Br	Bl	Blo	Br	Sr.	- 3	1852(73)
44.		1800	1782	6	6	2	2		166				Bl		Blo		18	1791(71)
45.	74	1880	1872	6	6	2	2	151	159	19,	4	Bl	Bl	Br	Br	So.	8	1876(74)
46.	71	1782	1818	6	6	2	2	130	135	17,	0	Bl	Bl	Br	Br	Fr.	-36	1800(71)
47.	68	1720	1721	6	6	2	2	155	159	22,	0	Gr	Bl	Br	Br	Jr.	- 1	1720(68)
48.*	63							110		21,	5	Br		Br		Sr.		
49.	68	1710	1708	6	6	2	2		199	38,	8	Br	Br	Bk	Bk	Gr.	2	1709(67)
50.	66	1690	1683	6	6	2	2	165	173	19,	8	Gr	Gr	Br	Br	So.	7	1686(66)
51.	72	1840	1814	6	6	2	2	178	182	20,	9	Br	Br	Bk	Bk	Sr.	26	1827(72)
52.	68	1740	1724	6	6	2	2	156	158	20,	11	Gr	Gy	Bk	Br	So.	16	1732(68)
53.	72	1783	1812	6	6	2	2	151	152	20,	2	Gr	Br	Br	Br	So.	-29	1798(71)
54.	74	1920	1885	6	6	2	2	213	217	19,	7	Bl	Bl	Br	Br	So.	35	1902(75)
55.	69	1750	1770	6	6	2	2	174	181	20,	3	Gy	Gy	Br	Br	Jr.	-20	1760(69)
56.	70	1800	1804	6	6	2	2	155	161	25,	5	Bl	Bl	Br	Br	Sr.	- 4	1802(71)
57.	68	1740	1750	6	6	2	2	165	168	24,	7	Gy	Bl	Br	Br	Gr.	-10	1745(69)
58.	68	1720	1721	6	6	2	2	115	116	20,	2	Br	Br	Br	Br	Jr.	- 1	1720(68)
59.	74	1901	1890	6	6	2	2	163	168	20,	2	Bl	Bl	Br	Br	Jr.	11	1896(75)
60.	69	1740	1750	6	6	2	2	157	163	22,	0	Bl	Bl	Br	Br	Sr.	-10	1745(69)
61.*	67	1740	1725	6	6	2	2	140	149	21,	9	Bl	Gy	Br	Br	Sr.	15	1732(68)
62.	71	1830	1821	6	6	2	2	135	132	19,	4	Bl	Bl	Br	Br	So.	9	1826(72)
63.	69	1740	1751	6	6	2	2	145	150	20,	2	Br	Br	Br	Br	Sr.	-11	1746(69)

Table VIII.1. (3) Continued

Student[a]	Own in.	mm (A)	mm (B)	ft. (A)	ft. (B)	yd. (A)	yd. (B)	Weight Own	Weight (B)	Age own (yr., mo.)	Eye color Own	Eye color (A)	Hair color Own	Hair color (A)	Student[b] Own	Hts. (mm) A-B	Hts. (mm) $\frac{A+B}{2}$ (in.)
64.	63	1620	1619	5	5	2	2	126	128	23, 1	Bk	Br	Bk	Bk	Sr.	1	1620(64)
65.		1790	1781	6	6	2	2		166			Br		Bk		9	1786(70)
66.	68	1730	1736	6	6	2	2	145	146	17, 9	Br	Br	Br	Br	Fr.	- 6	1733(68)
67.	67	1680	1703	6	6	2	2	165	176	20, 0	Br	Br	Br	Br	So.	-23	1692(67)
68.	70	1800	1797	6	6	2	2	126	131	19, 0	Bl	Bl	Br	Br	So.	3	1798(71)
69.	66	1700	1715	6	6	2	2	148	154	19, 3	Gr	Br	Br	Br	So.	-15	1708(67)
70.	68	1730	1748	6	6	2	2	145	150	20, 0	Bl	Bl	Br	Br	So.	-18	1739(68)
71.	67	1710	1708	6	6	2	2	153	153	20, 6	Br	Br	Br	Br	Sr.	2	1709(67)
72.*	63							110		21, 9	Br		Br		Sr.		
73.*	67	1701	1713	6	6	2	2		147	19, 8	Br	Br	Br	Br	So.	-12	1707(67)
74.	74							172		20, 2	Bl		Blo		So.		
75.	68							148		21, 11	Bl		Br		Sr.		
76.	69	1750	1748	6	6	2	2	135	142	20, 2	Gr	Br	Br	Br	So.	2	1749(69)
77.	71	1820	1795	6	6	2	2	172	179	19, 3	Gy	Br	Br	Br	So.	25	1808(71)
78.	68							184		20, 5					So.		
79.*	63	1620	1637	5	5	2	2	110	112	17, 10	Bl	Bl	Br	Br	Fr.	-17	1628(64)
80.*	65	1680	1660	6	6	2	2	125	125	18, 0	Gr	Br	Br	Br	Fr.	20	1670(66)
81.*	62	1600	1613	5	5	2	2	95	95	20, 2	Br	Br	Br	Br	Sr.	-13	1606(63)
82.	74	1900	1870	6	6	2	2	175	176	20, 1	Br	Br	Br	Br	Jr.	30	1885(74)
83.	68	1750	1744	6	6	2	2	215	218	23, 4	Gr	Br	Br	Br	Jr.	6	1747(69)
84.	72	1890	1893	6	6	2	2	240	252	21, 10	Br	Br	Br	Br	Sr.	- 3	1892(74)
85.	65	1640	1641	5	5	2	2	142	155	17, 3	Br	Br	Br	Bk	Fr.	- 1	1640(65)
86.	68	1720	1733	6	6	2	2	150	151	20, 1	Gy	Br	Br	Br	Jr.	-13	1726(68)
87.	71							170		20, 0	Br		Br		So.		
88.		1800	1801	6	6	2	2		144			Bl		Br		- 1	1800(71)
89.	71	1820	1794	6	6	2	2	143	141	20, 11	Bl	Bl	Br	Br	Jr.	26	1807(71)
90.	68	1680	1716	6	6	2	2	165	172	25, 5	Br	Br	Rd	Br	Jr.	-36	1698(67)
91.	68	1740	1737	6	6	2	2	160	170	20, 7	Gr	Bl	Br	Blo	Jr.	3	1738(68)
92.	72	1840	1840	6	6	2	2	184	188	22, 8	Bl	Bl	Br	Br	Sr.	0	1840(72)
93.	71							166		21, 0	Gr		Bk		Jr.		
94.	69	1720	1739	6	6	2	2	160	160	20, 0	Bl	Bl	Bk	Bk	Jr.	-19	1730(68)
95.	71							185		20, 6	Br		Br		Jr.		
96.	66	1710	1713	6	6	2	2	149	153	28, 0	Gr	Bl	Br	Br	Other	- 3	1712(67)
97.	73	1870	1856	6	6	2	2	213	215	20, 8	Gy	Br	Br	Br	Jr.	14	1863(73)
98.	72							184		18, 3	Gr		Br		Fr.		
Number	93	85	85	85	85	85	85	89	85	93	92	85	92	85	93		

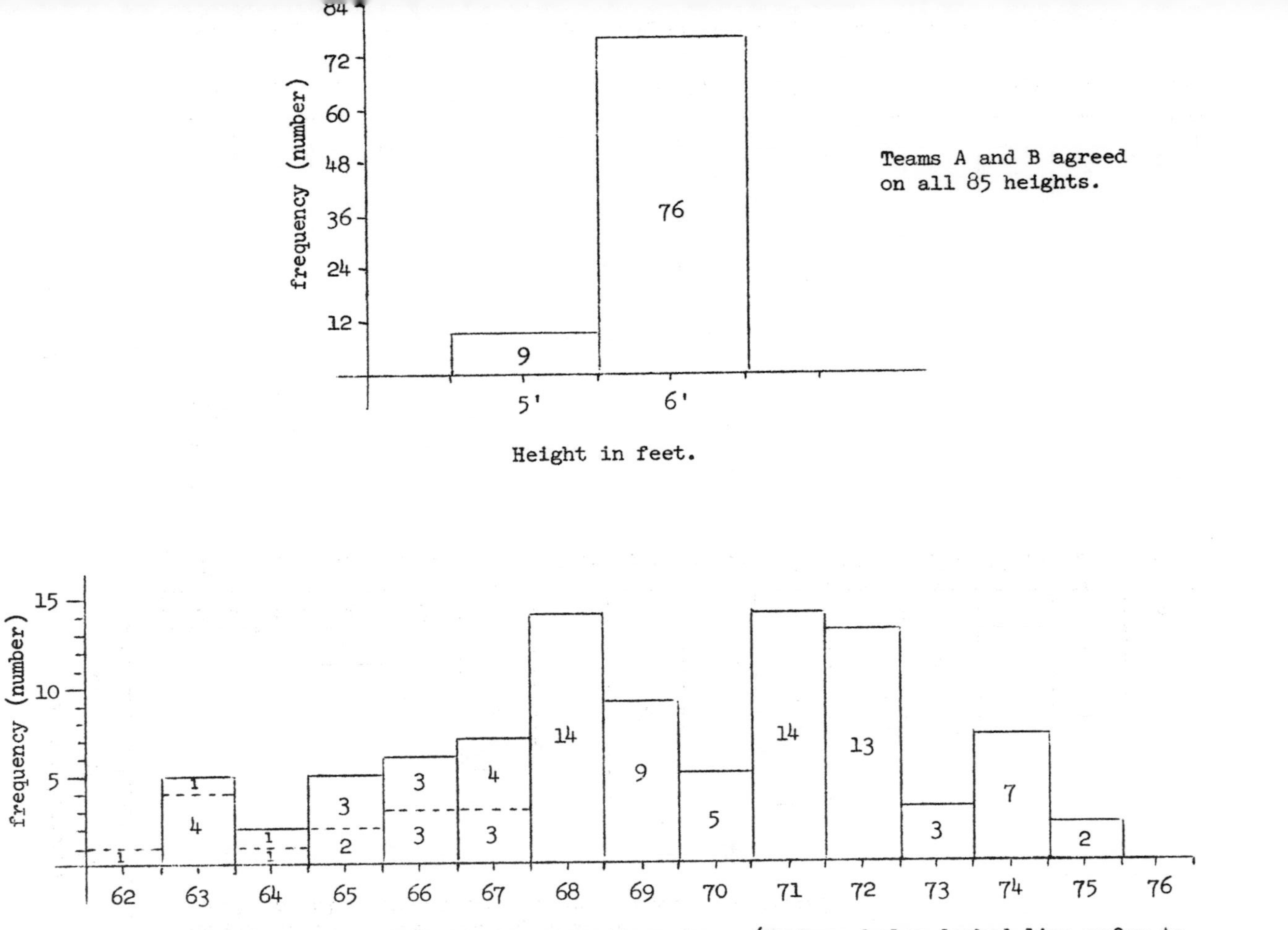

Height in inches as returned on class questionnaire. (Numbers below dashed line refer to number of girls out of 14.)

Figure VIII.1. Heights in feet as measured by teams A and B and in inches as recorded by the students themselves, Spring class, 1967.

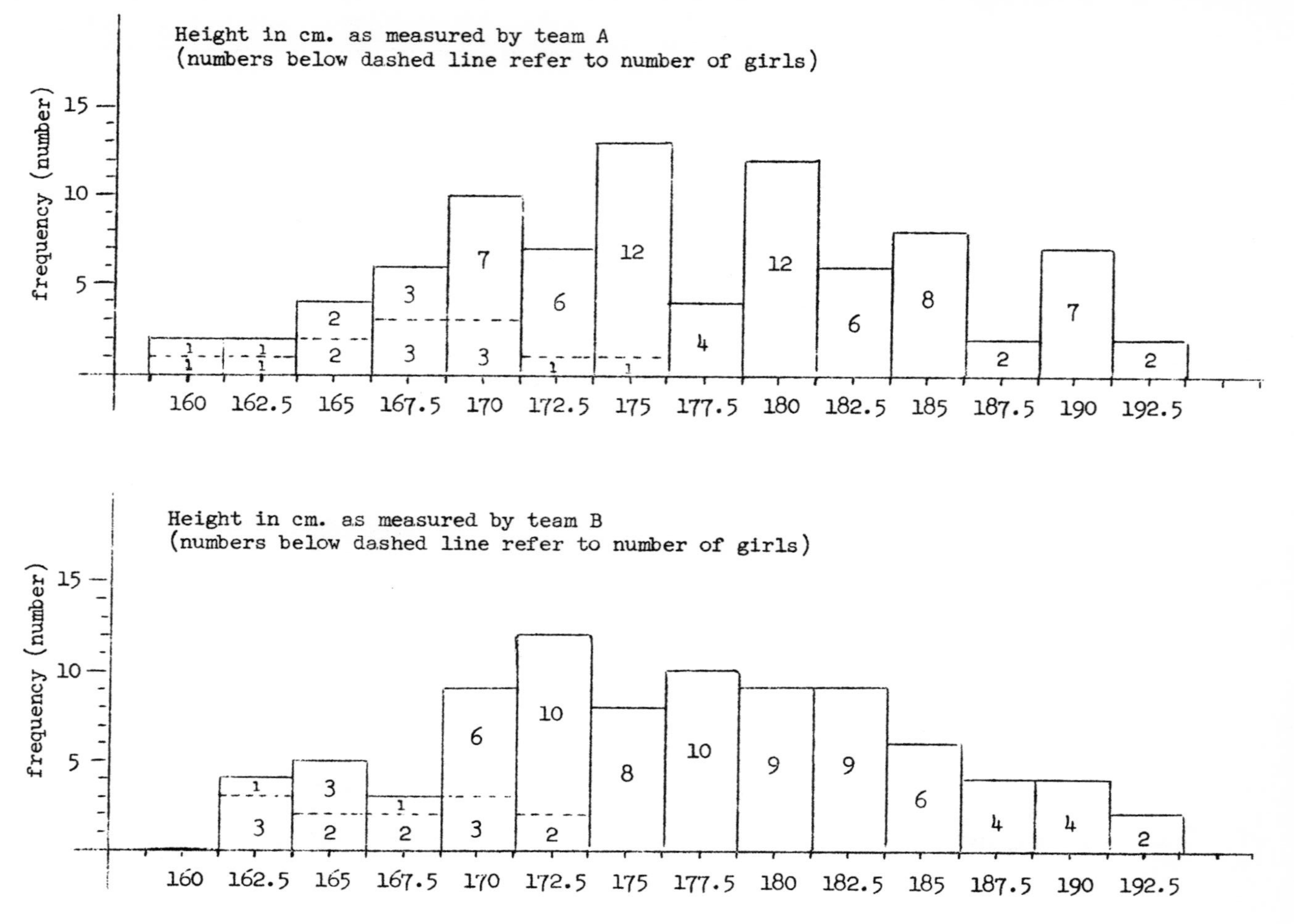

Figure VIII.2. Height in cm. as measured by teams A and B, Spring class, 1967.

Hence, we probably would conclude that team B obtained larger heights on the average than did team A (the short team).)

Instead of constructing a frequency distribution as a figure or graph, we may construct a table or chart. We may wish to study the two-way array of hair color and eye color as in table VIII.2 and to note the differences between own and A classifications. The most frequent combination is brown hair and blue eyes; the next most frequent is brown hair and brown eyes. Many combinations had zero frequency in our survey. These classes may also have zero frequency in the entire universe of individuals, as for instance, black eyes with naturally occurring red hair. Another manner of summarizing data is to present the results as proportions, as in the last row and last column of table VIII.2. Of the 92 individuals returning a questionnaire 1/92 had red hair, 7/92 had blond hair, 73/92 had brown hair, and 11/92 had black hair. Thus, approximately 3/4 of the class had brown hair. We have no idea of how representative this is of the student body at Cornell University because we have no idea of how individuals were allotted to this class with respect to the color of their hair. Perhaps this class could be considered as a representative sample of students at Cornell with respect to hair color or eye color. If so, then about 1/4 (25/92) of Cornellians in the Spring of 1967 reported that they had brown eyes and brown hair, and, as determined by Team A, about 1/3 (33/85) had brown eyes and brown hair.

In table VIII.3, we have given the frequencies of age and class standing. The most frequent group in the two-way array is of 20 year olds, plus or minus 6 months, in the sophomore class with 16 individuals. The most frequent class in the class standing classification is the sophomore group. The most frequent age class is the 20 year $\pm$ 6 months group with 29 individuals. In this table the class with zero frequencies was left blank to contrast with the previous table, table VIII,2, where zeros were included. Either form is acceptable, but table VIII.3 may be easier to read.

From the two-way array of weights and heights in table VIII.4, we may note that all heights and weights of the 13 girl students (there were 15, but one presented no data, and one did not report her own weight) in this class fall in the upper left-hand corner of the table and that there appears to be a relationship of higher weights with the taller people. The upper right-hand and the lower left-hand portions of the table are devoid of frequencies in this sample. We know that individuals in these categories, of the "5 x 5"

Table VIII.2. Frequency distribution of eye and hair color, Spring, 1967.

For Own

Eye color	Hair color: black	brown	blond	red	Total	Proportion
black	1	0	0	0	1	1/92
brown	6	25	2	1	34	17/46
green	2	13	0	0	15	15/92
blue	2	28	5	0	35	35/92
grey	0	7	0	0	7	7/92
Total	11	73	7	1	92	- -
Proportion	11/92	73/92	7/92	1/92	-	

For A

Eye color	Hair color: black	brown	blond	red	Total	Proportion
black	0	0	0	0	0	0
brown	6	33	1	0	40	8/17
green	0	1	0	0	1	1/85
blue	1	34	5	0	40	8/17
grey	0	4	0	0	4	4/85
Total	7	72	6	0	85	- -
Proportion	7/85	72/85	6/85	0	-	

Table VIII.3. Frequency distribution of age and class standing, Spring, 1967.

Age class*	Class standing						Total
	Freshman	Sophomore	Junior	Senior	Graduate	Other	
16,7 to 17,6	2						2
17,7 to 18,6	8	1					9
18,7 to 19,6	2	15					17
19,7 to 20,6	1	16	8	4			29
20,7 to 21,6		1	5	7			13
21,7 to 22,6			3	9			12
22,7 to 23,6			1	2			3
23,7 to 24,6							0
24,7 to 25,6			1	1	1		3
> 25,6					3	2	5
Total	13	33	18	23	4	2	93

* 16,7 means 16 years and 7 months of age; 17,6 means 17 years and 6 months of age.

Table VIII.4. Frequency distribution of weights and heights of class, Spring, 1967.*

Own

Height class (in.)	Weight class (pounds)															Total
	91-100	101-110	111-120	121-130	131-140	141-150	151-160	161-170	171-180	181-190	191-200	201-210	211-220	221-230	231-240	
62	g															1
63		3g		1	g											5
64				g	1											2
65			1	g		g,2										5
66				2g	g	2		1								6
67				g,1	g		2	1								6
68			1			4	4	2		1			1			13
69					1	3	3		2							9
70				2		1	1		1							5
71				1	1	5		3	2	1						13
72					1	1	3		4	2				1	1	13
73								1	1				1			3
74							1	1	3				1			6
75								1			1					2
Total	1	3	2	10	7	19	14	10	13	4	1		3	1	1	89

Team B

Height class (in.)	Weight class (pounds)															Total
	91-100	101-110	111-120	121-130	131-140	141-150	151-160	161-170	171-180	181-190	191-200	201-210	211-220	221-250	251-260	
64	g		g	1		g										4
65				2g		2	1									5
66					g	g			1							3
67			1	3g		g	4		1		1					11
68			1			g,1	5	3	1							12
69				1		3		2		1			1			8
70						2	2	2		2						8
71					1	3	1	2	1	2						10
72					2	3	2		1	2			1			11
73							1	1					1			3
74							1	2	1	1			1			6
75										1		1			1	3
76										1						1
Total	1	0	3	7	4	18	17	12	6	10	1	1	4		1	85

* g refers to girl student and number prefixing g refers to number of girls in the class; the plain numbers refer to the number of boys in the class.

and of the "beanpole" categories do exist in the population, even though infrequent. The relatively heavier concentration of individuals along the diagonal, sloping downward, of the table indicates a positive relationship between height and weight. In general, taller people do weigh more than shorter people. There is simply more of them!

Other methods of summarizing these data will be considered later. Before proceeding to another example, we should note that the size of the class interval was selected arbitrarily. This was also true of the class center. In general, we may note that if the class interval is too broad, all observations fall into one class. This was true of the height measurement made in yards. If the class interval is too fine, for instance, one millimeter, only zero, one, or a few individuals will fall in the class interval. To obtain some idea of the modal class or classes and of the shape of the frequency distribution, we need more than 93 observations as recorded here. 1000 to 5000 observations with 20 to 30 classes should suffice to indicate the form of the distribution and the modal class. Also, such a number gives a good idea of the range of the sample observations. The smaller the number of observations, the broader the class intervals will need to be in order to obtain observations. Measurements of height in feet resulted in 9 in the 5-foot class and 76 in the 6-foot class; those in inches, however, produced some observations from the 93 recorded in all classes from 62" to 75". Had we grouped by 2-inch intervals starting with 61-62, there would have been more individuals in each class.

VIII.3. A More Extensive Example of a Frequency Distribution

In order to obtain an idea of the frequency distribution of heights of men and women, data originally reported by Karl Pearson and A. Lee in volume 2 of Biometrika, 1903, and referring to heights of English people about 1900, are reproduced in table VIII.5. The class interval in daughters' and fathers' heights was one inch. The class center was on the half-inch mark. If a height was reported as 62 inches, one-half was put in the 61.5 inch class and one-half in the 62.5 class. Likewise, if a father 63 inches tall had a daughter 52 inches tall 1/4 of this observation was put into each of the four classes: (62.5,52.5), (62.5,53.5), (63.5,52.5), and (63.5,53.5). This splitting of observations accounts for the fractions of observations found

Table VIII.5. Heights of English people about 1900.[1]

Height of daughters in inches	Height of fathers in inches																		
	58.5	59.5	60.5	61.5	62.5	63.5	64.5	65.5	66.5	67.5	68.5	69.5	70.5	71.5	72.5	73.5	74.5	75.5	Total
52.5					.25	.25													.5
53.5					.25	.25													.5
54.5																			
55.5								1											1
56.5	.25	.25		.25	1.25	.5		1	.5	.5									4.5
57.5	.25	.25	.5	1.5	4.5	1	1.5	1.5	2.5		.5	.5							14.5
58.5	.25	.75	.5	.75	.75	1	1.75	1.25	5	2.75	.5	.25							15.5
59.5	.5	1	2		6	4.75	5	6.25	11.75	3.5	3.5	2	1.75	.5					48.5
60.5	.75	.75		2.5	8	6.25	12.5	18.25	20.25	11	9	4.75	2.5	1.25	1.25				99
61.5		.5	1.75	2	9.75	11.5	13	23.75	23.75	20.25	16.5	10.25	4.25	3	1.25				141.5
62.5		1	2.25	2	4.5	12	22.75	26	33	28.25	24.75	14.25	13.75	4.75	.75	.5			190.5
63.5			.25	2	6	8.25	11	27.25	35.75	37.25	31.5	26.25	16.25	7.75	1.5	.75	.25		212
64.5			.25	2.5	1.75	3.25	9.25	23	18.75	28.5	33	34.25	24.5	11.75	5.5	1	.25	1	198.5
65.5				.5	1	.5	11	12.25	9.25	19.75	30	26.5	22.25	15	4.75	3.75	2	1	159.5
66.5				.5	.5	1.5	3.25	7.25	8.75	16	26.25	26.75	20.5	18.5	7.75	4.25	.25	.5	142.5
67.5							1	5.75	7	4	14.25	13.25	12	11.25	4.5	3.75	.75		77.5
68.5					.25	.25	.25	.25	1.5	3	5.5	4.25	5.75	5.25	3.75	2.5	1.5	2	36
69.5					.25	.25	.25	.25	.25	.25	1	2.5	6.5	2.25	2.75	2	1		19.5
70.5											1.75	.25	4.5	.75	1.25	.75	.25		9.5
71.5											.5		.5	.5	1.5	.75	.25		4
72.5											1								1
Total	2	4.5	7.5	14.5	45	51.5	92.5	155	178	175	199.5	166	135	82.5	36.5	20	6.5	4.5	1376

[1] From table 31 of R. A. Fisher: Statistical Methods for Research Workers, published by Oliver and Boyd, Edinburgh and by permission of the author and publishers.

in the table. The most frequent class in the two-way classification is the one with 67.5" for fathers and 63.5" for daughters. The modal class for fathers' heights is 68.5" and for daughters is the 63.5" class. None of the 1376 observations are to be found in the upper right-hand corner or the lower left-hand corner of the table. Tall daughters did not have short fathers and _vice versa_.

One item of interest is the symmetrical bell-shaped form of the frequency distribution represented by a histogram of the frequencies for heights of the fathers and of the daughters. Where measurements follow a bell-shaped frequency distribution such as this, we say that the observations follow the form for a _normal frequency distribution_. Many measurements follow a normal frequency curve, but many do not. For example, weights of adult humans would tend to follow a frequency distribution which is not normal, since there would be too many weights in the right-hand portion of the frequency distribution.

VIII.4. Graphs and Charts

There are many, many ways in which to draw graphs, in which to present the basic results in a meaningful manner, and in which to disillusion the reader. The reader is referred to Huff [1954], chapters 3, 5, 6, 7 and 9, Moroney [1956], chapter 3, Schmid [1956], and Bevan [1968], chapter 5, for additional and interesting reading on graphs and charts.

To illustrate various ways of graphically representing a set of data, we shall consider an example given by Bevan [1968] wherein he states that "in an exclusive interview with the author" Yogi Bear stated that of $100 he received for television and film work during a given month, he gave the reservation manager $25 for board and lodging, spent $45 on extra honey, gave $10 to his friend Bubu, bribed someone with $15, and did not remember how he spent the remaining $5. We may represent these expenditures in an _ideograph_ or _ideogram_ form as follows:

Twenty \$5 bills spent by Yogi Bear

\$5 \$5 \$5 \$5 \$5	\$5 \$5 \$5 \$5 \$5 \$5 \$5 \$5 \$5	\$5 \$5	\$5 \$5 \$5	\$5
five \$5 bills were spent on board and lodging	nine \$5 bills were spent on extra honey	two \$5 bills were given to his friend Bubu	three \$5 bills were spent on bribing someone	one \$5 bill could not be accounted for

Another form for presenting the above data would be in the form of a pictogram or a pictograph using money bags of varying sizes (varying only one dimension such as height), replicas of \$5 bills, or pictures of various activities with the amount of money indicated. Still another type of presentation of the disappearance of Yogi's \$100 could be in the form of a pie chart or pie graph as shown in figure VIII.3.

Horizontal and vertical bar charts may be utilized to show the expenditures as shown in figure VIII.4, VIII.5, and VIII.6.

The above results may also be represented as a vertical or horizontal line chart or graph, respectively, as shown in figures VIII.7 and VIII.8.

For the example considered above, there is no scale for the expenditure items as there is for dollars spent. In figures VIII.1. and VIII.2, both axes had known scales; the vertical one or the ordinate represented number or frequency of occurrence, and the abscissa or horizontal scale represented units of length measurements. The graphs in these pictures were denoted as frequency histograms. Thus we may note that a histogram is a bar chart which measures the relative area, proportion, or frequency of occurrence in a given class interval which is measured on the abscissa in scaled units of measurement. This means that the space between bars in a histogram is not arbitrary, as it is for presenting data similar to Yogi Bear's expenditures.

If the frequency at the successive midpoints of class intervals are connected by straight lines and if the ordinate and abscissa are scaled units of measurement, the resulting figure is denoted as a frequency polygon. An example of four frequency polygons in one graph is given in figure VIII.9.

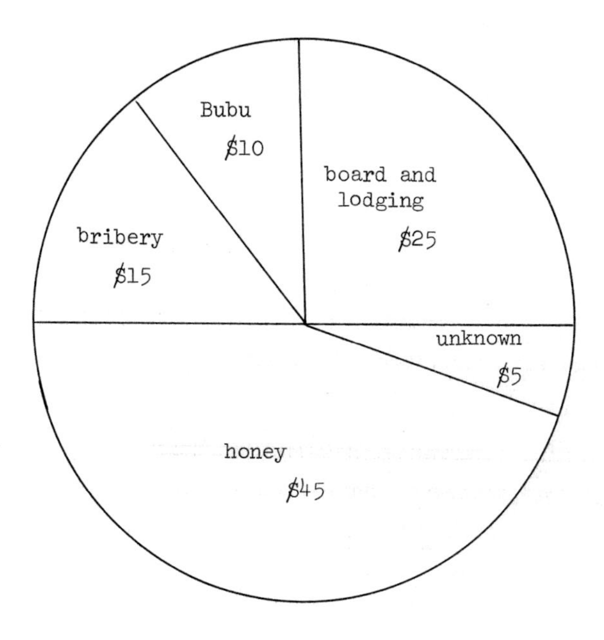

Figure VIII.3. Pie graph.

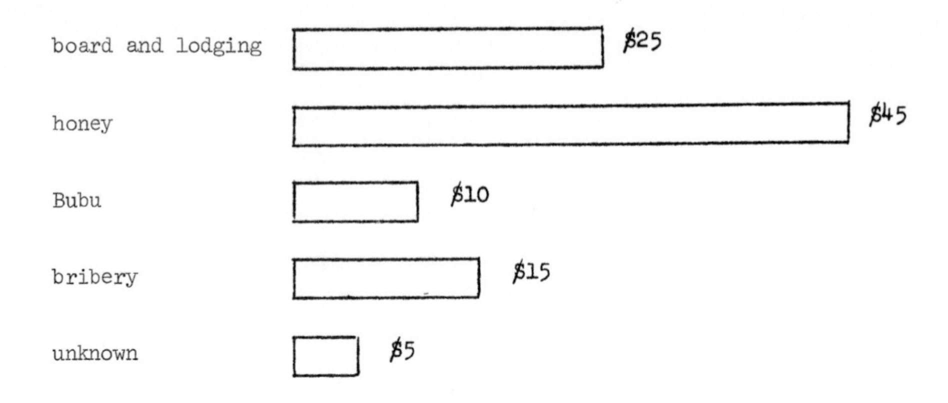

Figure VIII.4. Horizontal bar graph.

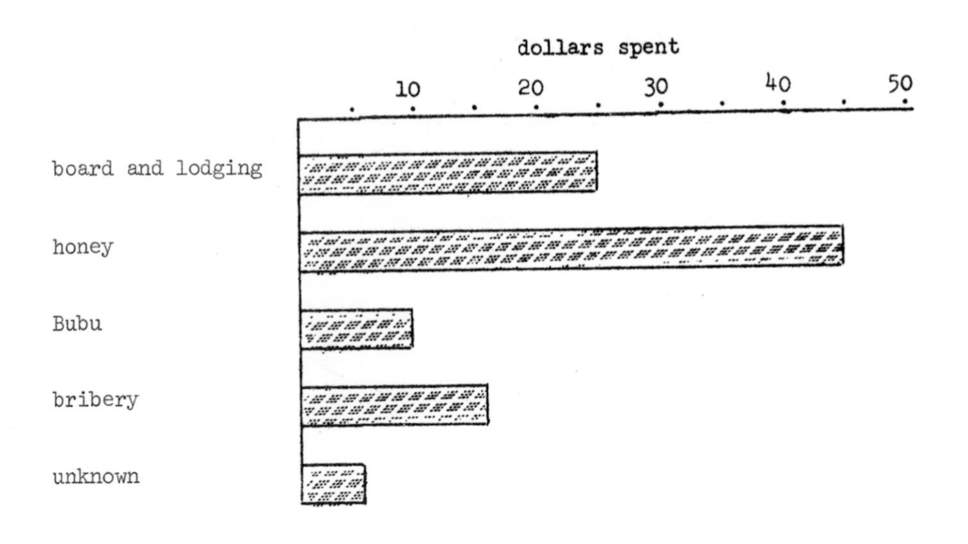

Figure VIII.5. Horizontal bar graph.

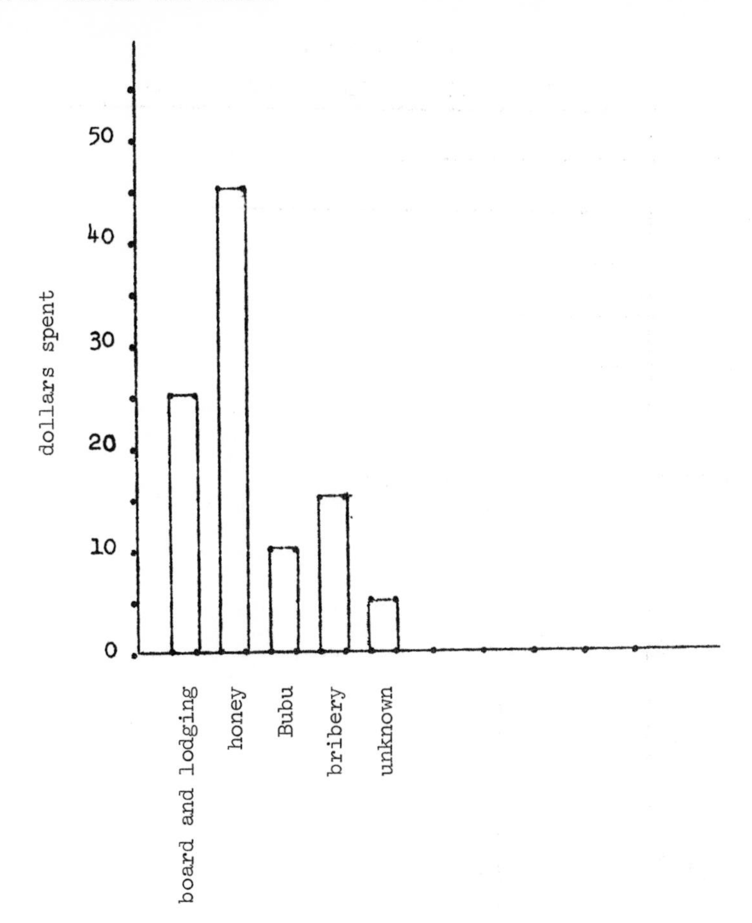

Figure VIII.6. Vertical bar graph.

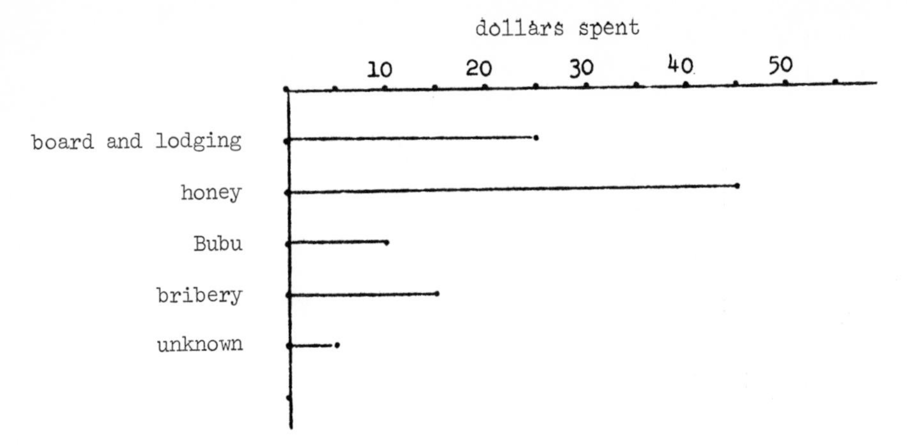

Figure VIII.7. Horizontal line graph.

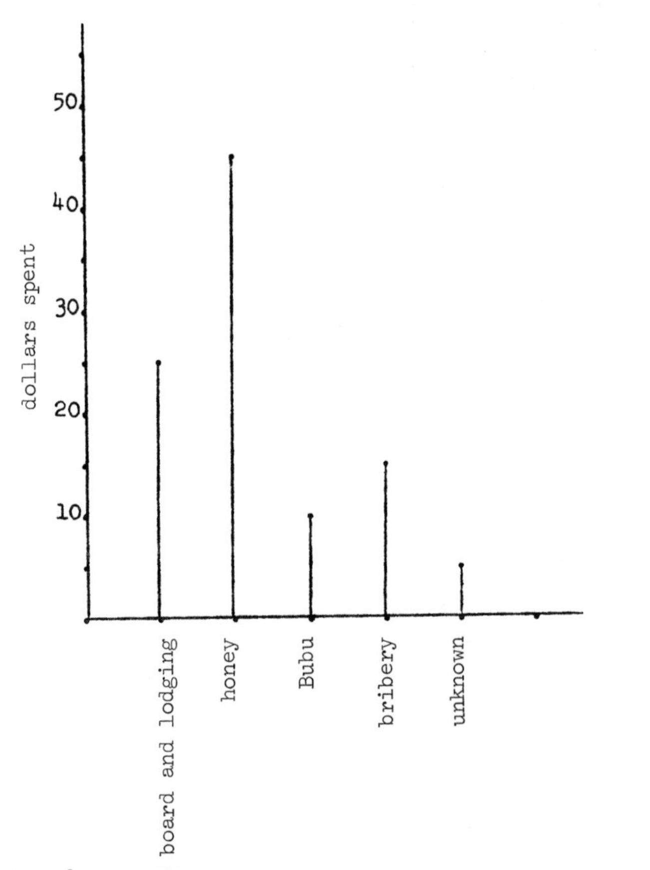

Figure VIII.8. Vertical line graph.

The number of deaths from motor vehicle accidents is given in terms of 10,000,000 born alive, to take into account the differences in proportions of various categories in the population. From the graph one can conclude that more men are killed in motor vehicle accidents than women. However, from these data *per se* it would be unwise to conclude that one should always go driving with a woman! It may be that women simply do not drive as much as men and that women are not in cars as frequently as men. Likewise, we should not conclude from these data in themselves that middle-aged white drivers are safer drivers than middle-aged non-whites. Cause and effect must be considered in light of all the facts and not just of one such set, as is done by persons who wish to "prove" (really to substantiate) their contention. For example, suppose that one were very much opposed to lowering the voting age to 18. One could use figure VIII.9 or, even better, a pooled estimate for all male and female data, to contend that immaturity is amply illustrated in the 15-20 and 20-25 age classes by the increased number of deaths, that instead of lowering the voting age to include more immature people we really should raise it, and that the data in figure VIII.3 "prove" one's point! (Perhaps the data substantiate the fact that women are more mature than men!) To determine whether or not the contentions are true would require considerable study and more precise definitions.

Instead of utilizing a frequency polygon, a researcher may wish to use cumulative frequencies in the form of an ogive or a cumulative frequency curve or graph. To illustrate, let us accumulate deaths over age classes for non-whites as in figure VIII.10. At each age, the number of deaths due to motor vehicle accidents per 10 million born alive has been accumulated at each 5-year age interval. Likewise, one could reverse the cumulative frequency curves and have "number-yet-to-die" instead of "number-died-to-age-class". Both forms are sometimes used in "intelligence quotients" $= \frac{\text{mental age}}{\text{actual age}} \times 100$ charts where it is desired to project "more than" or "less than" a given I.Q. score (see Moroney [1956], page 25, for an illustration).

Note that a smooth curve was drawn through the midpoints of the age classes. Such a curve can be misleading because it would appear that these results are arrived at from a much larger sample of data than is actually the case. It is recommended that for samples the cumulative frequency graph

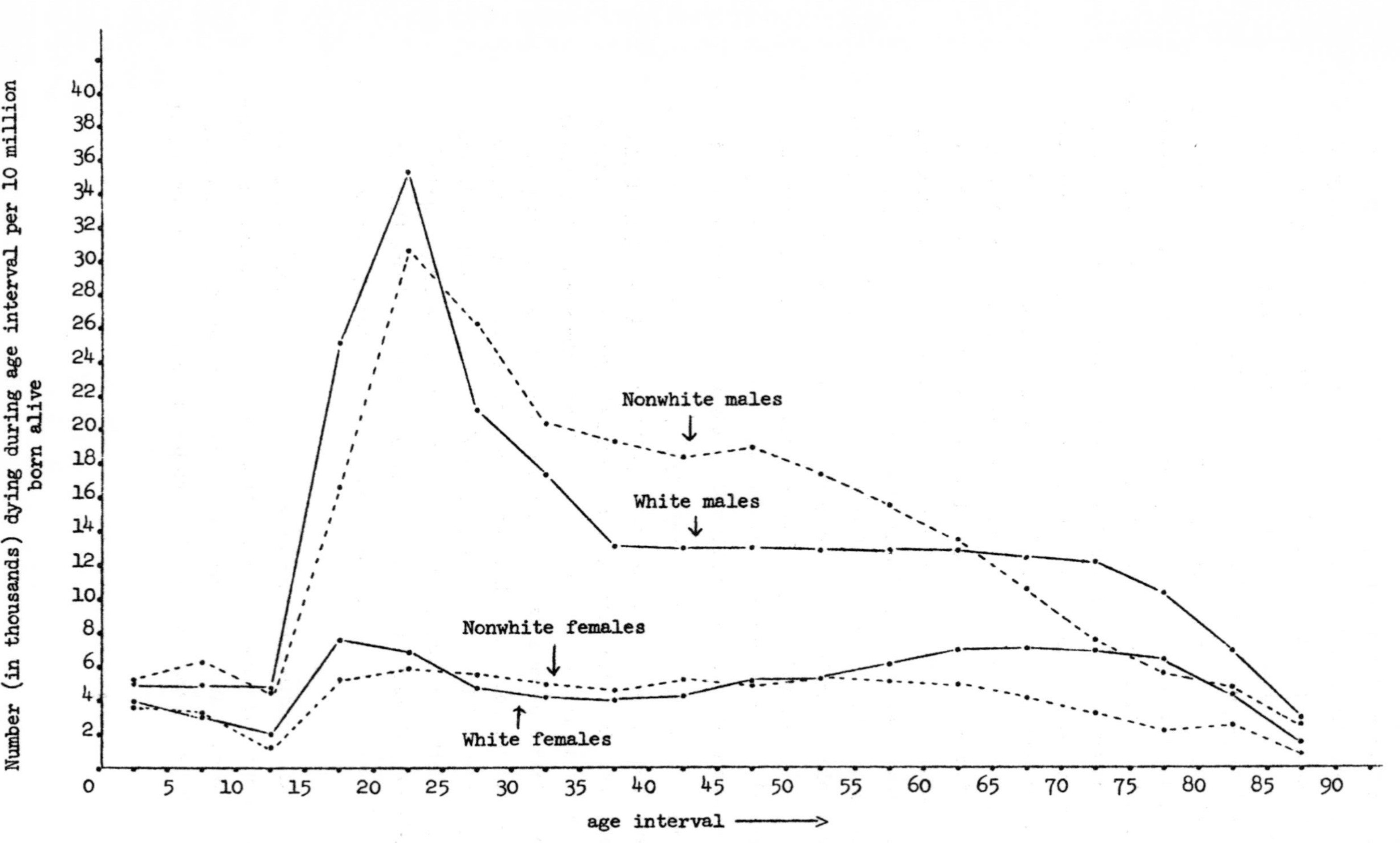

Figure VIII.9. Number dying in each 5-year age interval (midpoints given) from motor vehicle accidents, 1959-61. (Source: U.S. Life Tables by Causes of Death: 1959-61, volume 1, no. 6, U.S. Dept. of Health, Education, and Welfare, May, 1968.)

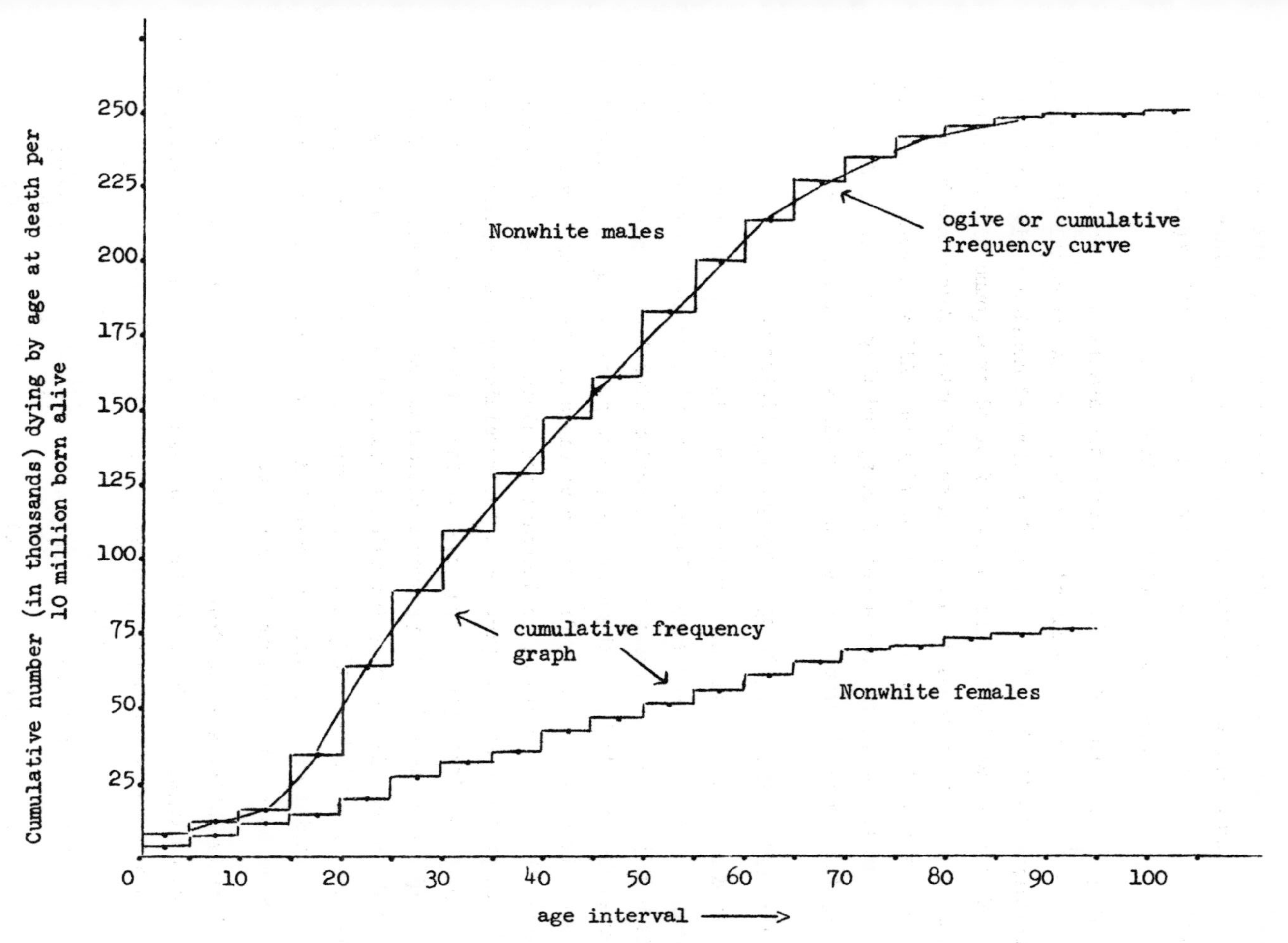

Figure VIII.10. Cumulative frequency of number of deaths due to motor vehicle accidents for nonwhite males and females over age classes. (Source: U.S. Life Tables by Causes of Death: 1959-61, volume 1, no. 6, U.S. Dept. of Health, Education, and Welfare, May, 1968.)

be used, instead of the ogive which should be reserved for the population. Also, one could connect the class midpoints with straight lines instead of curved ones if such a representation were desirable. With appropriate explanation, the reader should not be deceived.

Many other forms of graphs and charts are available to convey the pertinent facts from an experiment or survey. It is essential that a result be portrayed in its true form. Sometimes three-dimensional figures in the form of volumes, as for instance a loaf of bread or a pound of butter, are useful in portraying facts from a given set of data. Colors may be utilized to liven up a drab but important set of figures. All such methods are useful in summarizing facts from a set of data, and they are also useful in deceiving the reader about the facts, as we shall see in the next section.

VIII.5. Some Comments on Graphs, Charts, and Pictures.

Huff's [1954] delightful little book entitled How to Lie with Statistics contains a wealth of examples of misuses of statistics and deceptive tricks. It is highly recommended reading in connection with this chapter. As may be observed from Huff's book, there are many ways of deceiving the unsuspecting reader. The scales of the ordinate and abscissa can be changed relative to each other to increase or to decrease the slope of a curve. The scales can be chopped off and the picture of the remaining segment enlarged to make the slope of a curve appear steep when in fact it was almost flat. Class intervals can be selected to produce a histogram that may present a biased picture. In the class survey for heights measured in yards, the class interval was from 4'6" to 7'6"; all 85 students fell into this class, resulting in no variation in heights as presented. Class intervals can always be selected so as to exclude zero frequencies.

An interesting example from Huff [1954], pages 118-119, relates to the use of percentages and the selection of a base which can make prices appear to go up or down as the person pleases. The data given are:

Item	price: last year	price: this year	ratio: this year / last year	ratio: last year / this year
milk	20¢	10¢	50%	200%
bread	5¢	10¢	200%	50%
	arithmetic average		125%	125%

Now let us present the data in graphical form; in figure VIII.11, last year is represented by the base period (denominator) for the left graph and this year is the base for the graph on the right.

By seeing either of the graphs, the reader could easily be misled. The pertinent fact is that this year it costs 5¢ less for one loaf of bread plus one quart of milk than it did last year! The use of percentages did nothing to clarify the issue. Also, the selection of items and prices to be utilized in constructing percentages can have a considerable effect; in the above, an arithmetic average of prices was used, whereas a geometric average, $\sqrt{(50\%)(200\%)} = 100\%$, would show that prices did not change, that is, one doubled in price and the other item was one-half the price, and it does not matter which year is used as the base.

An example illustrating misuse of a picture is figure VIII.12, depicting consumption of cheese from various sources in the United Kingdom in 1953. This appeared in "Season at a Glance" in the New Zealand Dairy Board Annual Report (p. 13, 30th edition). The practice of presenting pictures such as this leads to the misconception that consumption of foreign cheese is only about 1/16 that of Commonwealth cheese, whereas in fact it is about 1/4; the height as well as the diameter of the cylinder was changed. A picture such as figure VIII.13 would be more appropriate, since here only the heights are changed. Another method of presenting the same idea is to use one cylinder of cheese cut into segments as shown in figure VIII.14, or into wedges as in figure VIII.15.

A procedure that may create misconceptions is a change from actual numbers to proportions or percentages. For example, suppose that politician K wishes to create the impression that he is doing better than politician E relative to number of people unemployed. Suppose that the number in the labor force under the administration of E has been 60 million, and under K it was 70 million, and that his aides prepared the two-bar graphs shown in figure VIII.15. Politician K might be tempted to use the right-hand bar graph. In fact, he might consider chopping off the graph below 4.2%, to indicate that he is doing twice as well as E had.

On the other hand, E could use the left-hand bar graph to show that there were 20,000 fewer individuals unemployed under him than under K. In situations like this, one should present both bar graphs and explain that

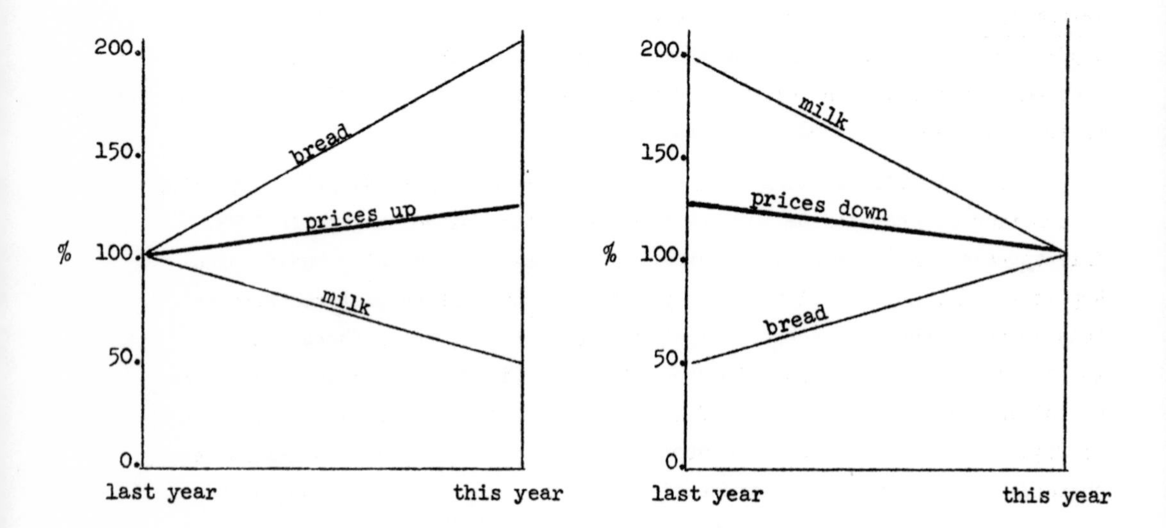

Figure VIII.11. Two ways of depicting the same data.

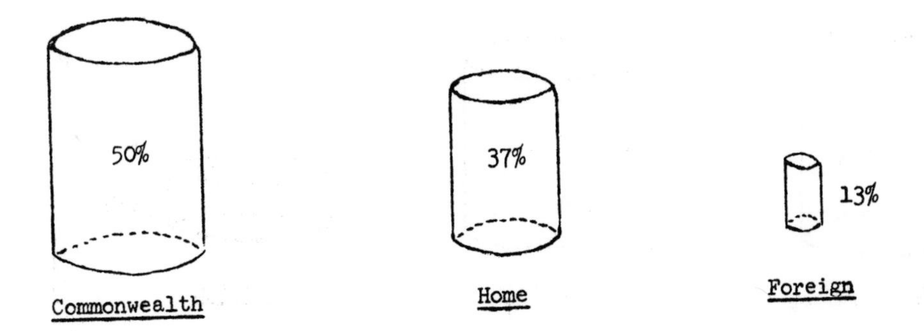

Figure VIII.12. Biased cylinder graph.

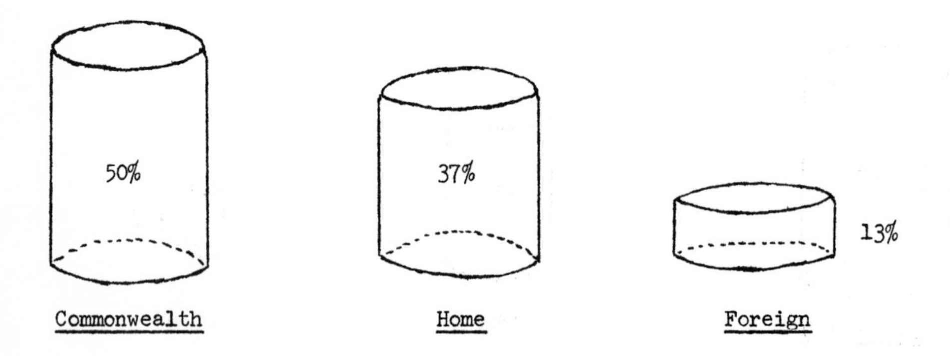

Figure VIII.13. Cylinder graph.

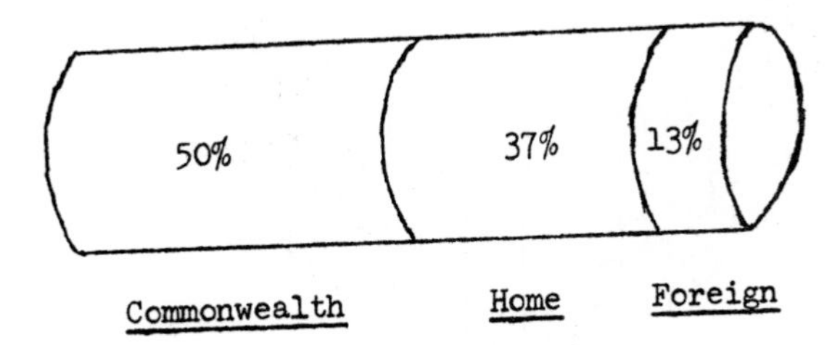

Figure VIII.14. Cylinder graph.

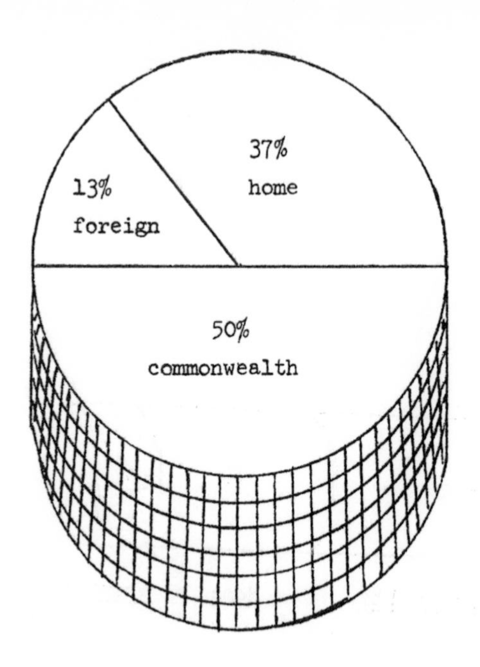

Figure VIII.15. Cylinder pie graph.

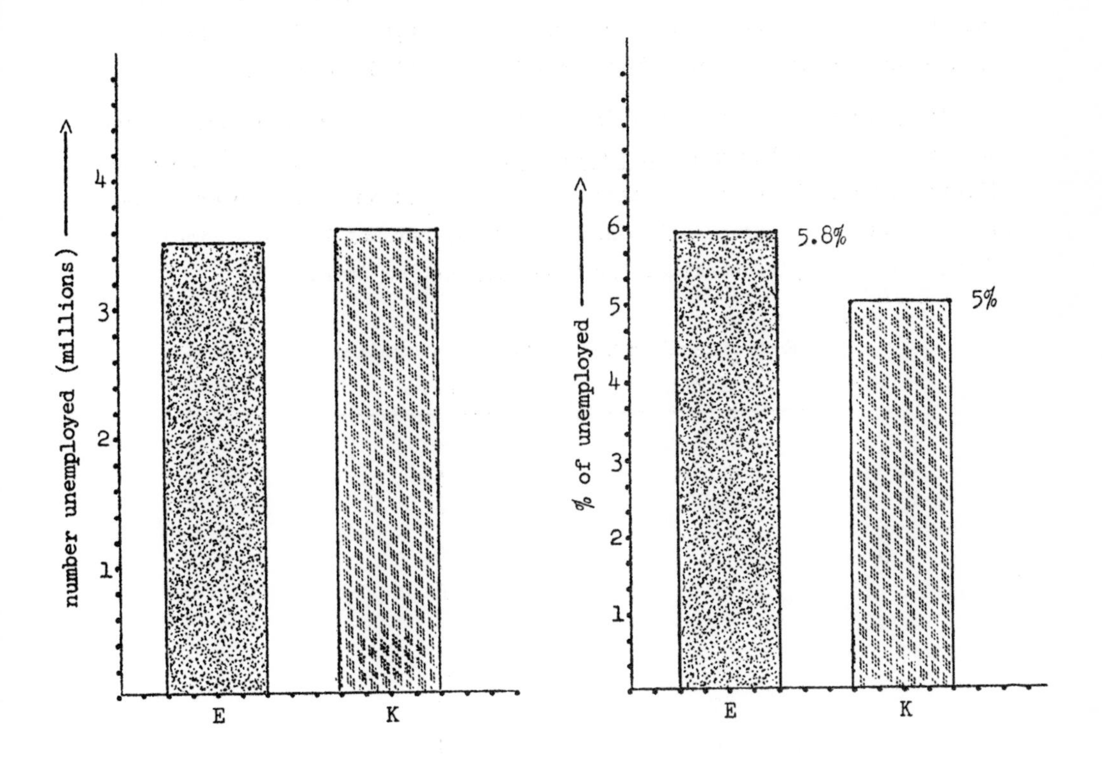

Figure VIII.16. Bar graphs.

the number in the labor force had changed from 60 to 70 million workers, that the number of unemployed had increased by 20,000, and that the percentage had decreased from 5.8% to 5% of the total labor force.

The process of reading off intermediate values on a graph between two observed values is known as interpolation. This may be all right in some situations, say where the variates on the abscissa and ordinate are continuous, that is, take on all possible values, but may be nonsensical in others. For example, consider the following set of data:

Number of children per family	Number of families observed
0	31
1	25
2	35
3	20
4	16
5	0
6	12
7	5
8	1
Total	145

The above results are presented in the following two forms in figure VIII. 17, one a histogram and the second as a linegraph or frequency polygon. It would be nonsensical to say that 10 families had 4.4 children (as seemingly indicated by the dotted lines)!

The process of extending a graph beyond the data points is known as extrapolation. Figure VIII.18 illustrates the decline in average annual death rate decade by decade from 1890 to 1930 (solid line). The estimated annual death rate (dashed line) for the 1930-1940 decade is under .05/1000 = 5/100,000. Since we have gone this far, why not also estimate the annual death rate for the 1940-1950 decade? This gives the unrealistic result that there is a return from death, that is, the death rate is about -2/100,000.

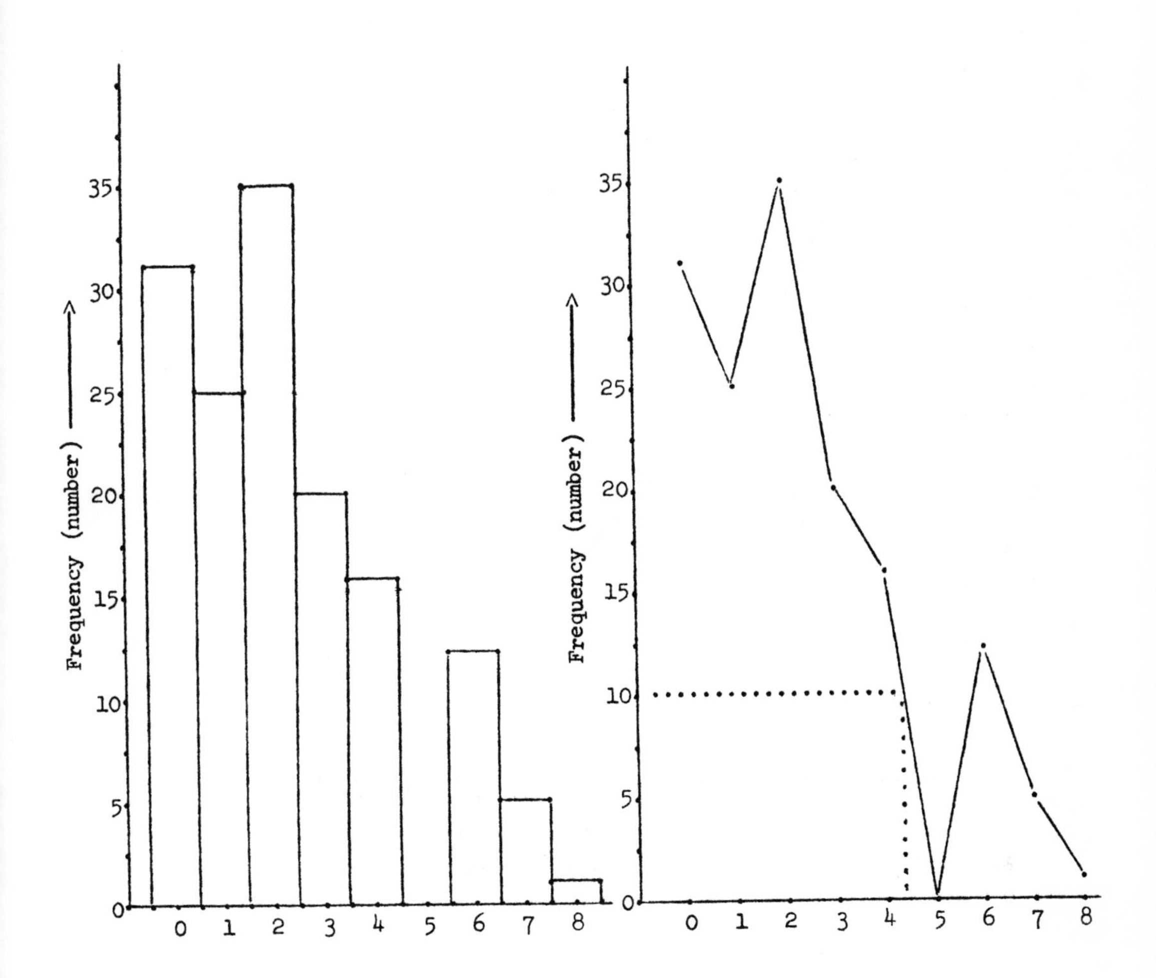

Figure VIII.17. Histogram and linegraph.

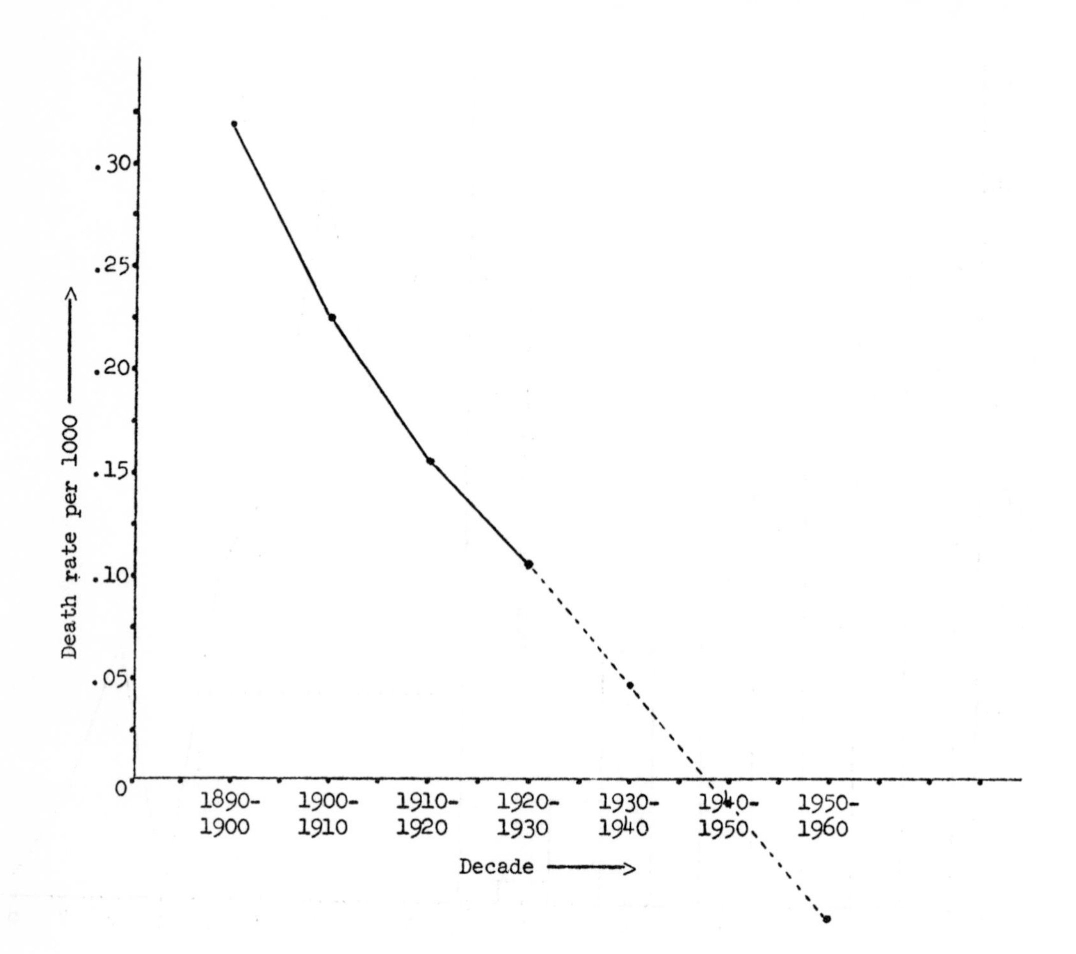

Figure VIII.18. Extrapolation of linegraph.

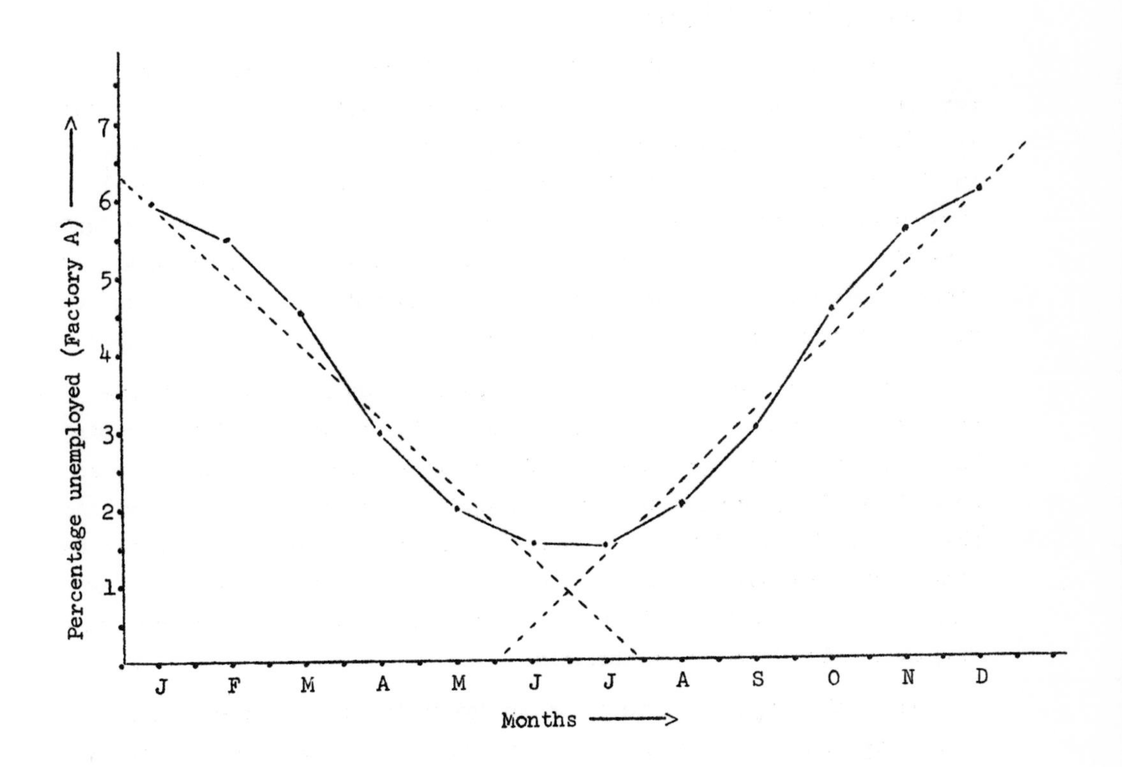

Figure VIII.19. Straight fitting to selected parts of a curved function.

By the 1950-1960 decade, the dead are returning by hordes! Of course, this is nonsense. We would suspect that the annual death rate would flatten out and remain nearly constant somewhere above the zero point. Here we see one of the dangers of extrapolation.

As a second example of the danger of extrapolation, suppose that the percentage of seasonally unemployed workers for factory A runs between 1½% and 6%. Suppose that the percentage unemployed by month may be graphed as in figure VIII.19 (solid line). Now suppose that only the data from the first 6 months are available to an investigator and that he is a "straight-line advocate". He fits a straight line to the data points by standard linear regression techniques (the dashed line) and gleefully predicts that there will be no unemployment by mid-July!

Suppose, on the other hand, that only the last 6 months' unemployed percentages were available to our "straight-line advocate". After drawing his standard linear regression (the dashed line) through the data points, he would woefully predict that over 7% would be unemployed in January! As plainly illustrated above, extrapolation can quickly lead to misleading and nonsensical results. Therefore, the experimenter must be wary of any extrapolations, especially when he does not know the form of the response function over all values of the abscissa.

As a third example, note the study by Professor U. Bronfenbrenner, Cornell University, on the response by girls and by boys to different degrees of severity of punishment. The "straight-line advocates" stated that girls should not be punished and that boys should be punished severely! The fact was that the maximum response to severity of punishment was reached much faster by girls than by boys and that both response equations were curved and not straight lines; this lead many writers to the above false conclusion.

From various sources, data are available on the number of passengers per car on highways for each year. For example, in 1940 a survey showed that the average number of passengers per car was 3.2, in 1950 the figure was 2.1, and in 1960 the average was only 1.4. If we plot these three pairs of figures as in figure VIII.20 and connect the adjacent points with straight lines, a sharp decline is to be noted. Suppose that our "straight-line advocate" draws a "best fitting" line through all three points and extends it to the year axis,

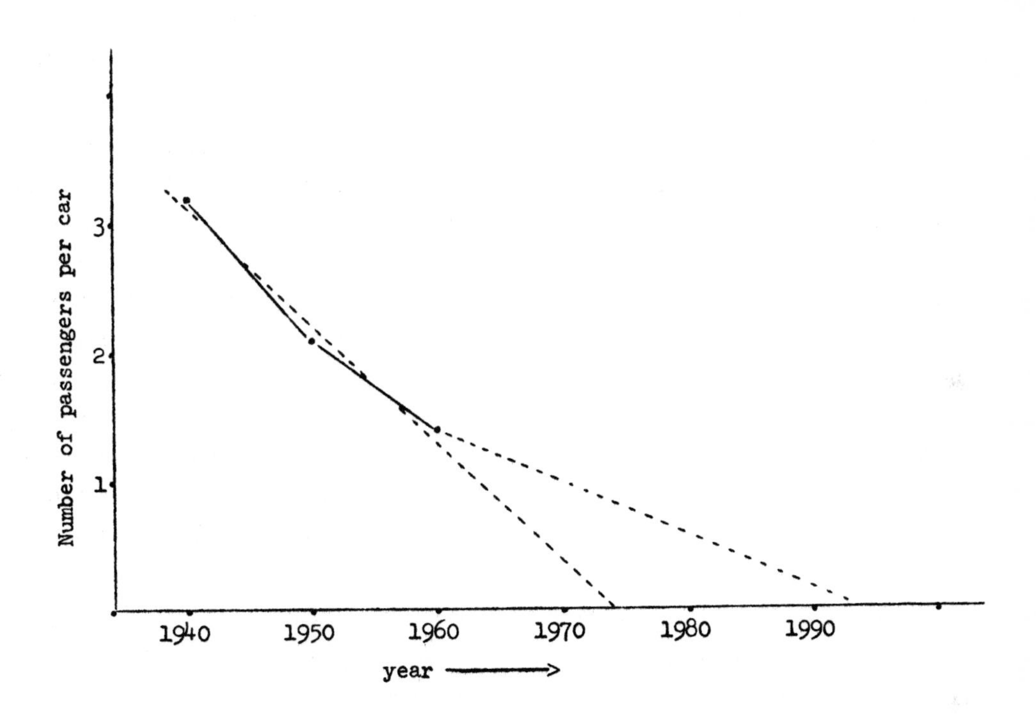

Figure VIII.20. Number of passengers per car on the highways by year.

the abscissa. We note that by 1974 cars will be going down the highways with no one in them! Even if the "straight-line advocate" is playing it "safe" and only uses a line with a smaller decrease, the dashed line on the right, it will still indicate that cars will contain no passengers by 1994. Probably the true situation is that this is a curved function which approaches one passenger per car as a limit. Although some of these examples may be amusing, they are intended to illustrate the dangers of extrapolation beyond the data available.

In presenting a graph for a set of data, one should present a scatter diagram or two-way representation of the data (see table VIII.4, for example) of the points as given in figure VIII.21. If a line is drawn through the points in the scatter diagram, the method used to plot the points on the line should be given. Frequently a person will draw an "eye-fitted" line through the points. This can be misleading, especially if the points are omitted. For example, suppose that one has the scatter diagram in figure VIII.22. Any straight line using linear regression methods will fit equally well, as there is no relationship of the Y_i and X_i values. The eye-fitter then can draw any line to suit his purpose. If he omits the points in the scatter diagram the reader could be mislead into thinking that the Y_i values are influenced by the value of the X_i, when in fact they were not.

Be wary of any graph or figure for which the scale has been omitted. Figure VIII.23 was adapted from Moroney [1956], page 29. Here we note that the two halves of the figure may, and quite probably do, have different scales. The figure would indicate that the return to "normal state" is much faster using "Snibbo" than was the increase in "Inter-Pocula Index". Also, note that no scales are given and that it would appear that the subjects were more than half intoxicated when the study began on the left-hand figure and that they were completely normal at the end of the study on the right-hand portion of the figure. Obviously, the advertiser wishes to sell "Snibbo"!

We have covered some of the honest and some of the dishonest ways of presenting results. There are others, but the general idea should be clear from what has been presented thus far. Charts, graphs, and figures can lead or mislead equally well. Be certain to read figures carefully and determine what they mean rather than what they appear to mean. Factual and mis-

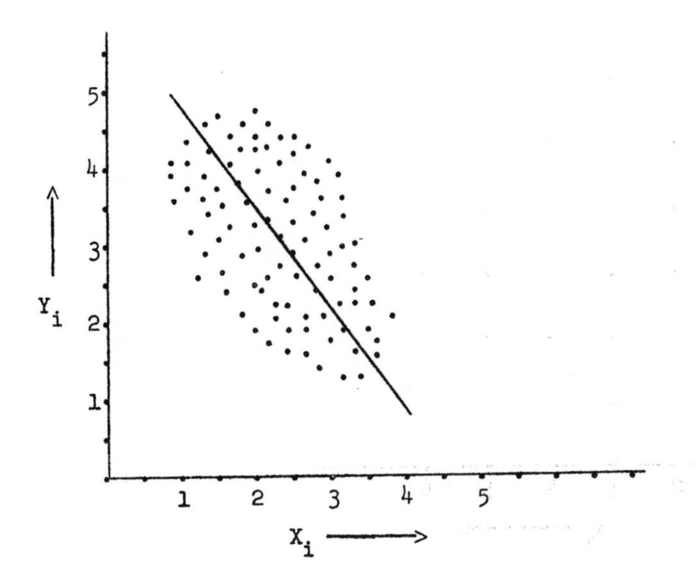

Figure VIII.21. Scatter diagram.

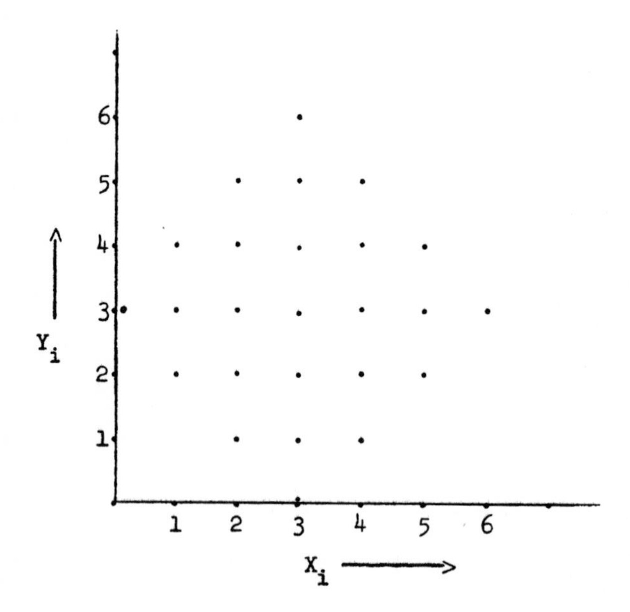

Figure VIII.22. Scatter diagram.

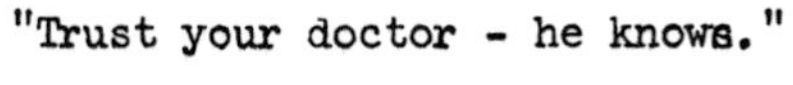

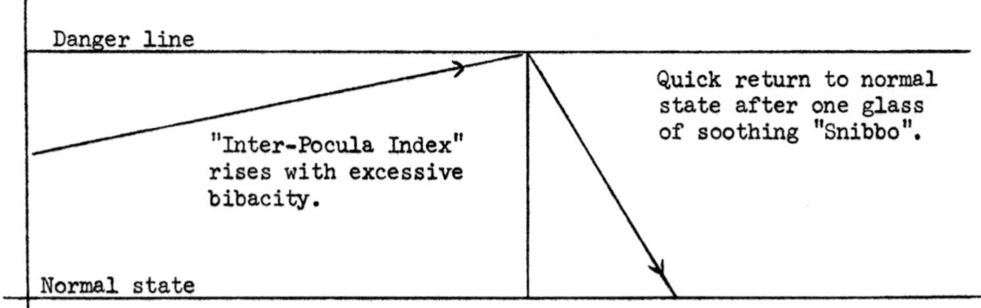

After consumption of alcohol your Inter-Pocula Index rises to what may prove a dangerous level, with serious risk to muscular atony. In such cases the taking of a therapeutic nostrum has untold effect as a sedative and restorative. There is no finer nostrum than quick-acting soothing Snibbo.

Figure VIII.23. Scaleless graph.

leading graphs and charts may be found in all types of literature, often side by side. They are of daily occurrence in newspapers and magazines. Frequently there is no attempt to mislead, since an uncritical writer may not even realize that he has "stretched" one scale, changed scales in the middle of a graph, omitted a scale, or otherwise mislead the reader. In any event, the reader must be wary and critical.

VIII.6. Problems

VIII.1. Find two or more examples from newspapers, magazines, or other published literature of misleading graphs, and show how to present the results in order not to mislead but to present the facts.

VIII.2. Prepare a scatter diagram of the 93 pairs of height and weight measurements as presented by the student from table VIII.1. Circle the dots which pertain to measurements for girls.

VIII.3. From the data in table VIII.1 complete the following table and comment on the results.

Frequency of last digit of heights in mm

	digit									
Team	0	1	2	3	4	5	6	7	8	9
A										
B										
expected	8.5	8.5	8.5	8.5	8.5	8.5	8.5	8.5	8.5	8.5

VII.4. From the data in table VIII.1 complete the following table:

Frequency of last digit for weights

	digit									
	0	1	2	3	4	5	6	7	8	9
Own - observed - expected	8.9	8.9	8.9	8.9	8.9	8.9	8.9	8.9	8.9	8.9
Team B - observed - expected	8.5	8.5	8.5	8.5	8.5	8.5	8.5	8.5	8.5	8.5

The expected number is computed as 89/10 or 85/10. Do people tend to select one digit over another in reporting weights, for instance, zeros and nines?

VIII.5. Eight matchboxes with the number of matches in each box as indicated are arranged as follows:

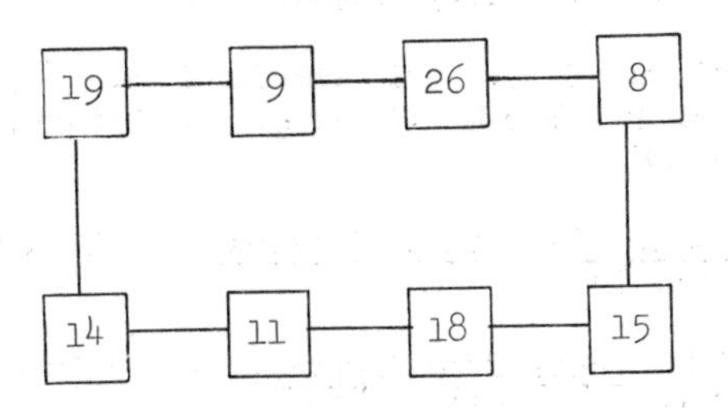

The first box contains 19 matches, the second 9 matches, etc. It is permissible to move any match from one box to another if these are connected by a line. The matches must be shifted in such a manner as to have an equal number of matches in each box. Show how this can be done to minimize the total number N of matches shifted and show graphically that your solution is a minimum.

VIII.6. Construct the graph for the following function using different colors for the different parts of the graph:

$$y = \frac{1}{x-1} + \frac{1}{x+1} \text{ for all values of x and y.}$$

VIII.7. We have illustrated methods of deceiving the reader with various graphical presentations. Words which in themselves are vague or convey misinterpretations, may be attached to graphs to make them even more misleading. A good source for finding such words is in literature circulated by campus or other groups promoting a cause. Such words or phrases as imperialism, working class, middle-class values, liberation groups, exploitation, anarchist, liberal, conservative, sexual liberation, and generation-gap, are often used to mislead. Find 10 examples of such words, and illustrate how you believe that the word has been used to mislead.

VIII.7. References and Suggested Reading

Bevan, J. M. [1968]. Introduction to Statistics. Philosophical Library, Inc., New York, pp. vii + 220, chapter five.

Gelfand, I. M., Glagoleva, E. G., and Schnol, E. E. [1969]. Functions and Graphs, Volume 2. The M.I.T. Press, Cambridge, Mass., and London, England, pp. v + 105.

Huff, D. [1954]. How to Lie with Statistics. W. W. Norton and Company, Inc., New York, pp. 142.

Moroney, M. J. [1956]. Facts from Figures. 3rd edition. Penguin Books, Baltimore, Maryland. pp. xiii + 472.

Schmid, C. F. [1956]. What price pictorial charts? Estadística 14:12-25.

CHAPTER IX. A LITTLE PROBABILITY

IX.1. Introduction

Throughout the first eight chapters we have alluded to several probabilistic concepts. We have relied on the reader's intuitive definition of odds, chance, or probability concepts rather than on a formal presentation, since all that was needed was a general notion of what is meant by probability. Few readers would be able to develop a precise, definite, and consistent meaning of the term which would withstand scholarly criticism; for many, it falls in the same vague, relatively undefined category as do a large number of other terms such as "imperialist", "reactionary", "liberal", "conservative", "radical", "substandard housing", "poverty", and other terms current in present day news media and in printed handouts of certain campus groups. To some, the term probability brings to mind thoughts of betting on the horses, Las Vegas, chances of survival from auto and airplane crashes, male-female ratio, flipping coins, chances of rain, throwing dice, something insurance companies and gambling institutions worry about, and so on; their interpretation may include such concepts as chance, random event, odds, a victim of Fate, personal belief, what happens to the other fellow, frequency or proportion, mechanistic, and terms learned by rote from the statistician's language. We shall briefly discuss three bases for definitions of probability:

1. relative frequency and empirical probability,

2. *a priori*, classical, or analytic probability, and

3. personal belief or subjective probability.

The relative frequency concept is the one often utilized; the controversy among a segment of the scholarly does not affect the practical use of the frequency concept of probability.

IX.2. Some Terminology

One of the first items to consider in establishing a definition of probability is the *scale* or *range* that probability values can take. So far, all persons dealing with probability have agreed to measure probability on a scale marked zero on one end and one at the other, with all possible values in-between and including the endpoints, that is, for a probability value p we may say $0 \leq p \leq 1$. A scale such as the one in figure IX.1 indicates the possible range of values with various kinds of happenings. If something simply cannot happen, then its probability of occurrence is given a value of zero. If something is certain to happen, then it has a probability value of one. Happenings which have the endpoint values of zero and one usually cause little difficulty. The in-between values, the uncertainties, cause many problems in everyday application, and this keeps the statistician busy.

The quantity to be measured on the above scale is the uncertainty of a future happening, such as the probability that it will rain next Sunday, or of a past occurrence, such as the probability that Sir Francis Bacon wrote the plays commonly attributed to William Shakespeare. In such situations there will be certain possible outcomes which we shall denote as events. For example, in coin tossing, we might categorize the possible events as follows:

E_h denotes a head on the side facing up,

E_t denotes a tail on the side facing up,

E_s denotes the coin standing on edge,

E_a denotes the coin remaining suspended in the air, and

E_v denotes the outcome when the coin vanishes.

We would most probably decide that E_s, E_a, and E_v could happen only with probability zero and are figments of the imagination rather than possible outcomes. This would leave us with the two possible events E_h and E_t. We would be concerned about probability values of occurrence for these two possible outcomes.

Many situations in nature have two possible outcomes; dead or alive, student or nonstudent, married or not married, go/no-go, if and if not, and either/or are examples of such a dichotomous classification. Many actions also fall into two possible categories: whether or not to sleep in class, to skip class, to carry an umbrella, to go to the baseball game, to vote in an election, to shave in the morning, to go to the Ice Follies, and so on. Despite the abundance of examples involving only two possible events, there is an even larger set of examples with more than two outcomes.

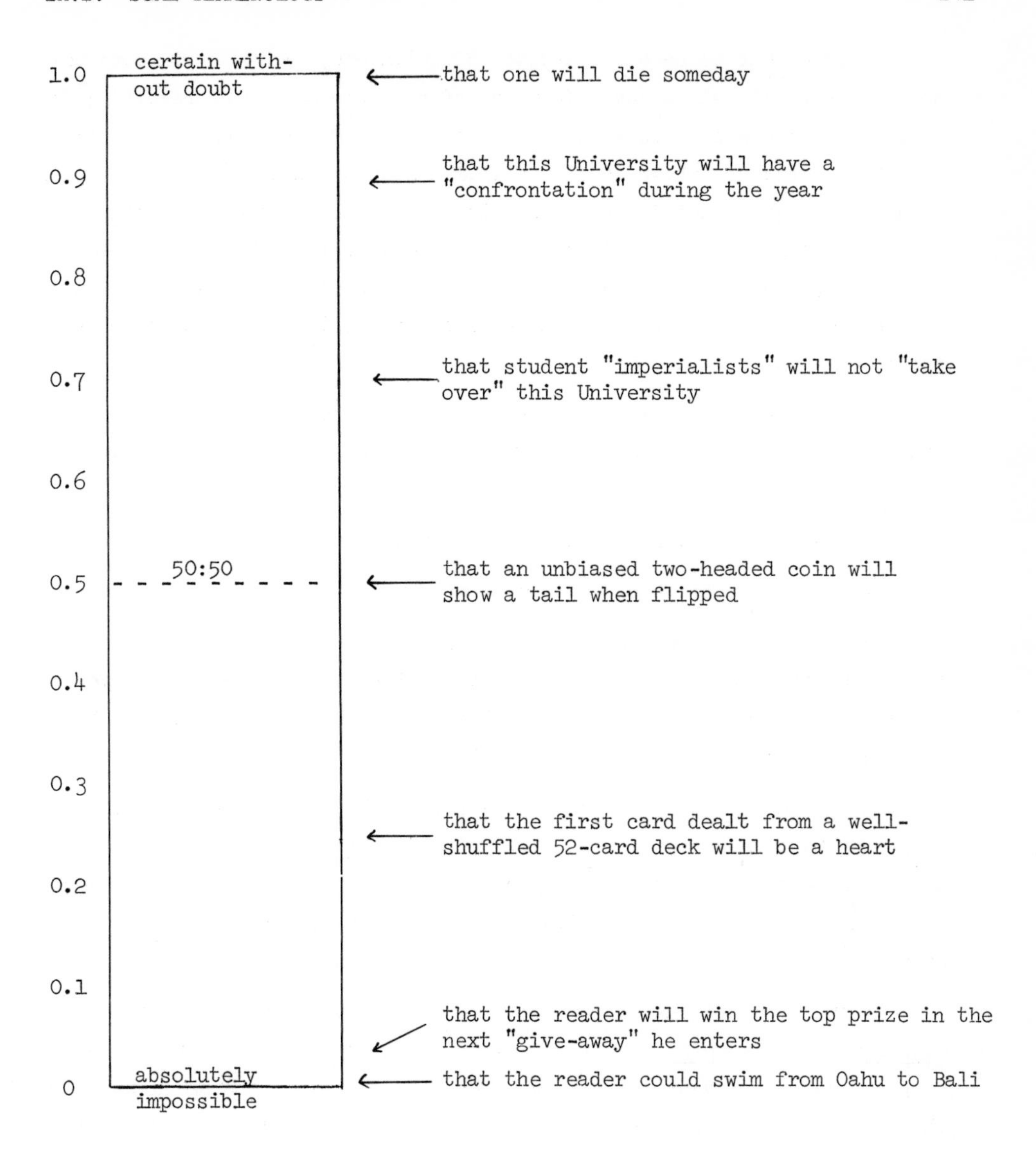

Figure IX.1 The probability scale of measurement.

Suppose that we have a six-sided die with one, two,..., six dots respectively on its six sides. In terms of possible outcomes, resulting from a roll of the die and observing the side that is facing upward, the six possible events are shown in figure IX.2. Each of these events is denoted as an <u>elementary event</u> in that there is no simpler or more basic manner of defining a happening of rolling a die. We assume that there are no additional happenings of relevance such as the die standing on an edge or a corner; hence the following six elementary events define the totality of all possible events for this activity.

Instead of considering the six possible events listed above, we might consider only the following two events as possible outcomes of the activity of casting a die:

E_6 denotes six dots showing on the top side of the die

$E_{\not 6}$ denotes other than six dots shown on the top side.

$E_{\not 6}$ (read E not 6) is composed of five elementary events, E_1, E_2, E_3, E_4, and E_5 and is denoted as a <u>compound event</u>. The event $E_{\not 6}$ happens whenever there are one, two, three, four, or five dots showing on the side of the die facing upward.

As another example, let us consider the activity of drawing one of ten ping-pong balls numbered 0, 1, 2,...,9 from a container; the ten possible elementary events are shown in figure IX.3. If our events were considered to be the compound events of odd numbers and even numbers, say E_o and E_e, each compound event would consist of five elementary events.

Another concept that is useful in probability contexts is that of <u>mutually exclusive events</u>: if one event in the set of possible outcomes happens, no other event can happen. Thus, in the coin tossing example with the elementary events E_h and E_t, the events are mutually exclusive, because either one or the other happens. On the other hand, suppose that the possible outcomes in the ping-pong example are the even numbers = E_e, the odd numbers E_o, and the numbers less than 5 = $E_{<5}$; events E_e and E_o are mutually exclusive, because these compound events contain no elementary events in common. However, $E_{<5}$ is not mutually exclusive of either E_e or E_o, because $E_{<5}$ has three elementary events in common with E_e and two with E_o.

Another useful idea which has been utilized previously in drawing

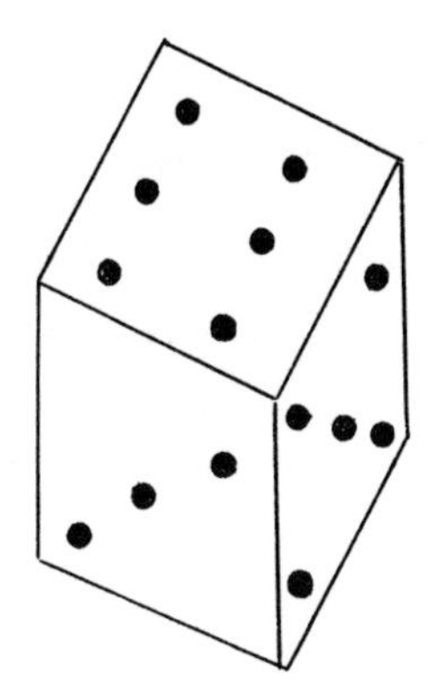

six-sided die

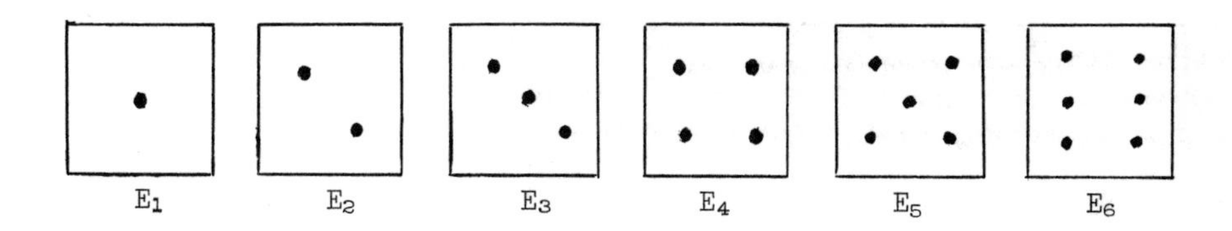

Figure IX.2. Six-sided die and six possible events.

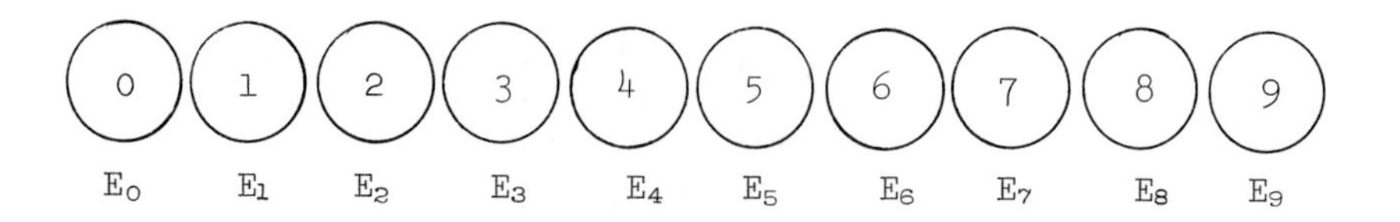

Ten elementary events

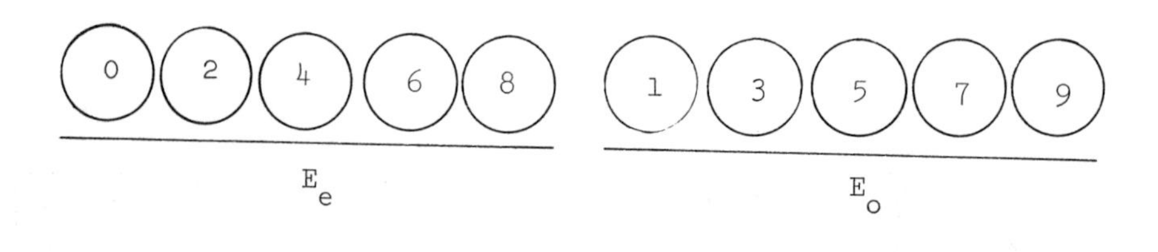

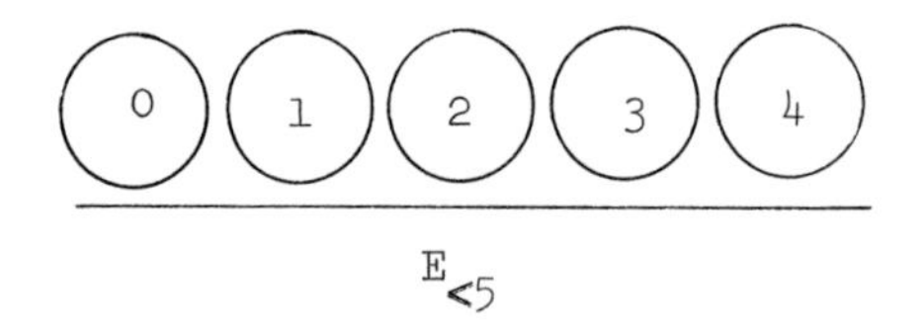

Three compound events

Figure IX.3. Elementary and compound events.

random samples is that of an <u>independent</u> <u>event</u>. In an activity, if the happening of one event does not influence the happening of a second event, the two events are said to be independent. To illustrate, suppose that the elementary events are represented by the selection of one of the six letters a, b, c, d, e, and f and suppose that three are to be selected to form a sample. Furthermore, suppose that our mechanistic process allows only the samples abc or def. The selection of a simultaneously selects b and c, and hence these are not independent events. On the other hand, if our mechanistic process is one which allows no influence of selection of one on the other then the events are independent. One such process is to place the letters on six identical ping-pong balls, thoroughly mix the balls in a container, and blindly draw three balls from the hat with no replacement after each draw. Such a process results in one of the 20 possible combinations of three letters, as listed in chapter V.

Another example is to consider the activity of rolling a die with the events $E_{\div 2}$ meaning that the number on the top face is divisible by 2, $E_{\not\div 2}$ meaning that the number on the top face is not divisible by 2, $E_{\div 3}$ meaning that the number on the top face is divisible by 3, and $E_{\not\div 3}$ meaning that the number on the top face is not divisible by 3. Clearly, event $E_{\div 2}$ and $E_{\div 3}$ are not independent, because 6 is divisible by 2 and by 3, 2 and 4 are not divisible by 3, and 3 is not divisible by 2. No pair of the compounds events given in this example is independent.

IX.3. <u>Empirical</u> <u>Probability</u>

Inherent in many practical applications of probability is the idea of a relative frequency of an outcome; a proportion or odds falls into this category. Associated with this is the idea of conceptually repeatable samplings or experiments. The samplings are conceived as being repeatable an infinite number of times. This would certainly fit a coin-tossing experiment or a die-casting experiment; conceptually, we could keep on throwing dice forever, not just 26,306 times as Mr. Weldon did (see table XI.1). He threw 12 dice at one time, which resulted in 12(26,306) = 315,672 elementary events, of which 106,602 showed either a five or a six. The proportion of times that a five or a six occurred was 106,602/315,672 = .3376986 which is greater than the predicted proportion of 1/3, resulting in the conclusion that these 12 dice were biased. We say, then, that the probability of obtaining a five or a six when these 12 dice are thrown is .3376986. More formally, we could write this as: $P(E_5$ or E_6 occurs) $= 0.3376986$. Also, $P(E_5$ or E_6 does not occur) $= P(E_1, E_2, E_3$ or E_4 occurs) $= 1 - 0.3376986$, since the sum of the probabilities for the possible events must equal unity, that is, we must account for all the happenings. This is another property of probability which we may write more formally, as follows: let p_i = probability of the ith event for $i=1,2,\ldots,n$ possible events which exhaust all possible events; then

$$\sum_{i=1}^{n} p_i = 1.$$

To illustrate this, suppose that we had an unbiased die and each of the six possible events had a probability of 1/6 of occurrence. Then we may write

$$P(E_1) + P(E_2) + P(E_3) + P(E_4) + P(E_5) + P(E_6) = \frac{1}{6} + \frac{1}{6} + \frac{1}{6} + \frac{1}{6} + \frac{1}{6} + \frac{1}{6} = 1.$$

When all events have the same probability of occurrence, they are referred to as <u>equally likely events</u>.

All the above implies that probability can be defined as "the ratio of the number of times an event occurs to the total number of possible occurrences" or "as the ratio of the number of successes to the total number of trials." This relative frequency definition is quite useful in many walks of life and in statistical applications. The ratio $a/(b+a)$ is often stated in terms of odds as $a:b$ or a to b.

To test a mechanism for drawing random samples, we would want to determine whether all possible samples of size n are equally likely events, that is, have equal chances of being selected. To do this, we would conduct an experiment to obtain a very large number of observations, and we would then compute the proportion of times each event occurred. If the proportions are considered to be equal, we could conclude that the process results in a random selection of samples.

To illustrate a familiar use of probabilities in the relative frequency sense, let us suppose that we have an alphabetical listing of the 15,000 students at University X, that the list contains 12,000 boys' names and

3,000 girls' names, and that we wish to draw a completely random sample of names from the list. The probability that a girl's name will be drawn is 3,000/15,000 = 1/5 = 0.2 and that a boy's name will be drawn is 12,000/15,000 = 4/5 = 0.8, provided a truly random procedure is utilized in drawing the name from the list. Suppose that in addition to the above, we know that there are 600 freshman, 500 sophomore, 500 junior, 500 senior, and 900 graduate girl students. We could then compute the various probabilities for randomly selecting a name for these categories as:

P(name belonging to a freshman girl) = 600/15,000 = 1/25,

P(name belonging to a sophomore girl) = 500/15,000 = 1/30,

P(name belonging to a junior girl) = 500/15,000 = 1/30,

P(name belonging to a senior girl) = 500/15,000 = 1/30, and

P(name belonging to a graduate girl) = 900/15,000 = 3/50.

A random sample of 200 names, say, containing no names of girls in any category would be considered a highly unlikely event, but we must remember that such an event does have a nonzero, albeit very low, probability of occurrence. Such an event should cause us to scrutinize the process utilized for drawing the sample. The probability that one of the above five categories would not be represented in the sample does not have such a low probability of occurrence. If we want all categories represented, we should use a form of stratified sampling as suggested in chapter V.

IX.4. Analytical, A Priori, or Classical Probability

In many situations we construct probabilities and a definition of probability from a theoretical consideration of an activity without any idea of conducting an experiment for collecting observations. From the very nature of the activity and of the associated possible events, we may be able to define the various probabilities of the various events. For example, we can observe that a coin has a head on one side and a tail on the other and that the coin is symmetrical; without any data from tossing coins, we state that $P(E_h) = \frac{1}{2} = P(E_t)$ where E_h = occurrence of a head and E_t = occurrence of a tail. Likewise, for the six-sided die we can observe by physical measurements that the die is symmetrical and unweighted, and hence we can state that the probability that any side will fall upward is 1/6, or that the six possible events are equally likely.

The use of theoretical knowledge of an activity to determine probability values such as those described above has been variously termed analytical, a priori, or classical probability. We actually used this basis previously when we considered an unbiased two-sided coin or six-sided die to state the expected probability values. This concept of probability is very useful for computing hypothetical probability values.

IX.5. Personal Belief, Subjective, or Psychological Probability

In the preceding two sections we have considered empirical and analytical bases for constructing probability values and concepts. Another basis is the "personal", "degree-of-belief", "psychological", "subjective", "egoistic", or "self-confidence" basis for proability. All relate to a person's belief. This approach has been formally studied by mathematicians, philosophers, and statisticians.

To illustrate the nature of the personal belief approach, consider such probability statements as:

P(my son's marriage will not end in divorce) = 0.75;

P(the next car I buy will be a "lemon") = 0.61;

P(a "bloodbath" in South Vietnam in the event of a North Vietnamese takeover) = 0.999;

P(that the President's "honeymoon" will last 6 months longer) = 0.002;

P(that the cute girl sitting next to me in my genetics class has a date for next Saturday night) = 0.80;

P(I will spend more than $100 for washer and dryer repairs this year) = 0.50.

Aside from some "phony statistics" terms because of undefined terms, for instance, "lemon", statements such as the above are based on the personal beliefs of one person; they would not be the same as those set by another person for the same statements. Hence the question of how such probabilities should be used in a scholarly sense is still in question.

IX.6. Addition Law of Probability

As we have illustrated in previous sections, the probability of a compound event is the sum of the probabilities of the elementary events making up the compound event. For example, in the last illustration in section IX.3, the sum of the probabilities for the girls' names from the various classes is equal to

$$\frac{1}{25} + \frac{1}{30} + \frac{1}{30} + \frac{1}{30} + \frac{3}{50} = \frac{5}{50} + \frac{3}{30} = \frac{2}{10} = 0.2,$$

which is equal to the probability of a girls' name being selected, that is 3000/15,000 = 0.2 In equation form we could say

$$P(\text{name belonging either to a freshman, sophomore, junior, senior, or graduate girl student}) = \frac{1}{25} + \frac{1}{30} + \frac{1}{30} + \frac{1}{30} + \frac{3}{50} = 0.2 = P(\text{name belonging to a girl}).$$

This illustrates the <u>addition law of probability</u>, which states that the probability for an event is the sum of the probabilities for the elementary events making up that event. Note the "either" and "or" words in the probability statement. Also note that we could say that the probability of a specified outcome is the sum of the probabilities of the mutually exclusive events making up the outcome.

As a second illustration from section IX.3, consider the example for E_6 = a six on the top face of a die and for $E_{\not 6}$ = not a six on the top face of the die. Then, for unbiased dice

$$P(E_6) = 1/6,$$

$$P(E_{\not 6}) = 5/6 = P(E_1) + P(E_2) + P(E_3) + P(E_4) + P(E_5),$$

and

$$P(E_6) + P(E_{\not 6}) = \frac{1}{6} + \frac{5}{6} = 1.$$

The last statement states that if a die is thrown, it is certain to show one of the six faces. We have ruled out the possibility of a die standing on an edge, that is, we have given this event zero probability.

IX.7. Multiplication Law of Probability

Suppose that we have two unbiased coins each with a head on one side and a tail on the other, that the coins are to be tossed simultaneously, and that we wish to calculate the probabilities for the various events which are:

coin 1	H	H	T	T
coin 2	H	T	H	T

where H represents the head on one side of a coin and T represents the tail on the other. Each of these four results is equally likely and each has a probability of occurrence of 1/4. Note that the results for two consecutive tosses of the same coin produce the same set of four results. If we are given the probabilities of the events for coin 1 and for coin 2, we can compute the probabilities of the individual coins as follows:

$$P(\text{H on coin 1 } \underline{\text{and}} \text{ H on coin 2}) = \frac{1}{2} \cdot \frac{1}{2} = \frac{1}{4} ,$$

$$P(\text{H on coin 1 } \underline{\text{and}} \text{ T on coin 2}) = \frac{1}{2} \cdot \frac{1}{2} = \frac{1}{4} ,$$

$$P(\text{T on coin 1 } \underline{\text{and}} \text{ H on coin 2}) = \frac{1}{2} \cdot \frac{1}{2} = \frac{1}{4} , \text{ and}$$

$$P(\text{T on coin 1 } \underline{\text{and}} \text{ T on coin 2}) = \frac{1}{2} \cdot \frac{1}{2} = \frac{1}{4} .$$

Note the word "and" in the probability statement. Also note that the probability <u>either</u> of a head on coin 1 and a tail on coin 2 <u>or</u> a tail on coin 1 and a head on coin 2 is computed as follows:

$$P(\underline{\text{either}} \text{ an H on coin 1 and T on the other coin } \underline{\text{or}} \text{ a T on coin 1 and an H on coin 2}) = \frac{1}{4} + \frac{1}{4} = \frac{1}{2} ,$$

which illustrates the addition law of probability. Thus, the "either-or" statement is associated with addition of probabilities, and the "and" statement is associated with multiplication of probabilities.

More formally, we can say that the probability of an event which is the result of the simultaneous occurrence of two or more independent elementary events is the product of the probabilities of the elementary events. This is called the <u>multiplication law of probability</u>.

As a second illustration of the above, consider that we have a pair of six-sided unbiased dice one of which is red and the other white. What is the probability of a pair of sixes resulting from a throw of the dice? That is, we want to know what is the probability of the events E_{r6} = a six on the red die and E_{w6} = a six on the white die both occurring from one throw of the two dice. Such a combination is known as "boxcars" in the vernacular of "crap-shooters", that is, dice-throwers. We compute this probability as

$$P(\text{boxcars}) = P(E_{r6}) \cdot P(E_{w6}) = \frac{1}{6} \times \frac{1}{6} = \frac{1}{36} .$$

Likewise, the probability of simultaneously obtaining a six on the red die, E_{r6}, and something other than 6 on the white die, $E_{w\not{6}}$, may be computed as follows:

$$P(E_{r6} \text{ and } E_{w\not{6}}) = P(E_{r6}) \cdot P(E_{w\not{6}}) = \frac{1}{6} \cdot \frac{5}{6} = \frac{5}{36} .$$

The multiplication law may be utilized in many situations. For example, it can be used "to prove" that a 15-year old boy named Arthur with a registered cocker spaniel named Princess Taffy has a dog that is one in a billion. To do this, let us suppose that there are one-half as many boys as girls at age 15, that the name Arthur is given to one in 50 boys, that registered cocker spaniels constitute only 0.1% of the dog population, and that the name "Princess Taffy" is given to only 0.01% of the dogs. Then P(15 year old boy named Arthur owns a registered cocker spaniel named Princess Taffy) = P(boy) · P(name Arthur) · P(a registered cocker spaniel) · P(name Princess Taffy)

$$= \frac{1}{2} \cdot \frac{1}{50} \cdot \frac{1}{1{,}000} \cdot \frac{1}{10{,}000} = \frac{1}{1{,}000{,}000{,}000} .$$

Another example that may have some financial consequences for the reader pertains to horse racing. We observe the odds posted on the board or given on racing forms, and if we believe these odds we could make various kinds of bets. One method would be to place bets on two horses in one race. Our chances of winning is the sum of the probabilities that each horse will win. Likewise, we might select an unpredictable horse who wins occasionally and place a show, a place, and a win bet on the horse. These kinds of bets make use of the addition law. We make use of the multiplication law for the "daily-double" or accumulator type bet. For the latter, we pick a horse to win in the first race and bet all our winnings on a horse to win in the second race. Our chance of winning is the product of the probabilities of each horse winning his own race.

IX.8. Conditional Probability

We need methods to compute a probability of an event happening given the condition that another event has already happened. These are available, as we shall illustrate with the red-white pair of dice example. Given that the red die has already been thrown and shows a six, E_{r6}, what is the probability that the white die will show a five, that is, event E_{w5}? We know that the outcome on the throw of the white die is independent of what appears on the red die, and hence $P(E_{w5}) = 1/6$. Also, since we are given that E_{r6} has occurred, its probability of occurrence after the fact is unity. The probability that event E_{w5} will occur given that event E_{r6} has already occurred may be written as follows:

$$P(E_{w5} \text{ occurs given that } E_{r6} \text{ has already occurred}) = P(E_{w5}|E_{r6})$$

$$= P(E_{w5}) \; .$$

$$P(E_{r6} \text{ has occurred}) = \frac{1}{6} \cdot 1 = \frac{1}{6} \; .$$

As a second example to illustrate conditional probabilities, suppose that we have an urn containing 5 black and 3 white balls, thus:

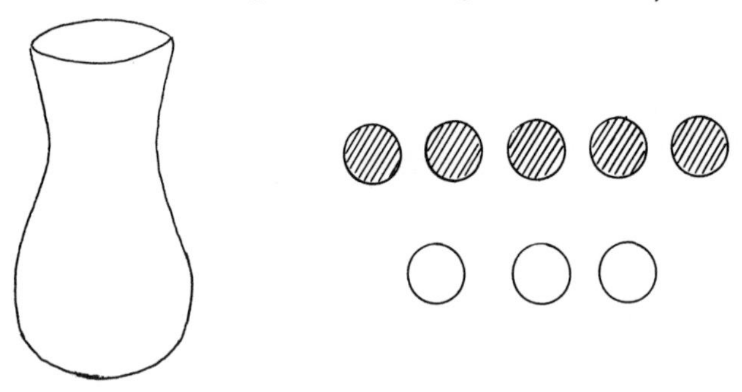

A random selection of a ball from the urn results in the following probabilities:

$$P(\text{ball is black}) = \frac{1}{8} + \frac{1}{8} + \frac{1}{8} + \frac{1}{8} + \frac{1}{8} = \frac{5}{8} \text{ and}$$

$$P(\text{ball is white}) = \frac{1}{8} + \frac{1}{8} + \frac{1}{8} = \frac{3}{8} .$$

Now suppose we wish to compute the probability of drawing a black ball given that the first ball is black and not replaced. The probability of drawing a black ball on the first draw is 5/8, but since there are only 4 black balls and 3 white balls remaining in the urn, the probability that the second ball drawn will be black is now 4/7. Hence the probability of the black ball on the first draw and a black ball on the second draw is

$$P(\text{black on 1st draw}) \cdot P(\text{black on 2nd draw/black on 1st}) = \frac{5}{8} \cdot \frac{4}{7} = \frac{20}{56} ,$$

whereas the probability of obtaining a black ball on the second draw given that a black ball was obtained on the first draw is 4/7 as given above. Confusion about these two types of statements can lead to erroneous computations of the probabilities. At this stage we begin to see the reasons for making very precise statements and for understanding what is meant by the probability statements; one word can change the probability statement completely.

A practical application of conditional probability using known facts or previous data is the following. Suppose one student is flipping a coin and a second student names the side of the coin that will turn face up and does this while the coin is in the air. If the coin matches his call, he wins and if not, he loses. Suppose that the coin came up heads five times in a row. What should the caller do on the next call of the coin? In order to maximize his winnings he should call heads, that is, he should adopt a "play-the-winner" strategy. The reason for this is possibly to bias the odds in his direction. If the coin has a head on one side and a tail on the other and is unbiased, it does not matter whether he calls heads or tails, so why not call heads? If the coin is not two-sided or is biased in favor of heads, his best call is heads in order to bias the odds in his favor. As a precautionary measure, one had better continue to keep track of the number of heads, because the coin could actually be biased in favor of more tails even though five heads were obtained on the first five throws.

IX.9. Permutations and Combinations

A teacher has five students and she wishes to set up a committee of three with a chairman = c, vice-chairman = v, and secretary = s. She wants all students to have an equal opportunity for all positions, but she does not want to be accused of bias. She rejects the idea of putting the five names into a hat,

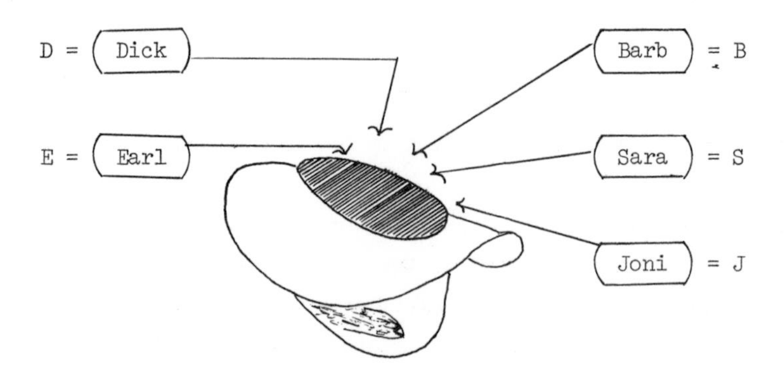

and blindly drawing sets of three names, because she knows about the vagaries of random sampling and that she would not be able to conduct the drawing for a long enough time. She comes up with the following scheme, where pos. = position on committee:

Committee

S B J	S B D	S B E	S J D	S J E	S D E	B J D	B J E	B D E	J D E
pos.	pos.	pos.	pos.	pos.	pos.	pos.	pos.	pos.	pos.
c v s	c v s	c v s	c v s	c v s	c v s	c v s	c v s	c v s	c v s
S B J	S B D	S B E	S J D	S J E	S D E	B J D	B J E	B D E	J D E
S J B	S D B	S E B	S D J	S E J	S E D	B D J	B E J	B E D	J E D
B S J	B S D	B S E	J S D	J S E	D S E	J B D	J B E	D B E	D J E
B J S	B D S	B E S	J D S	J E S	D E S	J D B	J E B	D E B	D E J
J S B	D S B	E S B	D S J	E S J	E S D	D B J	E B J	E B D	E J D
J B S	D B S	E B S	D J S	E J S	E D S	D J B	E J B	E D B	E D J

The top line gives the ten possible committees of size three and the six orderings under each committee give all possible orderings of the three individuals in the three positions on a committee. The teacher reasons that if she goes through all 60 orderings on 60 different days, then she should not be accused of prejudice.

We note that this scheme becomes somewhat unmanageable when there are more than n=5 students. Therefore, we might wish to observe what happens in general for any n and to determine if some mathematical formula is available for schemes such as the above. In order to count or to compute the number of possible orderings or permutations, observe that any one of the n=5 names may appear in the first position of chairman; then, given that a person is chairman, any one of the remaining n-1=4 names may appear in the second position of vice-chairman. Thus all possible orderings of 5=n names for the two positions yields n(n-1) = 5(4) = 20 possible orderings. When we consider all three positions, we proceed for the first two positions as above and note that with a given chairman and vice-chairman there are only 3=n-2 names remaining from which to select a secretary. Therefore there are n(n-1)(n-2) = 5(4)(3) = 60 possible orderings of n=5 individuals into committees of r=3 names each as, indeed, there are 60 orderings as written down by the teacher.

The question now arises concerning a formula for the number of orderings or permutations. If we use n instead of 5 and r instead of 3, we can write down the number of permutations as follows:

$$n(n-1)(n-2)(n-3)\cdots(n-r+1) = \frac{n(n-1)(n-2)\cdots(2)(1)}{(n-r)(n-r-1)\cdots(2)(1)} = \frac{n!}{(n-r)!}$$

where $n! = n(n-1)(n-2)\cdots(2)(1)$. Thus, if ${}_nP_r$ is the symbol used to express the number of permutations of n items taken r at a time, we can write the following simple mathematical formula:

$${}_nP_r = \frac{n!}{(n-r)!} ,$$

where 0! is defined to be equal to one; for the example above for n=5,

$${}_5P_3 = \frac{(5)(4)(3)(2)(1)}{(5-3)(5-4)} = (5)(4)(3) = 60$$

possible permutations of 5 names into 3 committee positions.

To determine the total number of possible committees of n=5 items into committees of size r=3, we note that to form a committee or combination, the order of the r=3 names is unimportant and that there are 3(2)(1) = 3! =r! permutations for any given combination of r=3 items. If we divide the total number of permutations = 60 by the number of permutations of 3 names, we obtain the total number of combinations as follows:

$$\frac{{}_5P_3}{3!} = \frac{5!}{(5-3)!3!} = \frac{5(4)(3)(2)(1)}{(2)(1)(3)(2)(1)} = 10 \text{ committees} = {}_5C_3 = \binom{5}{3}$$

where ${}_5C_3 = \binom{5}{3}$ is the symbol used to denote the total number of combinations of 5 items taken 3 at a time. For any n, we may proceed to justify in a similar manner as for n=5 and r=3 and to write

$$\frac{{}_nP_r}{r!} = \frac{n!}{(n-r)!r!} = {}_nC_r = \binom{n}{r}$$

which is a simple and compact mathematical formulation for the total number of combinations of n items taken r at a time.

Some additional useful relations for permutations and combinations are the following. From

$$\binom{n}{r} = \frac{n!}{(n-r)!r!} = \binom{n}{n-r},$$

we see that we obtain the same number of combinations taking n items n-r at a time as we do taking n items r at a time. For the above example, we obtained ten committees of size three for the five students. We would also obtain ten committees of size 5-3=2. Another relation is

$$r!\binom{n}{r} = \frac{r!n!}{(n-r)!r!} = {}_nP_r$$

for n different items taken r at a time. Again using the above example for n=5 and r=3, we note that there were ten combinations and that 3!(10)

= 1(2)(3)(10) = 60 = the number of permutations that were obtained.

In another situation, suppose that there are n items such that p are of one kind and are identical, q are of a second kind and are identical, r are of a third kind and are identical, and the remaining items each occur once. For example, in the word "statistics", s and t each occur 3 times, i occurs twice, and the remaining two letters a and c each occur once, such that 3 + 3 + 2 + 1 + 1 = 10 letters. Suppose that we wish to know the number of different permutations of the 10 letters taken 10 at a time. The total number of permutations is

$$_{10}P_{10} = \frac{10!}{(10-10)!} = 10!$$

but not all permutations are different. Since there are p! = 3! permutations of s which are identical, q! = 3! permutations of t which are identical, and k! = 2! permutations of these 10 letters taken 10 at a time is

$$\frac{10!}{3!3!2!} = \frac{1(2)(3)(4)(5)(6)(7)(8)(9)(10)}{1(2)(3)(1)(2)(3)(1)(2)} = 2(5)(7)(8)(9)(10) = 50{,}400.$$

By the same reasoning, we may write the general formula for p items of one kind, q of a second kind, k of a third kind, and n-p-q-k items which are of only one kind and for the total number of different permutations of n items taken n at a time as

$$\frac{n!}{p!q!k!} .$$

Furthermore, if we have n items such that p are of one kind, q of a second kind, k of a third kind, and n-p-q-k different single items, and if we wish to compute the number of permutations of r different items, we may use the following formula

$$\frac{[n-(p-1)-(q-1)-(k-1)]!}{[n-(p-1)-q-1)-(k-1)-r]!} .$$

Many more such relations need to be constructed to handle all types of combinations and permutations, but this is the subject of more advanced courses in statistics.

For the binomial distribution in chapter XI we note the term $n!/(n-k)!k!$ which may be thought of in terms of combinations or in terms of permutations. For the latter we want to know the number of permutations of n items taken n at a time with k of one kind and n-k of a second kind. For the example given, suppose that we have five unbiased two-sided coins which are tossed simultaneously. The total possible different permutations are:

event	coin 1	coin 2	coin 3	coin 4	coin 5	
5 tails	T	T	T	T	T	} 1
4 tails, 1 head	H	T	T	T	T	
	T	H	T	T	T	
	T	T	H	T	T	} 5
	T	T	T	H	T	
	T	T	T	T	H	
3 tails, 2 heads	H	H	T	T	T	
	H	T	H	T	T	
	⋮					} 10
	T	T	T	H	H	
2 tails, 3 heads	H	H	H	T	T	
	⋮					} 10
	T	T	H	H	H	
1 tail, 4 heads	H	H	H	H	T	
	⋮					} 5
	T	H	H	H	H	
5 heads	H	H	H	H	H	} 1
Total						32

For all tails or all heads there is only one permutation of 5 of one kind: $5!/5!$. For 4 tails and one head, or conversely 4 heads and one tail, we compute the number of permutations of 5 items with 4 of one kind and only one of the second kind as $5!/4!1! = 5$. Likewise, for n=5 items with k=3 of one kind and n-k=2 of a second kind, the number of permutations is computed as $5!/2!3! = 10$ permutations. This illustrates the usefulness of the computational formulas for permutations and combinations in computing the coefficients for the binomial distribution and consequently the probabilities of the various events. The theoretical probability of occurrence of 5 heads or 5 tails from the toss of 5 unbiased coins is 1/32, the probability of either 4 heads and one tail, or conversely one head and 4 tails, is 5/32, and the probability of occurrence of 3 heads and 2 tails, or *vice versa*, is 10/32. Although it was easy to write out the 32 permutations for n=5, it becomes more difficult as n increases. With the above formulas we need only compute the various probabilities as was done in tables XI.1 and XI.2 to obtain the expected proportions or probabilities, say p_i. Then the various p_i were multiplied by the total number of observations to obtain the expected numbers for any given event.

This short introduction should be sufficient to acquaint the reader with some elementary notations of probability and to form a basis for some of the material in the next chapters. For further insight into more complex applications, the reader is directed to references listed below.

IX.10. Problems

IX.1. (Adapted from example 330B of Wallis and Roberts [1956].) Suppose that license plates are numbered from 1 to 500 and suppose that the first digit is obtained for randomly drawn license numbers. What is your guess as to the frequency of the digits 0, 1, 2, 3, 4,...,9? Compute the actual frequency for each digit for this set of numbers. What is the frequency of the last digit of numbers? Suppose that there had been 1000 license plates in your population. What would have been the frequency of the first and last digits in the license numbers?

IX.2. (Adapted from example 331 of Wallis and Roberts [1956].) The following parlor game may be used to increase your financial resources at the expense of your gullible friends. Take any book of statistical tables such as the *Statistical Abstract of the United States*. Open to a page in any haphazard manner. Then play the following game. Gullible (your friend) pays Sharpie (you) one dime every time a 1, 2, 3, or 4 appears and Sharpie pays Gullible a dime every time a 5, 6, 7, 8 or 9 appears. Why are the odds in favor of Sharpie?

IX.3. A game with which the author has not had much success is a game called chuck-a-luck and is often played at carnivals and fairs. If you wish a carnival affair at your next party and perhaps want to make a little money, obtain a cage with see-through sides, and place three large six-sided dice into it. Then build a board, with numbers 1, 2, 3, 4, 5 and 6, divide it into six equal parts, and mark. You are now ready to play. The rules are as follows. Bets are placed on the numbers, and pay-offs are made if the dice numbers (of dots) correspond to the player's number. Otherwise, the operator keeps the amount wagered. If two of the dice show the same number, the pay-off is doubled, and if all three dice show the same number, the pay-off is tripled. The player often reasons that the odds are in his favor since his chances of winning are 50-50 when he plays three numbers and, in addition, he receives a double pay-off when two dice show the same number and a triple pay-off when all three dice show the same number. He erroneously believes that his chances of winning are greater than 50-50. Show that the odds of a player winning are 91:125.

IX.4. (Adapted from example 330A, Wallis and Roberts [1956].) A young Romeo wanted to give each of four girls, Miss East, Miss West, Miss North, and Miss South, equal time. He took a commuter train to see each of the girls. He reasoned that he would be fair and it would be more fun if he would randomly select the minute to arrive at the station and then take the train leaving next after his arrival. He kept track of the number of times that he saw each girl and to his chagrin he was not seeing them an equal proportion of the time. The train schedules were:

Eastbound	every	20 minutes	starting	on the hour,
Westbound	"	"	"	"	one minute after the hour,
Northbound	"	"	"	"	ten minutes after the hour,
Southbound	"	"	"	"	twelve minutes after the hour.

What proportion of the time was he seeing each girl and why? Could you devise a selection of trains which would yield equal time with each girl?

IX.11. References and Suggested Reading

Chapman, D. G. and Schaufele, R. A. [1971]. Elementary Probability Models and Statistical Inference. Ginn and Blaisdell, Waltham, Mass., pp. xix + 358.

Goldberg, S. [1963]. Probability. An Introduction, 4th printing. Prentice-Hall, Inc., Englewood Cliffs, New Jersey, pp. ix + 322.

McCarthy, P. J. [1957]. Introduction to Statistical Reasoning. McGraw-Hill Book Company, Inc., New York, Toronto, London, pp. xiii + 402.

(Read chapter 7, giving particular attention to the examples.)

Moroney, M. J. [1956]. Facts from Figures. Penguin Books, Baltimore and London, pp. viii + 472.

(Read chapter 2 and associated problems.)

Mosteller, F. [1965]. Fifty Challenging Problems in Probability with Solutions. Addison-Wesley Publishing Company, Reading, Mass., viii + 82.

Mosteller, F., Rourke, R. E. K. and Thomas, G. B. [1961]. Probability with Statistical Applications. Addison-Wesley Publishing Company, Reading, Mass., xv + 478 pp.

Wallis, W. A. and Roberts, H. V. [1956]. Statistics: A New Approach. The Free Press, Glencoe, Illinois, pp. xxxviii + 646.

(Read chapter 10.)

CHAPTER X. STATISTICAL SUMMARIZATION OF DATA

X.1. Arithmetic Mean, Median, Mode, Harmonic Mean, Geometric Mean, and Proportion or Percentage.

Reference to the arithmetic mean and to the median was made in previous chapters. Further discussion of these measures of central tendency and of other measures is presented in this chapter. In addition, several measures of variation are discussed.

The calculation of the arithmetic mean, the median and the mode is described below and is illustrated with two examples.

Example 1. Suppose the following sets of data are available:

Set 1: 5, 7, 8, 9, 3, 2, 4, 5, 6, 5

The sum of the 10 numbers is 54.

Set 2: 8, 3, 5, 7, 4, 7, 4, 5, 6

The sum of the 9 numbers is 49.

Rearranging the numbers in lineal order, we obtain:

Set 1: 2, 3, 4, $\overbrace{5, 5, 5}^{\text{mode}}$ 6, 7, 8, 9 (median ↑ between the 5th and 6th numbers)

Set 2: 3, $\overbrace{4, 4}^{\text{mode}}$, $\overbrace{5, 5}^{\text{mode}}$, 6, $\overbrace{7, 7}^{\text{mode}}$, 8 (median ↑ at the second 5)

The mode is defined as the most frequent (or popular) number. In set 1 the number 5 occurs 3 times and no other number occurs more than once. In the second set of data there are three modes, 4, 5, and 7.

The sample median is that value which lies midway between the n/2 and (n+2)/2 ordered variates for n an even number and is the value of the (n+1)/2 ordered observations when n is odd. The population median is the value which divides the totality of observations such that 50% fall below the median and 50% fall above the median. For set 1, there are 10 observations, and the median falls between the 5th and 6th observations when they are arrayed in lineal order. Since both the 5th and 6th observations have the number 5, the median value is 5. If the values had been different, say 5 and 8, the median would have been $(5+8)/2 = 6.5$. For the second set of data, the number of observations is odd and the median is the value associated with the $(n+1)/2$th observation which is the fifth observation. Since the value of the 5th observation is 5, this is the median for these data.

The arithmetic mean or average is defined to be the sum of the n observations divided by the number of observations. The averages of the observations for the two sets are:

$$\frac{5+7+8+9+3+2+4+5+6+5}{10} = \frac{54}{10} = 5.4 \;;$$

$$\frac{8+3+5+7+4+7+4+5+6}{9} = \frac{49}{9} = 5\frac{4}{9} \doteq 5.44 \;.$$

Example 2. Suppose that we have a camp of 99 workers whose salaries are all equal, say \$2000 per year, and that their leader, a political appointee, receives \$100,000 per year. The mode and the median are both \$2000, but the arithmetic mean is:

$$\frac{99(\$2000) + \$100{,}000}{100} = \frac{\$298{,}000}{100} = \$2980 \;.$$

Here we note that there is only one individual above the arithmetic average, and there are 99 below. The formula expressing this mean of n observations, with the ith $(i=1,2,\cdots,n)$ observation being defined as Y_i, is:

$$\frac{Y_1+Y_2+Y_3+\cdots+Y_n}{n} = \frac{1}{n}\sum_{i=1}^{n} Y_i = \frac{Y_.}{n} = \bar{y}_. \;.$$

The symbol $\cdots$ indicates that all values of Y_i from Y_3 to Y_n are to be included. The symbol $\sum_{i=1}^{n}$ is a summation sign indicating that the sum is over all observations from 1 to n. Likewise, the symbol $Y_.$ indicates a summation over all values of i in Y_i. These two symbols are simply shorthand notations for the left side of the above equation. The arithmetic average may be likened to a center of gravity or to a fulcrum for a scaled board on which the data are placed in their respective positions relative to their values. Thus for the set 1 data, where each datum is represented by a square and the mean by a fulcrum, we can picture the result as shown in figure X.1. When the fulcrum is placed at 5.4 the board balances and is level.

For the data of example 2, the fulcrum would be placed at $2980 as shown in figure X.2. Since there are 99 values of $2000 the fulcrum is set at $2980 to balance the board which has one value at $100,000.

The example illustrates that the three averages arithmetic mean, median, and mode do not yield the same values, and that the proportion of the observations below the arithmetic mean depends upon the symmetry of the relative frequencies of the data. Since these different averages yield different values, it is imperative that the type of average be specified and understood by the user. All averages are useful for specific but different purposes.

If the distribution of values of observations is not symmetrical, that is, it is skewed, the relative locations of the mode, median and mean are shown in figure X.3. One half of the area under the curve is to the left of the median (unshaded area) and one half is to the right of the median (shaded area).

A situation wherein the arithmetic average leads to the wrong answer is the following: The speeds on 4 sides of a square of side 100 miles are 100 mi./hr., 200 mi./hr., 300 mi./hr., and 400 mi./hr. respectively. The arithmetic average of the four speeds is $(100 + 200 + 300 + 400)/4 = 250$ m.p.h. That using the arithmetic average is not correct is easily seen from the following:

Time to travel first side = 1 hour = 60 minutes

Time to travel second side= $\frac{1}{2}$ hour = 30 minutes

Time to travel third side = $\frac{1}{3}$ hour = 20 minutes

Time to travel fourth side= $\frac{1}{4}$ hour = 15 minutes

Total time 25/12 hours= 125 minutes

Therefore, total distance/total time = 400 miles/25/12 hours = 192 m.p.h.,

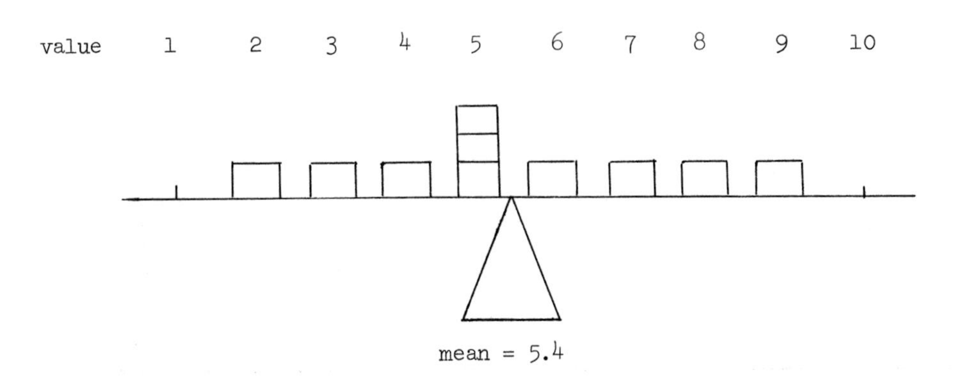

Figure X.1. Arithmetic mean represented as a center of gravity.

Figure X.2. Arithmetic mean represented as a center of gravity.

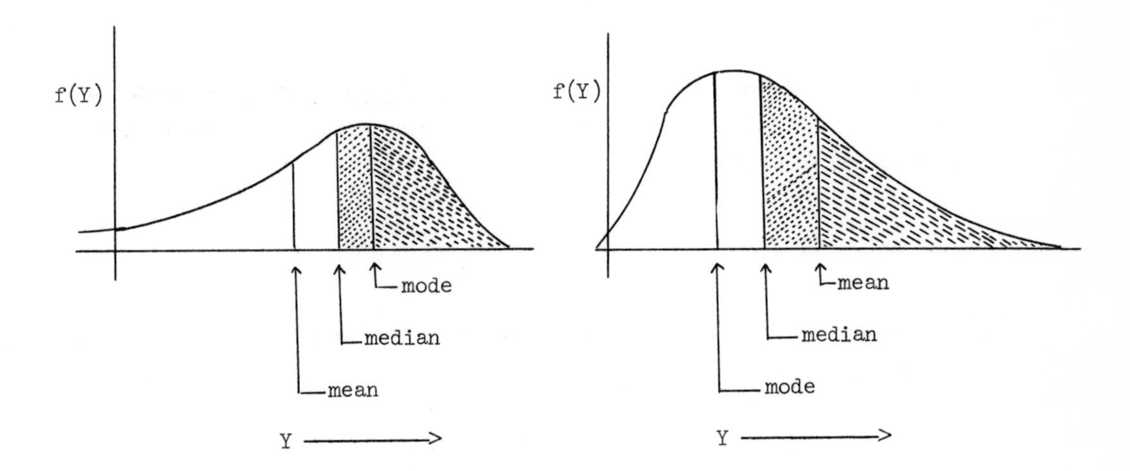

Figure X.3. Relative locations of mean, median, and mode.

which is the correct result. There are other ways to obtain the correct answer; for instance, one could use a weighted mean as follows:

$$\frac{1(100)+\frac{1}{2}(200)+\frac{1}{3}(300)+\frac{1}{4}(400)}{1+\frac{1}{2}+\frac{1}{3}+\frac{1}{4}=25/12} = \frac{12(400)}{25} = 192 \text{ m.p.h.}$$

$$\frac{60(100)+30(200)+20(300)+15(400)}{60+30+20+15=125} = \frac{24000}{125} = 192 \text{ m.p.h.}$$

In lieu of these, one could use a <u>harmonic mean or average</u>, which is defined as the reciprocal of the mean of the reciprocals. The formula for computing the harmonic mean is:

$$\bar{y}_h = 1/\sum_{i=1}^{n} 1/Y_i/n = n/\sum_{i=1}^{n} 1/Y_i .$$

This type of average is appropriate when dealing with rates or prices. For the example on speeds

$$\bar{y}_h = 4/\left(\frac{1}{100}+\frac{1}{200}+\frac{1}{300}+\frac{1}{400}\right) = \frac{4(1200)}{12+6+4+3} = 192 \text{ m.p.h.},$$

which, we may note, is the correct average. If the times had been constant and the distances variable, then the arithmetic average for m.p.h. would have been correct; however, with the distances constant and the times variable, the harmonic average is the correct one.

Another type of average is the <u>geometric mean or average</u>, which is defined to be the nth root of the product of n observations. The formula for computing the geometric mean is:

$$\bar{y}_g = (Y_1 \cdot Y_2 \cdot Y_3 \cdot \cdots \cdot Y_n)^{1/n} = \left(\prod_{i=1}^{n} Y_i\right)^{1/n} ,$$

where the symbol $\prod_{i=1}^{n}$ indicates a product of n variables. This type of average is appropriate for situations where the Y_i values increase in a geometric or exponential manner. The growth data for chick embryos from 6 to 12 days in chapter VII is an example of data following an exponential increase. Suppose that we had the 6-day weight = Y_6 = 0.029 and the 16-day weight

$= Y_{16} = 2.812$ and wished to estimate the 11-day weight. The geometric mean is:

$$\bar{y}_g = \sqrt{0.029(2.812)} = 0.286$$

whereas the observed Y_{11} was 0.261. The arithmetic mean would be $(0.029 + 2.812)/2 = 1.420$ which is considerably different from $Y_{11} = 0.261$ or from $\bar{y}_g = 0.286$. Another example wherein a geometric mean would be appropriate is the case of the change in population of a city when the birth and death rates stay constant and when there is no migration in or out of the city.

Another type of average is one called an index number. Moroney [1956], for example, gives an introduction to index numbers through the use and construction of a simple Cost of Living Index.

A percentage or ratio is an average in the same sense that the preceding ones are. Some of us are acquainted with the batting averages of baseball players. We know that this average is a ratio of the number of hits, say h, to the total times at bat, say b. The batting average is $[h(1)+(b-h)(\text{zero})]/b$ = ratio expressed as a 3-decimal number such as .341. This means that $h/b = 34.1\%$ of the times at bat = b, the batter scored a hit = h, and $(b-h)/b = 65.9\%$ of the time he did not get a hit or got zero hits.

We utilized proportions to summarize some of the data from the class survey as, for instance, in table VIII.2. The results in table VIII.3 could have been similarly treated.

X.2. Means from Orthogonal Experimental, Treatment, and Sample Survey Designs

In the completely randomized design with v treatments each repeated r times, or in a sample survey design with a random selection of the clusters and of the r elements in each cluster, data of the following nature are available:

Treatment or cluster number

	1	2	3	...	v
	Y_{11}	Y_{21}	Y_{31}	...	Y_{v1}
	Y_{12}	Y_{22}	Y_{32}		Y_{v2}
	Y_{13}	Y_{23}	Y_{33}		Y_{v3}
	$\vdots$	$\vdots$	$\vdots$	$\ddots$	$\vdots$
	Y_{1r}	Y_{2r}	Y_{3r}	...	Y_{vr}
Total	$Y_{1.}$	$Y_{2.}$	$Y_{3.}$		$Y_{v.}$
Mean	$\bar{y}_{1.}$	$\bar{y}_{2.}$	$\bar{y}_{3.}$		$\bar{y}_{v.}$

Grand total = sum of all the observations

$$= \sum_{i=1}^{v} \sum_{j=1}^{r} Y_{ij} = Y_{..}$$

overall mean = $Y_{..}/rv$

$= \bar{y}_{..}$

The treatment or cluster means, $Y_{i.}/r = \bar{y}_{i.}$, and the overall mean $\bar{y}_{..}$ are the means which summarize the information about arithmetic averages for these data when the yield equation is of the form:

Y_{ij} = true population mean + true treatment or cluster effect + deviation of observation from true cluster mean

= estimated population mean + estimated treatment or cluster effect + estimated random deviation

$$= \bar{y}_{..} + (\bar{y}_{i.} - \bar{y}_{..}) + \hat{e}_{ij}.$$

In a stratified sample survey design with c randomly selected clusters per stratum and with r randomly selected elements per cluster, data of the following nature are available:

Stratum number

	1				2				...	s			
	cluster mean				cluster mean					cluster mean			
	1	2	...	c	1	2	...	c		1	2	...	c
	Y_{111}	Y_{121}		Y_{1c1}	Y_{211}	Y_{221}		Y_{2c1}		Y_{s11}	Y_{s21}		Y_{sc1}
	Y_{112}	Y_{122}		Y_{1c2}	Y_{212}	Y_{222}		Y_{2c2}		Y_{s12}	Y_{s22}		Y_{sc2}
	$\vdots$												
	Y_{11r}	Y_{12r}		Y_{1cr}	Y_{21r}	Y_{22r}		Y_{2cr}		Y_{s1r}	Y_{s2r}		Y_{scr}
Total	$Y_{11\cdot}$	$Y_{12\cdot}$		$Y_{1c\cdot}$	$Y_{21\cdot}$	$Y_{22\cdot}$		$Y_{2c\cdot}$		$Y_{s1\cdot}$	$Y_{s2\cdot}$		$Y_{sc\cdot}$
Mean	$\bar{y}_{11\cdot}$	$\bar{y}_{12\cdot}$		$\bar{y}_{1c\cdot}$	$\bar{y}_{21\cdot}$	$\bar{y}_{22\cdot}$		$\bar{y}_{2c\cdot}$		$\bar{y}_{s1\cdot}$	$\bar{y}_{s2\cdot}$		$\bar{y}_{sc\cdot}$
Stratum total				$Y_{1\cdot\cdot}$				$Y_{2\cdot\cdot}$					$Y_{s\cdot\cdot}$
Stratum mean				$\bar{y}_{1\cdot\cdot}$				$\bar{y}_{2\cdot\cdot}$					$\bar{y}_{s\cdot\cdot}$

The ij^{th} cluster mean is equal to the ij^{th} total divided by r, that is, $Y_{ij\cdot}/r = \bar{y}_{ij\cdot}$. The ith stratum mean is equal to $Y_{i\cdot\cdot}/rc = \bar{y}_{i\cdot\cdot}$, where $Y_{i\cdot\cdot}$ is the total for the ith cluster and rc is the number of observations in a stratum total.

For a randomized complete block design with v treatments in r different complete blocks, the data rearranged in a systematic fashion may be represented as:

Treatment	Block 1	2	3	...	r	Treatment Total	Mean
1	Y_{11}	Y_{12}	Y_{13}	...	Y_{1r}	$Y_{1\cdot}$	$\bar{y}_{1\cdot}$
2	Y_{21}	Y_{22}	Y_{23}	...	Y_{2r}	$Y_{2\cdot}$	$\bar{y}_{2\cdot}$
3	Y_{31}	Y_{32}	Y_{33}	...	Y_{3r}	$Y_{3\cdot}$	$\bar{y}_{3\cdot}$
⋮	⋮	⋮	⋮	⋱	⋮	⋮	⋮
v	Y_{v1}	Y_{v2}	Y_{v3}	...	Y_{vr}	$Y_{v\cdot}$	$\bar{y}_{v\cdot}$
Block total	$Y_{\cdot 1}$	$Y_{\cdot 2}$	$Y_{\cdot 3}$	...	$Y_{\cdot r}$	$Y_{\cdot\cdot}$	-
Block mean	$\bar{y}_{\cdot 1}$	$\bar{y}_{\cdot 2}$	$\bar{y}_{\cdot 3}$	...	$\bar{y}_{\cdot r}$	-	$\bar{y}_{\cdot\cdot}$

The arithmetic means are computed as the total divided by the number of observations. The block means $\bar{y}_{\cdot j} = Y_{\cdot j}/v$, the treatment means $\bar{y}_{i\cdot} = Y_{i\cdot}/r$, and the overall mean $\bar{y}_{\cdot\cdot} = Y_{\cdot\cdot}/rv$ are the means required to summarize the information in a randomized complete block design when the yield datum is represented by the linear additive equation

$$Y_{ij} = \mu + t_i + b_j + e_{ij}$$

= true overall mean + true treatment effect + true block effect + true random error.

$$Y_{ij} = \hat{\mu} + \hat{t}_i + \hat{b}_j + \hat{e}_{ij}$$ = estimated mean + estimated treatment effect + estimated block effect + estimated random error

$$= \bar{y}_{\cdot\cdot} + (\bar{y}_{i\cdot} - \bar{y}_{\cdot\cdot}) + (\bar{y}_{\cdot j} - \bar{y}_{\cdot\cdot}) + (Y_{ij} - \bar{y}_{i\cdot} - \bar{y}_{\cdot j} + \bar{y}_{\cdot\cdot}) .$$

The various totals in terms of true effects are

$$Y_{\cdot\cdot} = rv\,\mu + r\sum_{i=1}^{v} t_i + v\sum_{j=1}^{r} b_j + \sum_{i=1}^{v}\sum_{j=1}^{r} e_{ij}$$

$$Y_{i.} = r(\mu+t_i) + \sum_{j=1}^{r} b_j + \sum_{j=1}^{r} e_{ij}$$

$$Y_{.j} = v(\mu+b_j) + \sum_{i=1}^{v} t_i + \sum_{i=1}^{v} e_{ij}$$

Here we note that the difference between any two $\bar{y}_{i.}$ contains only the difference between two treatment effects and some error terms. The same is true for any two $\bar{y}_{.j}$, and, hence, the design is orthogonal by definition

Example X.1. To illustrate the above consider the following numerical example with v=3 and r=4 for a randomized complete block design:

	Block				Treatment	
Treatment	1	2	3	4	Total	Mean
1	5	13	7	3	28	7
2	9	10	8	9	36	9
3	1	7	6	6	20	5
Block total	15	30	21	18	84	-
Block mean	5	10	7	6	-	7

In the latin square design, there are three different arithmetic means, the row mean, the column mean, and the treatment mean. These, together with the overall arithmetic mean constitute all the means necessary to summarize the information from a k × k latin square design when the yield datum for a single observation is of the following nature,

$$Y_{hij} = \mu + t_i + b_j + c_h + e_{hij}$$

= true overall mean + true treatment effect + true block or row effect + true column effect + true random error,

$$= \hat{\mu} + \hat{t}_i + \hat{b}_j + \hat{c}_h + \hat{e}_{hij} = \bar{y}_{...} + (\bar{y}_{.i.} - \bar{y}) + (\bar{y}_{..j} - \bar{y})$$

$$+ (\bar{y}_{h..} - \bar{y}) + (Y_{hij} - \bar{y}_{h..} - \bar{y}_{.i.} - \bar{y}_{..j} + 2\bar{y}_{...})$$

= estimated overall mean + estimated treatment effect + estimated row effect + estimated column effect + estimated random error.

The various totals in terms of effects are:

$$Y_{...} = k^2\mu + k\sum_{h=1}^{k} c_h + k\sum_{i=1}^{k} t_i + k\sum_{j=1}^{k} b_j + \sum_{h=1}^{k}\sum_{j=1}^{k} e_{hij}$$

$$Y_{h..} = k(\mu + c_h) + \sum_{i=1}^{k} t_i + \sum_{j=1}^{k} b_j + \sum_{j=1}^{k} e_{hij}$$

$$Y_{.i.} = k(\mu + t_i) + \sum_{h=1}^{k} c_h + \sum_{j=1}^{k} b_j + \sum_{h=1}^{k} e_{hij}$$

$$Y_{..j} = k(\mu + b_j) + \sum_{h=1}^{k} c_h + \sum_{i=1}^{k} t_i + \sum_{h=1}^{k} e_{hij}$$

The difference between any two treatment means, say $\bar{y}_{.1.}$ and $\bar{y}_{.2.}$, contains only the difference due to the two treatment effects, t_1 and t_2, and some error terms. The same is true for row and column means. Therefore, by definition, treatments are orthogonal to rows and columns, and rows and columns are orthogonal to each other.

The yields = Y_{hij}, totals, and means in a latin square design may be of the following form depending upon the particular randomization utilized:

Row	Column 1	2	3	...	k	Row total	Row mean
1	Y_{111}	Y_{221}	Y_{331}	...	Y_{k41}	$Y_{..1}$	$\bar{y}_{..1}$
2	Y_{122}	Y_{232}	Y_{342}	...	Y_{k52}	$Y_{..2}$	$\bar{y}_{..2}$
3	Y_{133}	Y_{243}	Y_{353}	...	Y_{k63}	$Y_{..3}$	$\bar{y}_{..3}$
⋮	⋮	⋮	⋮	⋱	⋮	⋮	⋮
k	Y_{1kk}	$Y_{2,k-1,k}$	$Y_{3,k-2,k}$	...	Y_{k1k}	$Y_{..k}$	$\bar{y}_{..k}$
Column total	$Y_{1..}$	$Y_{2..}$	$Y_{3..}$	...	$Y_{k..}$	$Y_{...}$	-
Column mean	$\bar{y}_{1..}$	$\bar{y}_{2..}$	$\bar{y}_{3..}$	...	$\bar{y}_{k..}$	-	$\bar{y}_{...}$

	Treatment number 1	2	3	...	k	Sum
Treatment total	$Y_{.1.}$	$Y_{.2.}$	$Y_{.3.}$	...	$Y_{.k.}$	$Y_{...}$
Treatment mean	$\bar{y}_{.1.}$	$\bar{y}_{.2.}$	$\bar{y}_{.3.}$	...	$\bar{y}_{.k.}$	$k\bar{y}_{...}$

Example X.2. As a small numerical example to illustrate the above general form, suppose that k=3 and that the following data and arrangement of treatments A, B, and C are available from an experiment designed as a latin square:

Row	Column 1	Column 2	Column 3	Row Total	Row Mean
1	B 23	A 17	C 29	69	23
2	A 16	C 25	B 16	57	19
3	C 24	B 18	A 12	54	18
Column Total	63	60	57	180	-
Column Mean	21	20	19	-	20

	Treatment A	Treatment B	Treatment C	Sum
Treatment Total	45	57	78	180
Treatment Mean	15	19	26	60 = 3(20)

Another orthogonal design is the simple change-over design. The treatment, row, column, and grand totals are obtained in much the same manner as for the latin square design. The schematic layout for v treatments in vs columns and v rows is, for yield = Y_{hij}:

Row	Column 1	2	...	v	v+1	...	vs	Row Total	Row Mean
1	Y_{111}	Y_{221}	...	$Y_{v,v,1}$	$Y_{v+1,1,1}$	...	$Y_{vs,v,1}$	$Y_{..1}$	$\bar{y}_{..1}$
2	Y_{1v2}	Y_{212}	...	$Y_{v,v-1,2}$	$Y_{v+1,2,2}$	...	$Y_{vs,v-1,2}$	$Y_{..2}$	$\bar{y}_{..2}$
⋮	⋮	⋮	⋱	⋮	⋮	⋱	⋮	⋮	⋮
v	Y_{12v}	Y_{23v}		$Y_{v,1,v}$	$Y_{v+1,v,v}$	...	$Y_{vs,1,v}$	$Y_{..v}$	$\bar{y}_{..v}$
Column total	$Y_{1..}$	$Y_{2..}$	...	$Y_{v..}$	$Y_{v+1..}$	...	$Y_{vs..}$	$Y_{...}$	-
Column mean	$\bar{y}_{1..}$	$\bar{y}_{2..}$	...	$\bar{y}_{v..}$	$\bar{y}_{v+1..}$	...	$\bar{y}_{vs..}$	-	$\bar{y}_{...}$

	Treatment number					
	1	2	3	...	v	Sum
Treatment total	$Y_{\cdot 1\cdot}$	$Y_{\cdot 2\cdot}$	$Y_{\cdot 3\cdot}$		$Y_{\cdot v\cdot}$	$Y_{\cdot\cdot\cdot}$
Treatment mean	$\bar{y}_{\cdot 1\cdot}$	$\bar{y}_{\cdot 2\cdot}$	$\bar{y}_{\cdot 3\cdot}$		$\bar{y}_{\cdot v\cdot}$	$v\bar{y}_{\cdot\cdot\cdot}$

The yield equation for a single datum is the same as for the latin square design.

Example X.3. A small numerical example for v=2 and s=3 is used to illustrate the computation of the row, column, treatment, and grand or overall means. The example is given below where the letter A or B designates the treatment and the number designates the yield data:

	Column						Row	
Row	1	2	3	4	5	6	Total	Mean
1	A 10	B 6	B 7	A 11	A 10	B 4	48	8
2	B 8	A 12	A 9	B 9	B 6	A 16	60	10
Column total	18	18	16	20	16	20	108	-
Column mean	9	9	8	10	8	10	-	9

	Treatment letter		
	A	B	Sum
Treatment total	68	40	108
Treatment mean	$11\frac{1}{3}$	$6\frac{2}{3}$	18 = 2(9)

From the definition of orthogonality in chapter VI we note that if the difference between two treatment means (or other means) does not contain any effects other than those due to the two specific treatments and random error, then the other effects are orthogonal to the treatments, and the experimental design is an orthogonal one. Since this is true the arithmetic means and linear combinations of these means summarize the information from the data.

As may be noted from chapter VII, the factorial treatment design is one in which the various effects of factors and the corresponding interactions are orthogonal to each other. The various arithmetic means suffice to estimate the various effects. This is not true for most fractional replicates of a factorial treatment design nor for the balanced incomplete block design.

X.3. Treatment Means Adjusted for Block Effects in a Balanced Incomplete Block Design

The estimation of treatment and block effects and adjusted treatment means in a balanced incomplete block (bib) design are included here to illustrate the fact that arithmetic means for treatments do not suffice to estimate the differences between treatments in a nonorthogonal design. Also, the bib design was selected because the computations remain relatively simple.

A particular systematic layout for a bib design with $v=4$ treatments in $b=6$ blocks of size $k=2$ each and with the various totals and means used in the analysis is given below:

	Block number						
	1	2	3	4	5	6	
	Y_{11}	Y_{12}	Y_{13}	Y_{24}	Y_{25}	Y_{36}	
	Y_{21}	Y_{32}	Y_{43}	Y_{34}	Y_{45}	Y_{46}	
Block total	$Y_{.1}$	$Y_{.2}$	$Y_{.3}$	$Y_{.4}$	$Y_{.5}$	$Y_{.6}$	$Y_{..}$ = grand total
Block mean	$\bar{y}_{.1}$	$\bar{y}_{.2}$	$\bar{y}_{.3}$	$\bar{y}_{.4}$	$\bar{y}_{.5}$	$\bar{y}_{.6}$	$\bar{y}_{..}$ = grand mean

	Treatment				
	1	2	3	4	Sum
Treatment total	$Y_{1.}$	$Y_{2.}$	$Y_{3.}$	$Y_{4.}$	$Y_{..}$

Note that a particular yield is written as

$$Y_{ij} = \mu + b_j + t_i + e_{ij}$$

= true overall mean + true j^{th} block effect + true i^{th} treatment effect + a true random error.

Then,

$$Y_{1.}/3 = \bar{y}_{1.} = (3\mu + b_1 + b_2 + b_3 + 3t_1 + e_{11} + e_{12} + e_{13})/3 ,$$

$$Y_{2.}/3 = \bar{y}_{2.} = (3\mu + b_1 + b_4 + b_5 + 3t_2 + e_{21} + e_{24} + e_{25})/3 ,$$

$$Y_{3.}/3 = \bar{y}_{3.} = (3\mu + b_2 + b_4 + b_6 + 3t_3 + e_{32} + e_{34} + e_{36})/3 ,$$

$$Y_{4.}/3 = \bar{y}_{4.} = (3\mu + b_3 + b_5 + b_6 + 3t_4 + e_{43} + e_{45} + e_{46})/3 ,$$

$$Y_{.1}/2 = \bar{y}_{.1} = (2\mu + 2b_1 + t_1 + t_2 + e_{11} + e_{21})/2 ,$$

$$Y_{.2}/2 = \bar{y}_{.2} = (2\mu + 2b_2 + t_1 + t_3 + e_{12} + e_{32})/2 ,$$

$$Y_{.3}/2 = \bar{y}_{.3} = (2\mu + 2b_3 + t_1 + t_4 + e_{13} + e_{43})/2 ,$$

$$Y_{.4}/2 = \bar{y}_{.4} = (2\mu + 2b_4 + t_2 + t_3 + e_{24} + e_{34})/2 ,$$

$$Y_{.5}/2 = \bar{y}_{.5} = (2\mu + 2b_5 + t_2 + t_4 + e_{25} + e_{45})/2 ,$$

$$Y_{.6}/2 = \bar{y}_{.6} = (2\mu + 2b_6 + t_3 + t_4 + e_{36} + e_{46})/2, \text{ and}$$

$$Y_{..}/12 = \bar{y}_{..} = \hat{\mu} = (12\mu + 2\sum_{i=1}^{4} b_j + 3\sum_{i=1}^{4} t_i + (\text{sum of 12 error terms}))/12.$$

From the above means we note that any difference between treatment means

$\bar{y}_{i.}$ contains block effects, and any difference between block means $\bar{y}_{.j}$ contains treatment effects. Hence, by the definition of orthogonality, the treatment and block effects are not orthogonal.

Now how do we estimate treatment effects $= \hat{t}_i$, which is an estimate of t_i? If we set up values denoted as $Q_{i.} = Y_{i.}$ - (sum of means of blocks in which treatment i occurs), and omit error terms from the equations, we obtain simple solutions as follows:

$$\hat{t}_1 = \left(Y_{1.} - (\bar{y}_{.1} + \bar{y}_{.2} + \bar{y}_{.3})\right)/2 = Q_{1.}/2$$

$$\hat{t}_2 = \left(Y_{2.} - (\bar{y}_{.1} + \bar{y}_{.4} + \bar{y}_{.5})\right)/2 = Q_{2.}/2$$

$$\hat{t}_3 = \left(Y_{3.} - (\bar{y}_{.2} + \bar{y}_{.4} + \bar{y}_{.6})\right)/2 = Q_{3.}/2$$

$$\hat{t}_4 = \left(Y_{4.} - (\bar{y}_{.3} + \bar{y}_{.5} + \bar{y}_{.6})\right)/2 = Q_{4.}/2$$

Likewise, $\bar{y}_{..} = \hat{\mu}$ is a solution for μ in the above. Therefore, a treatment mean adjusted for block effects is computed as $\bar{y}_{..} + \hat{t}_i = \bar{y}_{..} + Q_{1.}/2 = \bar{y}'_i$. The difference between any two adjusted treatment means or effects is for treatments 1 and 2, say,

$$\bar{y}'_{1.} - \bar{y}'_{2.} = \bar{y}_{..} + \hat{t}_1 - \bar{y}_{..} - \hat{t}_2 = \hat{t}_1 - \hat{t}_2 = (Q_{1.} - Q_{2.})/2 \ .$$

The solutions $\hat{b}_j$ of the b_j effects may be obtained by setting the error terms equal to zero (that is, omitting them) and substituting in the solutions for the $\hat{t}_i$. For example,

$$\hat{b}_1 = (Y_{.1} - 2\hat{\mu} - \hat{t}_1 - \hat{t}_2)/2 = (Y_{.1} - 2\bar{y}_{..} - Q_{1.}/2 - Q_{2.}/2)/2 \ .$$

In general, the solution for any $\hat{t}_i$ in a bibd with v treatments in b blocks of size k is given by

$$\hat{t}_i = Q_{i.}k/v\lambda \ ,$$

where λ is the number of times any given pair of treatments occurs together in the b blocks. In the example above, $\lambda = 1$ and $r = 3$ = the number of times each treatment occurs in the experimental design.

Example X.4. We shall approach the illustration a little differently this time in order to obtain deeper insight into the nature of the yield data in terms of the effects. Suppose that someone tells us that he obtained the following solutions from an experiment designed as a bib design with v=4, r=3, b=6, k=2, and λ=1 as described above:

$$\hat{\mu} = 10 \qquad \hat{t}_3 = 5 \qquad \hat{b}_2 = -3 \qquad \hat{b}_5 = 1$$
$$\hat{t}_1 = -1 \qquad \hat{t}_4 = 0 \qquad \hat{b}_3 = 1 \qquad \hat{b}_6 = -2$$
$$\hat{t}_2 = -4 \qquad \hat{b}_1 = 3 \qquad \hat{b}_4 = 0$$

$$\hat{e}_{11} = -1 \qquad \hat{e}_{32} = 0 \qquad \hat{e}_{24} = 1 \qquad \hat{e}_{45} = 2$$
$$\hat{e}_{21} = 1 \qquad \hat{e}_{13} = 1 \qquad \hat{e}_{34} = -1 \qquad \hat{e}_{36} = 1$$
$$\hat{e}_{12} = 0 \qquad \hat{e}_{43} = -1 \qquad \hat{e}_{25} = -2 \qquad \hat{e}_{46} = -1$$

Note that the yield in a bib design is

$$Y_{ij} = \hat{\mu} + \hat{t}_i + \hat{b}_j + \hat{e}_{\cdot j}$$

which for i=1 and j=1 is

$$Y_{11} = 10 - 1 + 3 - 1 = 11 .$$

Continuing this process for all 12 observations, we obtain the following reconstruction of the experimenter's yields:

	Blocks						
	1	2	3	4	5	6	
	Y_{11}=11	Y_{12}= 6	Y_{13}=11	Y_{24}= 7	Y_{25}= 5	Y_{36}= 9	
	Y_{21}=10	Y_{32}= 7	Y_{43}=15	Y_{34}= 9	Y_{45}=18	Y_{46}=12	
$Y_{\cdot j}$	21	13	26	16	23	21	120 = $Y_{\cdots}$

The treatment totals are:

$$Y_{1.} = 28 \ , \ Y_{2.} = 22 \ , \ Y_{3.} = 25 \ , \text{ and } Y_{4.} = 45 \ .$$

The process the experimenter went through to give us the above solutions was to compute the various totals above; then he computed the $Q_{i.}$ values as follows:

$$Q_{1.} = Y_{1.} - \tfrac{1}{2}(Y_{.1}+Y_{.2}+Y_{.3}) = 28 - \tfrac{1}{2}(21+13+26) = -2 \ ,$$
$$Q_{2.} = Y_{2.} - \tfrac{1}{2}(Y_{.1}+Y_{.4}+Y_{.5}) = 22 - \tfrac{1}{2}(21+16+23) = -8 \ ,$$
$$Q_{3.} = Y_{3.} - \tfrac{1}{2}(Y_{.2}+Y_{.4}+Y_{.6}) = 25 - \tfrac{1}{2}(13+16+21) = 0 \ , \text{ and}$$
$$Q_{4.} = Y_{4.} - \tfrac{1}{2}(Y_{.3}+Y_{.5}+Y_{.6}) = 45 - \tfrac{1}{2}(26+23+21) = 10 \ .$$

Then,

$$\hat{t}_1 = Q_{1.}/2 = -1 \ ,$$
$$\hat{t}_2 = Q_{2.}/2 = -4 \ ,$$
$$\hat{t}_3 = Q_{3.}/2 = 0 \ ,$$
$$\hat{t}_4 = Q_{4.}/2 = 5 \ , \text{ and}$$
$$\hat{\mu} = \bar{y}_{..} = 120/12 = 10 \ .$$

From these values we compute the $\hat{b}_j$ values as:

$$\hat{b}_1 = \big(21 - 2(10) - (-1) - (-4)\big)/2 = 3 \ ,$$
$$\hat{b}_2 = \big(13 - 2(10) - (-1) - 0\big)/2 = -3 \ ,$$
$$\hat{b}_3 = \big(26 - 2(10) - (-1) - 5\big)/2 = 1 \ ,$$
$$\hat{b}_4 = \big(16 - 2(10) - (-4) - 0\big)/2 = 0 \ ,$$
$$\hat{b}_5 = \big(23 - 2(10) - (-4) - 5\big)/2 = 1 \ , \text{ and}$$
$$\hat{b}_6 = \big(21 - 2(10) - 0 - 5\big)/2 = -2 \ .$$

Likewise, the $\hat{e}_{ij}$ may be computed as:

$$\hat{e}_{ij} = Y_{ij} - \hat{\mu} - \hat{t}_i - \hat{b}_j$$

which for i=1 and j=1 is:

$$\hat{e}_{11} = 11 - 10 - (-1) - 3 = -1 \ .$$

The remaining $\hat{e}_{ij}$ are computed similarly.

The adjusted treatment means are computed as follows:

$$\bar{y}'_{1.} = \hat{\mu} + \hat{t}_1 = 10 - 1 = 9 \ ,$$

$$\bar{y}'_{2.} = \hat{\mu} + \hat{t}_2 = 10 - 4 = 6 \ ,$$

$$\bar{y}'_{3.} = \hat{\mu} + \hat{t}_3 = 10 + 0 = 10 \ , \text{ and}$$

$$\bar{y}'_{4.} = \hat{\mu} + \hat{t}_4 = 10 + 5 = 15 \ .$$

The formulae gave us the same values we started with. This should increase our confidence in the formulae and give us some insight into the structure of the yields, of the effects, and of the computations.

X.4. Measures of Variation

As we have observed thus far, not all values fall at the average, regardless of what kind of average is used. There will be a scattering of values on both sides of an average. In addition to obtaining an average for a set of data, it is often of interest to have a statistic for summarizing information about scatter. One of the simplest measures of scatter is the range, which is simply the difference between the largest observed value and the smallest observed value in the sample, experiment, or population. For the class survey data, the range in heights (2nd column of table VIII·1) was 75" - 62" = 13", and the range in weights (9th column in table VIII·1) was 240 lbs. - 95 lbs. = 145 lbs.

Associated with the range is the idea of percentiles. For example, 10 percentiles or 25 percentiles (or quantiles) are frequently used to describe the scatter of grades on an examination or in a course. We often hear the term "in the upper 10%" or "in the upper 25%" of the class. Associated with this idea would be presentation of data from a frequency distribution as percents. We essentially did this for the data of table VIII·2 by presenting the proportion of individuals in the various hair and eye color classes. The percentage of individuals in the eye color classes black, brown, green, blue, and grey was 2%, 37%, 15%, 37%, and 9%, respectively. If there is a gradation of color from black to grey, we could speak of the lower 2 percentile, the lower 39%, and so forth.

Another measure of variation is <u>absolute error</u> or <u>absolute mean deviation</u>, which is the average of the absolute deviations from the mean, or

$$\sum_{i=1}^{n} |Y_i - \bar{y}|/n \ ,$$

where the vertical pair of lines means that the sign of the deviation is ignored. For example, suppose that $Y_1=1$, $Y_2=3$, $Y_3=5$, $Y_4=7$, and $Y_5=9$ are the sample values. Then $\sum_{i=1}^{5} Y_i = 25$ = sum of the n=5 observations, and $\bar{y} = 25/5 = 5$ = arithmetic mean. The deviations from the mean are $Y_1-\bar{y} = -4$, $Y_2-\bar{y} = -2$, $Y_3-\bar{y} = 0$, $Y_4-\bar{y} = 2$, and $Y_5-\bar{y} = 4$. The absolute mean deviation is

$$\frac{1}{5}\sum_{i=1}^{5} |Y_i - \bar{y}| = \frac{1}{5}\left\{|-4| + |-2| + |0| + |2| + |4|\right\} = \frac{12}{5} = 2.4 \ .$$

This measure has not been used to any extent for statistical summarization of data.

Another statistic that is useful in summarizing information about scatter is the <u>variance</u> which for a sample of size n is defined to be:

$$s^2 = \sum_{i=1}^{n} (Y_i - \bar{y})^2/(n-1) \ .$$

This statistic is an estimate of the population variance usually denoted by σ^2. The sample variance is an average of the squared distances from the arithmetic mean and refers to the scatter of single observations. For the above example,

$$s^2 = \left\{(-4)^2 + (-2)^2 + 0^2 + 2^2 + 4^2\right\}/(5-1) = \frac{40}{4} = 10 \ .$$

In referring to the scatter of means from a sample of size n, the <u>variance of a sample mean of n observations</u> is defined to be:

$$s_{\bar{y}}^2 = \frac{s^2}{n} = \sum_{i=1}^{n} (Y_i-\bar{y})^2/n(n-1) \ .$$

Here we note that as the sample size increases, the scatter decreases.

The <u>standard deviation of a single observation</u> is the square root of the sample variance, or $s = \sqrt{\sum_{i=1}^{n} (Y_i-\bar{y})^2/(n-1)}$; the <u>standard deviation of a mean of n observations</u>, or more commonly the <u>standard error of a mean</u>, is equal to

$$s_{\bar{y}} = \sqrt{\sum_{i=1}^{n} (Y_i-\bar{y})/n(n-1)} = \sqrt{s^2/n} \ .$$

Also, since we have been talking about differences between two independent sample means such as might occur in a completely randomized experiment or in a stratified random sample survey, we need to define another term, the <u>standard error of a difference between two treatment (or stratum) means</u>. This is equal to

$$s_{\bar{y}_1-\bar{y}_2} = \sqrt{\sum_{j=1}^{n_1} \frac{(Y_{1j}-\bar{y}_{1.})^2}{n_1(n_1-1)} + \sum_{j=1}^{n_2} \frac{(Y_{2j}-\bar{y}_{2.})^2}{n_2(n_2-1)}} \ ,$$

where n_1 observations are used to estimate the mean and variance for treatment (or stratum) one and n_2 observations are used to estimate the mean and variance for treatment (or stratum) two. If the deviations $(Y_{1j}-\bar{y}_{1.})$ and $(Y_{2j}-\bar{y}_{1.})$ are of the same degree of variability, that is, they both come from a population with a population variance of σ^2, then the variance of a difference between the two means becomes:

$$s^2_{\bar{y}_{1\cdot}-\bar{y}_{2\cdot}} = \left\{ \frac{\sum_{j=1}^{n_1} (Y_{1j}-\bar{y}_{1\cdot})^2 + \sum_{j=1}^{n_2} (Y_{2j}-\bar{y}_{2\cdot})^2}{(n_1-1) + (n_2-1)} \right\} / \left\{\frac{1}{n_1} + \frac{1}{n_2}\right\} ,$$

which for $r = n_1 = n_2$ becomes:

$$s^2_{\bar{y}_{1\cdot}-\bar{y}_{2\cdot}} = \frac{2}{r} \sum_{i=1}^{2} \sum_{j=1}^{r} (Y_{ij}-\bar{y}_{i\cdot})^2/2(r-1) .$$

If all treatments in a completely randomized design of v treatments with r replicates on each treatment have the same degree of scatter, then the estimated variance of a single observation is equal to

$$s^2_e = \sum_{i=1}^{v} \sum_{j=1}^{r} (Y_{ij}-\bar{y}_{i\cdot})^2/v(r-1) = \sum_{i=1}^{v} \sum_{j=1}^{r} \hat{e}^2_{ij}/v(r-1) .$$

The subscript e is placed on s^2 to denote that s^2_e , the error variance, refers to the scatter of the $\hat{e}_{ij}$ = estimated random errors = nonassignable errors. The standard error of a single mean of r replicates in a completely randomized design is simply s^2_e/r , and the standard error variance of a difference between two treatment means, say 1 and 2, each with r replicates, is $2s^2_e/r$; the standard error of a difference of two means is $s_e\sqrt{2/r}$.

With the above formulation we can proceed to obtain the error variance, s^2_e and other related quantities for other orthogonal designs such as the randomized complete block design, the simple change-over design, and the latin square design. We simply obtain the sum of squares of the random errors = $\hat{e}_{ij}$ or $\hat{e}_{hij}$ and divide by the appropriate constants, which are denoted as degrees of freedom, to obtain the estimated error variance. For the various designs these are computed as follows:

Completely randomized design (v treatments; r replicates)

$$s_e^2 = \sum_{i=1}^{v} \sum_{j=1}^{r} (Y_{ij}-\bar{y}_{i.})^2/v(r-1) = \sum_{i=1}^{v} \sum_{j=1}^{r} \hat{e}_{ij}^2/(\text{degrees of freedom} = v(r-1)).$$

$s^2_{\bar{y}_1-\bar{y}_2} = 2s_e^2/r$, where $\hat{e}_{ij} = Y_{ij}-\bar{y}_{i.} = Y_{ij}$ - estimated treatment mean = estimated random error.

Randomized complete block design (v treatments; r replicates)

$$s_e^2 = \sum_{i=1}^{v} \sum_{j=1}^{r} (Y_{ij}-\bar{y}_{i.}-\bar{y}_{.j}+\bar{y}_{..})^2/(r-1)(v-1) = \sum_{i=1}^{v} \sum_{j=1}^{r} \hat{e}_{ij}^2/\Big((r-1)(v-1) = \text{degrees of freedom}\Big).$$

$s^2_{\bar{y}_{1.}-\bar{y}_{2.}} = 2s_e^2/r$, where $\hat{e}_{ij} = Y_{ij}-\bar{y}_{i.}-\bar{y}_{.j}+\bar{y}_{..} = Y_{ij}$ - estimated treatment mean - estimated block mean + estimated grand mean = estimated random error.

Latin square design (k treatments; k replicates)

$$s_e^2 = \sum_{h=1}^{k} \sum_{j=1}^{k} \frac{(Y_{hij}-\bar{y}_{h..}-\bar{y}_{.i.}-\bar{y}_{..j}+2\bar{y}_{...})^2}{(k-1)(k-2) = \text{degrees of freedom}} = \sum_{h=1}^{k} \sum_{j=1}^{k} \frac{\hat{e}_{hij}^2}{(k-1)(k-2)}.$$

$s^2_{\bar{y}_{.1.}-\bar{y}_{.2.}} = 2s_e^2/k$, where $\hat{e}_{hij} = Y_{hij}-\bar{y}_{h..}-\bar{y}_{.i.}-\bar{y}_{..j}+2\bar{y}_{...} = Y_{hij}$ - estimated column mean - estimated treatment mean - estimated row mean + 2 (estimated grand mean) = estimated random error.

Simple change-over design (v treatments; vs replicates)

$$s_e^2 = \sum_{h=1}^{v} \sum_{j=1}^{vs} \frac{(Y_{hij}-\bar{y}_{h..}-\bar{y}_{.i.}-\bar{y}_{..j}+2\bar{y}_{...})^2}{(v-1)(vs-2)=\text{degrees of freedom}} = \sum_{h=1}^{v} \sum_{j=1}^{vs} \frac{\hat{e}_{hij}^2}{(v-1)(vs-2)}$$

$s^2_{\bar{y}_{.1.}-\bar{y}_{.2.}} = 2s^2_e/vs$, where $\hat{e}_{hij} = Y_{hij} - \bar{y}_{h..} - \bar{y}_{.i.} - \bar{y}_{..j} + 2\bar{y}_{...} = Y_{hij}$ - estimated column mean - estimated treatment mean - estimated row mean + 2(estimated grand mean) = estimated random error.

The computation of the sum of squares of the deviations is similar for all designs; the only difference involved is in the computation of the estimated error deviations and in the degrees of freedom. There is a mathematical reason for the latter, but we shall not go into that in this text.

We proceed in the same manner for a bib design with v treatments in b incomplete blocks of size k each. Here we cannot use arithmetic means but have to use the effects or the adjusted means instead. Since $Y_{ij} = \mu + t_i + b_j + e_{ij} = \hat{\mu} + \hat{t}_i + \hat{b}_j + \hat{e}_{ij}$, then $\hat{e}_{ij} = Y_{ij} - \hat{\mu} - \hat{t}_i - \hat{b}_j$. The error variance in a bib design is:

$$s^2_e = \sum_{\text{all } i,j} \frac{(Y_{ij} - \hat{\mu} - \hat{t}_i - \hat{b}_j)^2}{bk-v-b+1 = \text{degrees of freedom}} = \sum_{\text{all } i,j} \frac{\hat{e}^2_{ij}}{(bk-v-b+1)}$$

Since the design is not orthogonal, the variance of a difference of two means or effects adjusted for block effects is different from orthogonal designs. The variance of a difference between two adjusted means from a bib design is:

$$s^2_{\bar{y}'_{1.}-\bar{y}'_{2.}} = s^2_{\hat{t}_1-\hat{t}_2} = 2ks^2_e/v\lambda = \frac{2s^2_e}{r}\left\{1 + \frac{v-k}{v(k-1)}\right\} .$$

Let us now apply these formulae to the four numerical examples above. In example X.1, the $\hat{e}_{ij}$ from the randomized complete block design of v=3 treatments in r=4 replicates are:

Treatment	$\hat{e}_{ij}$ values, Block 1	2	3	4	Sum
1	0	3	0	-3	0
2	2	-2	-1	1	0
3	-2	-1	1	2	0
Sum	0	0	0	0	0

$$\sum_{i=1}^{3} \sum_{j=1}^{4} \hat{e}_{ij}^2 = 38$$

$$(r-1)(v-1) = (4-1)(3-1) = 6$$

$$\hat{s}_e^2 = 38/6 = 19/3 .$$

$s^2_{\bar{y}_{1.}-\bar{y}_{2.}} = 2(19)/3(4) = 19/6$, and $s^2_{\bar{y}_{i.}} = 19/3(4) = 19/12$.

For the latin square design as given in example X.2 with k=3, the various computations are:

Row	$\hat{e}_{hij}$ values, Column 1	2	3	Sum
1	B 0	A -1	C 1	0
2	A 1	C 0	B -1	0
3	C -1	B 1	A 0	0
Sum	0	0	0	0

$$\hat{e}_{2A1} + \hat{e}_{1A2} + \hat{e}_{3A3} = 0$$

$$\hat{e}_{1B1} + \hat{e}_{3B2} + \hat{e}_{2B3} = 0$$

$$\hat{e}_{3C1} + \hat{e}_{2C2} + \hat{e}_{1C3} = 0$$

$$\hat{s}_e^2 = 6/(3-1)(3-2) = 3$$

$$s^2_{\bar{y}_{.A.}-\bar{y}_{.B.}} = s^2_{\bar{y}_{.A.}-\bar{y}_{.C.}} = s^2_{\bar{y}_{.B.}-\bar{y}_{.C.}} = \frac{2(3)}{k=3} = 2 \text{ ; } s^2_{\bar{y}_{.i.}} = 3/3 = 1 .$$

For the simple change-over design in example X.3, the error deviations are:

Row	Column 1	2	3	4	5	6	Sum
1	$-\frac{1}{3}$	$\frac{1}{3}$	$2\frac{1}{3}$	$-\frac{1}{3}$	$\frac{2}{3}$	$-2\frac{2}{3}$	0
2	$\frac{1}{3}$	$-\frac{1}{3}$	$-2\frac{1}{3}$	$\frac{1}{3}$	$-\frac{2}{3}$	$2\frac{2}{3}$	0
Sum	0	0	0	0	0	0	0

$$\sum_{h=1}^{6} \sum_{j=1}^{2} \hat{e}_{hij}^2 / (2-1)(6-2) = 240/9(4) = 20/3 = s_2^2 \ .$$

$$s^2_{\bar{y}_{.A.}-\bar{y}_{.B.}} = \frac{2}{6}\left(\frac{20}{3}\right) = \frac{20}{9} \text{ and } s^2_{\bar{y}_{.i.}} = \frac{1}{6}\left(\frac{20}{3}\right) = \frac{10}{9} \ .$$

For the bib design in example X.4, the $\hat{e}_{ij}$ are already computed. Therefore,

$$s_e^2 = \sum_{i,j} \frac{\hat{e}_{ij}^2}{(12-4-6+1)} = \frac{16}{3} \ ; \ s^2_{\bar{y}'_{1.}-\bar{y}'_{2.}} = \frac{2(2)}{4(1)}\left(\frac{16}{3}\right) = \frac{16}{3} = s^2_{\hat{t}_1-\hat{t}_2}$$

X.5. Problems

X.1. For the class survey data in table VIII.1, for the 12 girls who were in class, what are the median values for height (4th column), weight (10th column), and age (11th column)? What are the modal values? Compute the arithmetic means if a computing machine is available; otherwise, show how to compute the arithmetic mean.

X.2. A man travels by car from town X to town Y, and he gets only 20 miles to a gallon of gasoline. On the return journey, he gets 30 miles to a gallon. Then, by assuming that the distance from X to Y is 60 miles, verify that the harmonic mean is the correct average to calculate. Find the arithmetic average for comparison.

X.3. Groups of boys and girls are tested for reading ability. The forty boys make an average score of 80%. The sixty girls have an average score of 40%. Calculate the arithmetic average for boys and girls combined. Would you use a weighted or unweighted mean? Why, or why not?

X.4. On March 1st a baby weighed 10 lbs. On June 1st it weighed 22.5 lbs. Use the geometric mean to estimate its weight on April 15th. Also compute the arithmetic mean and compare it with the geometric mean.

X.5. The following data (about 1936-38) represent percentages of individuals enrolled in various types of schools. The base is the number of individuals in the age bracket, and the numerator is the number of individuals enrolled. Thus more than 100% can be enrolled by this method since, for instance, children under 6 and over 13 are in elementary school.

Country	6-13 years % enrolled in elementary schools	14-17 years % enrolled in secondary schools	18-21 years % enrolled in college and universities
United States	119.5	63.6	14.6
Canada	113.0	55.8	6.2
Australia	119.6	21.1	8.5
New Zealand	111.0	27.9	5.6
EUROPE			
Belgium	92.5	55.7	3.6
Czechoslovakia	83.8	43.3	2.5
Denmark	102.0	44.3	3.0
France	114.8	33.5	2.6
Germany	83.4	43.5	3.0
Great Britain	104.3	50.8	3.6
Netherlands	88.8	40.9	2.0
Russia	99.0	32.9	6.0
Switzerland	92.3	42.6	3.8

Data from Appendix in Education - America's Magic by R. M. Hughes and W. H. Lancelot.

Find the median, range, and mode (if one exists) for each set of the above data. Prepare bar graphs of percentages with the countries arranged by increasing order of percentage. Where would you expect to find current data of this nature?

X.6. The following data were constructed for ease of computation to illustrate the computations for a completely randomized design. There are v=4 treatments and r=3 replicates on each treatment. Compute the arithmetic means ($\bar{y}_{i.}$), the sum of squares of the error deviations ($\Sigma\Sigma\, \hat{e}^2_{ij} = 172$), the error variance for all treatments ($s^2_e = 172/8 = 21.5$), and the variance of difference between two treatment means $\left(s^2_{\bar{y}_{i.}-\bar{y}_{i'.}} = 2(21.5)/3 = 43/3\right)$.

Treatments 1	2	3	4	
Y_{11} 11	Y_{21} 13	Y_{31} 21	Y_{41} 10	
Y_{12} 4	Y_{22} 9	Y_{32} 18	Y_{42} 4	
Y_{13} 6	Y_{23} 14	Y_{33} 15	Y_{43} 19	
$Y_{1.}$	$Y_{2.}$	$Y_{3.}$	$Y_{4.}$	$Y_{..}$ =
$\bar{y}_{1.}$	$\bar{y}_{2.}$	$\bar{y}_{3.}$	$\bar{y}_{4.}$	$\bar{y}_{..}$ =

Show all computations, as for example, $Y_{1.} = 11 + 4 + 6 = 21$, $\bar{y}_{1.} = 21/3 = 7$, etc., including computation of $\hat{e}_{ij}$.

X.7. Professor Snedecor, in example 11.2 of his text, constructed a set of data which resulted in relatively simple computations for a randomized complete block design. Here v=3 treatments were arranged in r=3 complete blocks. The yields arranged in a systematic fashion are:

	Blocks			Treatment	
Treatment	1	2	3	Total	Mean
1	Y_{11} 6	Y_{12} 5	Y_{13} 4	$Y_{1.}$	$\bar{y}_{1.}$
2	Y_{21} 15	Y_{22} 10	Y_{23} 8	$Y_{2.}$	$\bar{y}_{2.}$
3	Y_{31} 15	Y_{32} 15	Y_{33} 12	$Y_{3.}$	$\bar{y}_{3.}$
Block total	$Y_{.1}$	$Y_{.2}$	$Y_{.3}$	$Y_{..}$	-
Block mean	$\bar{y}_{.1}$	$\bar{y}_{.2}$	$\bar{y}_{.3}$	-	$\bar{y}_{..}$

Compute the various block, treatment, and grand totals and means. Verify that $\sum_{i=1}^{3} \sum_{j=1}^{3} \hat{e}_{ij}^2 = 10$. Show how to compute (do not actually compute) the variance of a treatment mean, variance of a difference between two treatment means, and standard error of a difference between two treatment means.

X.8. The following 3 X 3 latin square design was constructed for ease of computation:

Row = week	Column = Store 1	2	3	Week Total	Mean
1	Y_{11B} 13	Y_{12A} 7	Y_{13C} 19	$Y_{1..}$	$\bar{y}_{1..}$
2	Y_{21C} 14	Y_{22B} 8	Y_{23A} 2	$Y_{2..}$	$\bar{y}_{2..}$
3	Y_{31A} 6	Y_{32C} 15	Y_{33B} 6	$Y_{3..}$	$\bar{y}_{3..}$
Store total	$Y_{.1.}$	$Y_{.2.}$	$Y_{.3.}$	$Y_{...}$ 90	-
Store mean	$\bar{y}_{.1.}$	$\bar{y}_{.2.}$	$\bar{y}_{.3.}$	-	$\bar{y}_{...}$

				Sum	
Treatment total	$Y_{..A}$	$Y_{..B}$	$Y_{..C}$	$Y_{...}$	90
Treatment mean	$\bar{y}_{..A}$	$\bar{y}_{..B}$	$\bar{y}_{..C}$	$3(\bar{y}_{...})$	30

Show computations necessary to obtain s_e^2 the error variance, the standard error of a difference between any two treatment means, and complete the computations in the above tables.

X.6. References and Suggested Reading

Bevan, J. M. [1968]. Introduction to Statistics. Philosophical Library, Inc., New York, pp. vii + 220.

(Chapters 4 and 5.)

Campbell, R. C. [1967]. Statistics for Biologists. Cambridge University Press, London, pp. xii + 242.

(Chapter 2.)

Fisher, R. A. [1950]. Statistical Methods for Research Workers. Eleventh edition, Oliver and Boyd Ltd., Edinburgh and Hafner Publishing Company, New York.

(Chapter 3.)

Huff, D. [1954]. How to Lie with Statistics. W. W. Norton and Company, Inc., New York, pp. 142.

(Chapters 1, 2, and 4.)

McCarthy, P. J. [1957]. Introduction to Statistical Reasoning. McGraw-Hill Book Company, Inc., New York, Toronto, and London, pp. xiii + 402.

(Chapters 3 and 4.)

Moroney, M. J. [1956]. Facts from Figures, 3rd edition. Penguin Books, Baltimore and London, pp. viii + 472.

(Chapters 7, 8, and 9.)

Pearce, S. C. [1965]. Biological Statistics: An Introduction. McGraw-Hill Book Company, New York, St. Louis, San Francisco, Toronto, London, Sydney, pp. xiii + 212.

(Chapters 1 and 2.)

Snedecor, G. W. and Cochran, W. G. [1967]. Statistical Methods, 6th edition. The Iowa State University Press, Ames, Iowa, pp. xiv + 593.

(Chapters 1, 2, 3, and 8.)

Steel, R. G. D. and Torrie, J. H. [1960]. Principles and Procedures of Statistics. McGraw-Hill Book Company, New York, Toronto, and London, pp. xvi + 481.

CHAPTER XI. ORGANIZED OR PATTERNED VARIABILITY

XI.1. Introduction

So far, we have considered two types of statistics for summarizing information about a set of data arranged in a frequency distribution: an average and a measure of scatter. We have emphasized two particular statistics, the arithmetic mean and variance for single samples and for a number of experimental designs. In certain situations, we can go further and define the variability completely when we know that the observations follow a prescribed pattern, that is, the variability is organized into a pattern rather than being completely unknown. When this pattern is known, we say that the distribution of all observations or data in the population follows a pattern or organization. Many types of organized variation have been described mathematically. Probability forms the basis for distribution theory. Some concepts of probability were discussed in chapter IX.

XI.2. Binomial Distribution

If an event either occurs or does not occur in a series of n trials, and if these two events are mutually exclusive, then the random variable Y is said to have a binomial distribution. The probability that there are k occurrences of an event in n trials (that is, the random variable Y takes the value k for $k = 0,1,2,\ldots,n$ occurrences of the event) is $P(Y = k) = n!p^{n-k}q^k/(n-k)!k!$; this is the mathematical equation for the binomial distribution. To illustrate, suppose that we toss 5 unbiased coins and observe the number of heads which occur on any one throw. This number could be k=0, k=1, k=2, k=3, k=4, k=5 heads. The probabilities of the various outcomes = number of heads for $p = \frac{1}{2}$ would be as follows:

$$k=0: \quad \frac{5!}{(5-0)!0!} \left(\tfrac{1}{2}\right)^5 \left(\tfrac{1}{2}\right)^0 = \frac{1}{32}$$

$$k=1: \quad \frac{5!}{(5-1)!1!} \left(\tfrac{1}{2}\right)^4 \left(\tfrac{1}{2}\right)^1 = \frac{5}{32}$$

$$k=2: \quad \frac{5!}{(5-2)!2!} \left(\tfrac{1}{2}\right)^3 \left(\tfrac{1}{2}\right)^2 = \frac{10}{32}$$

$$k=3: \quad \frac{5!}{(5-3)!3!} (\tfrac{1}{2})^2 (\tfrac{1}{2})^3 = \frac{10}{32}$$

$$k=4: \quad \frac{5!}{(5-4)!4!} (\tfrac{1}{2})^1 (\tfrac{1}{2})^4 = \frac{5}{32}$$

$$k=5: \quad \frac{5!}{(5-5)!5!} (\tfrac{1}{2})^0 (\tfrac{1}{2})^5 = \frac{1}{32}$$

where $0! = 1$, $5! = (1)(2)(3)(4)(5)$, and $n! = (1)(2)(3)\ldots(n-1)(n)$. The total of the probabilities adds to unity, since $(p+q)^5 = (\tfrac{1}{2} + \tfrac{1}{2})^5 = 1^5 = 1$.

The data in table XI.1 illustrate occurrences which follow the binomial distribution for $p = .3376986$. They do not follow the binomial for $p = \frac{1}{3}$ as one would expect from unbiased dice, that is, $\frac{1}{3}$ of the outcomes obtained from throwing a six-sided die should have a 5 or 6 face up on the average. Many other illustrations of the binomial distribution may be found in statistical texts. The data in table XI.2 illustrate data on proportions which do not follow a binomial distribution.

XI.3. Poisson Distribution

In the binomial distribution we are concerned with success or failure of an event, whereas in the Poisson distribution we are concerned with the number of times an event occurs and not at all with the number of times it fails. Both of these distributions are concerned with discrete or discontinuous data ordinarily known as enumeration data. The probabilities of $k=0,1,2,\ldots$ events are given by the following equation denoting the Poisson distribution:

$$P(Y=k) = \frac{m^k e^{-m}}{k!} \quad ,$$

where m is the arithmetic mean of the population, k is the number of times the event occurs, and e is the base of natural logarithms. The probabilities for the various events are:

Table XI.1.* Weldon's data on 26,306 throws of 12 dice.

Number of dice with 5 or 6	Observed frequency	Expected true dice	Expected biased dice	Measure of divergence $\frac{x^2}{m}$	
				True dice	Biased dice
0	185	202.75	187.38	1.554	.030
1	1149	1216.50	1146.51	3.745	.005
2	3265	3345.37	3215.24	1.931	.770
3	5475	5575.61	5464.70	1.815	.019
4	6114	6272.56	6269.35	4.008	3.849
5	5194	5018.05	5114.65	6.169	1.231
6	3067	2927.20	3042.54	6.677	.197
7	1331	1254.51	1329.73	4.664	.001
8	403	392.04	423.76	.306	1.017
9	105	87.12	96.03	3.670	.838
10	14	13.07	14.69		
11	4	1.19	1.36	.952	.222
12	...	.05	.06		
	26306	26306.02	26306.00	35.491 n = 10	8.179 n = 9

* Taken from R. A. Fisher, Statistical Methods for Research Workers, 11th edition, page, 64, published by Oliver & Boyd, Edinburgh and by permission of the author and publishers.

Table XI.2.* Geissler's data on distribution of boys in 53,680 German families of eight children each.

Number of boys	Number of families observed	Expected p=.514677	Excess (x)= observed minus expected	$\frac{x^2}{m}$
0	215	165.22	+ 49.78	14.998
1	1485	1401.69	+ 83.31	4.952
2	5331	5202.65	+128.35	3.166
3	10649	11034.65	-385.65	13.478
4	14959	14627.60	+331.40	7.508
5	11929	12409.87	-480.87	18.633
6	6678	6580.24	+ 97.76	1.452
7	2092	1993.78	+ 98.22	4.839
8	342	264.30	+ 77.70	22.843
	53680	53680.00		91.869

* Taken from R. A. Fisher, Statistical Methods for Research Workers, 11th edition, page 67, published by Oliver & Boyd, Edinburgh and by permission of the author and publishers.

Occurrence of event	Probability of occurrence
0 time	e^{-m}
1 time	me^{-n}
2 times	$m^2e^{-m}/2$
3 times	$m^3e^{-m}/3!$
4 times	$m^4e^{-m}/4!$
⋮	⋮

The sum of the probabilities of all occurrences of events is unity.

An example of data following a Poisson distribution is described by R. A. Fisher in his _Statistical Methods for Research Workers_. The data were obtained by Bortkewitch from records of 10 army corps for 20 years; they refer to k = number of deaths per year per army corps resulting from horse-kicks. The data are:

Deaths per year per army corps	Observed	P(Y=k)	NP(Y=k)= expected
0	109	0.543	108.67
1	65	0.331	66.29
2	22	0.101	20.22
3	3	0.021	4.11
4	1	0.003	0.63
> 4	0	0.003	0.08
Total	200 = N	1.000	200.00

The mean is computed as

$$\{0(109) + 1(65) + 2(22) + 3(3) + 4(1)\}/200 = 0.61 .$$

The parameter m in the Poisson distribution is set equal to 0.61 and then the various $P(Y=k)$ are computed for $k=0,1,2,3$, and 4. The last value of 0.003 is obtained by subtraction; likewise, the value 0.08 is obtained by subtraction.

There is close agreement between observed and expected values in the above table. An example of data with poor agreement between observed and expected values and not following the Poisson distribution, even when the zero class is omitted (see Rao and Chakravarti, Biometrics 12:264-282, 1956), is given below. The data represent number of decayed teeth in boys.

Number of decayed teeth	0	1	2	3	4	5	6	7	8	9	10	11	12
Frequency of boys	61	47	43	35	28	15	20	5	5	2	1	2	1

XI.4. Normal Distribution

The normal frequency distribution may be represented as a bell-shaped curve and is expressed mathematically by the following equation:

$$f(Y \text{ given } \mu \text{ and } \sigma^2) = \frac{1}{\sqrt{2\pi\sigma^2}} e^{-(Y-\mu)^2/2\sigma^2} ,$$

where Y is the particular value of the random variable from a normal distribution, μ is the center of gravity or arithmetic mean of the entire distribution, and σ^2 is the population variance. (This is strictly analagous to the square of the radius of gyration in the theory of moments of inertia in physics.) The mean and the variance are the only parameters associated with the normal frequency distribution. The development of this function is due independently to three persons: a French immigrant to England, De Moivre, a Frenchman, LaPlace, and a German, Gauss.

A large proportion of statistical theory is based on the normal frequency distribution, and statistical techniques or procedures utilizing this theory are frequently used to summarize data from surveys and experiments. As has been amply demonstrated with examples for the binomial and Poisson distributions, there are many instances of nonnormal or anormal distributions. In fact, there are many types of distributions of various statistics: Student's t distribution, Snedecor's F or variance-ratio distribution, Fisher's z distribution, Wishart's distribution of sum of squares, chi-square distribution, uniform or rectangular distribution, triangular distribution, Cauchy distribution, negative binomial, Weibull distribution, contagious distributions, distribution of mixtures of binomials, distribution of mixtures of exponential distributions, exponential distribution, distribution of the correlation coefficient, distribution of confidence intervals, hypergeometric distribution, to name only a few. In many cases a simple transformation of yield data following one of the above distributions (say from Y to log Y or from Y to $\sqrt{Y}$) will allow normal theory to be utilized. This feature makes normal theory all the more popular among statisticians and users of statistics.

If a random variable Y is normally distributed with mean μ and variance σ^2, then the mean of a random sample of size n is normally distributed with the same mean μ and variance σ^2/n. The usefulness of this result is increased by the fact that if the original distribution of the random variable is not normal and if the variance σ^2 is finite, the distribution of means from samples of size n tends to normality as the sample size n increases. (These results are embodied in the "Central Limit Theorem", a very important theorem in statistics from both the theoretical and applied points of view.) This result allows use of normal theory for practical applications in situations where the exact form of the sampled population is unknown. Even for relatively small values of n, say from 5 to 10, the approximation is usually close enough for practical applications. Since many summarizations of data involve the arithmetic mean, normal theory receives considerable attention is statistical textbooks.

The bell-shaped figure in figure XI.1. illustrates the form of the normal frequency curve and the parameters as they relate to the curve. The point $Y=\mu$ centers and locates the curve on the Y axis. The parameter σ^2 determines the shape of the curve. For example, three different frequency distributions centered at μ with three different population variances, say $\sigma_1^2 \neq \sigma_2^2 \neq \sigma_3^2$, would look something like figure XI.2. depending upon the actual values of the population variances.

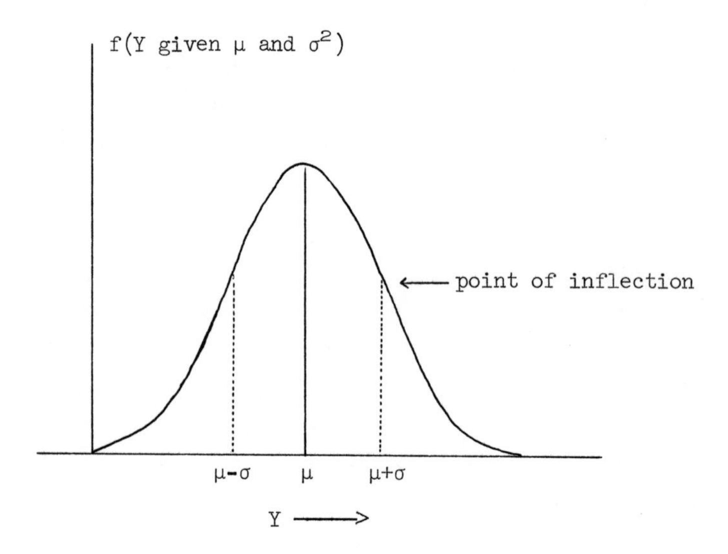

Figure XI.1. Normal frequency curve.

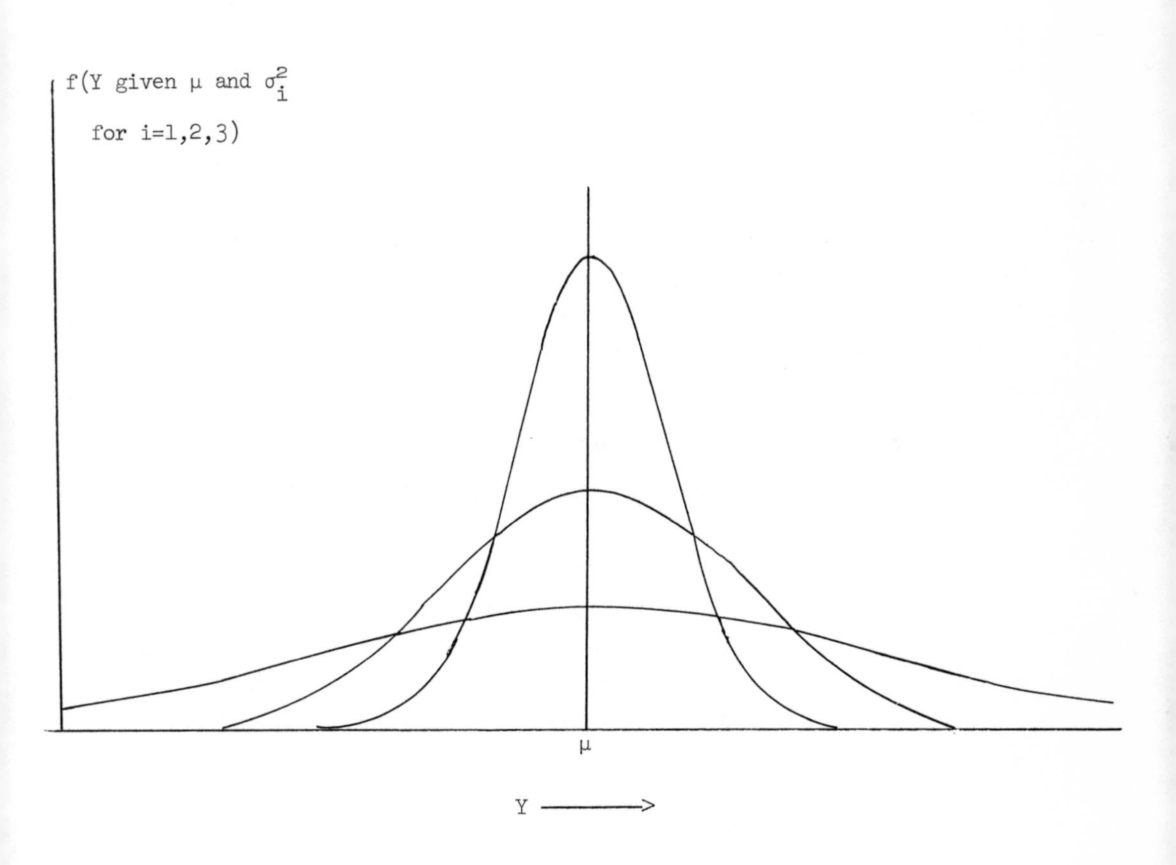

Figure XI.2. Normal frequency curves.

The area under the normal curve has been tabulated for many values of $(Y-\mu)/\sigma$. The distribution of $(Y-\mu)/\sigma$ for Y having a normal distribution is again a normal distribution with a new mean of zero and a new variance of unity. This is called the unit normal distribution since it has unit variance.

Figure XI.3 indicates a use of the unit normal curve and its relation to an actual distribution of the variable Y = I.Q. Here we note the average I.Q. is 100 and its variance is $13^2 = 169$. The proportion of the area under the curve between -1 and +1 is about 68% (the shaded area); the proportion of the area between 0 and +2 is approximately $47\frac{1}{2}$%; and the area between -2 and +2 is about 95%. Likewise, the area on the right of +2 and on the left of -2 is approximately 5%. The probability that a randomly selected observation Y falls in the interval between a = -2 and b = +2 is $P(a \leq Y \leq b) = P(-2 \leq Y \leq +2)$ is approximately 95%.

XI.5. Interval Estimate of a Parameter -- Confidence Interval

Instead of obtaining a point estimate like the arithmetic mean as a measure of central tendency or of a variance as a measure of scatter or variability, it is often desirable to have an interval estimate. Instead of saying that the point described by the arithmetic mean is an estimate of the population mean, we say that the population mean is included in an interval and that the interval is an estimate of the population mean. We can even say what proportion of the time we would expect the estimated confidence interval to contain the population mean. Since we have "confidence" from theory that the population mean is contained in the estimated interval $1-\alpha$ percent of the time and is not contained in the estimated intervals α percent of the time, then we speak of the interval as a $1-\alpha$ confidence interval (with a specified error rate base).

To illustrate the computation of a confidence interval, let us consider that a random sample of n = 5 observations is available as follows:

$$Y_1 = 2,\ Y_2 = 3,\ Y_3 = 6,\ Y_4 = 7, \text{ and } Y_5 = 11.$$

The median value is $Y_3 = 6$. Consider the range from 2 to 11. What is the probability that all five sample values fall below the population median? Since any single Y_i has a probability of $\frac{1}{2}$ of falling below the population median, and since all five values were independently obtained, the multi-

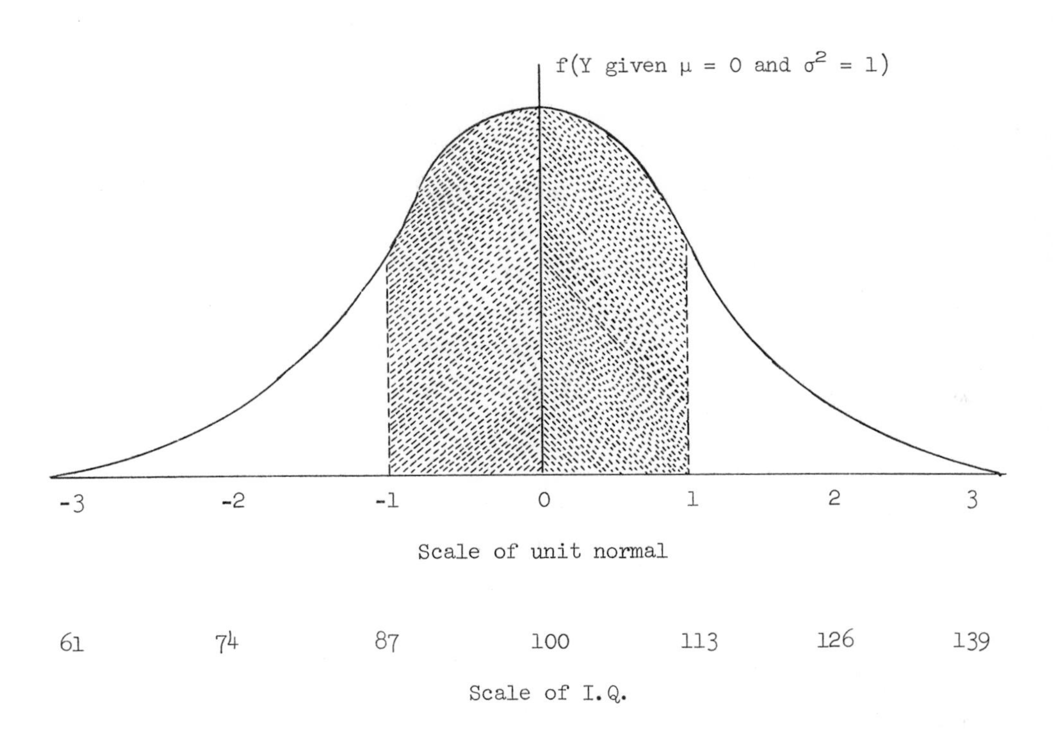

Figure XI.3. Relating unit normal deviates to I.Q. scores.

plication law of probability holds, that is, P(all 5 sample values fall below the population median) = $(\frac{1}{2})^5$ = 1/32. Likewise, the probability that all five sample values fall above the population median is 1/32. Therefore the probability that the interval 2 to 11 contains the population median is 1 - 2(1/32) = 15/16. We would have a $1 - \alpha$ = 15/16 = 93.75% confidence that the sample interval contains the population median.

To further illustrate the construction of confidence intervals, suppose that the range from the second smallest to the second largest sample value above was used as the confidence interval, that is, from $Y_2 = 3$ to $Y_4 = 7$. From the binomial distribution with $p = \frac{1}{2}$, the probability that all five values are below the population median is 1/32, and the probability that four values are below and one is above the median is 5/32. Likewise, the probability that all five values are above the population median is 1/32, and the probability that four values are above and one is below the population median is 5/32. Therefore the probability that the interval $Y_2 = 3$ to $Y_4 = 7$ contains the population median is equal to 1 - (1/32 + 5/32 + 1/32 + 5/32) = 1 - 12/32 = 5/8 = 62.5%.

In order to compute a confidence interval for means of samples drawn from a normal distribution, we need to have Student's t-distribution tabulated in the same manner as was described for the unit normal case. ($t = (\bar{y}-\mu)/s_{\bar{y}}$ where $\bar{y}$ = sample mean, $s_{\bar{y}}$ = standard error of a sample mean, and μ is the population mean.) Fortunately, this has been done. The t-distribution is a symmetric distribution with mean zero which depends only on the degrees of freedom. As degrees of freedom become large, the t-distribution approaches the unit normal distribution and is equal to it for infinite degrees of freedom.

The tabulated values of t in table XI.3. are taken from table IV of Fisher: Statistical Methods for Research Workers (published by Oliver & Boyd, Edinburgh and by permission of the author and publishers). As stated before, the t-distribution has been tabulated for many values of the degrees of freedom and of the probabilities. The tabulated t value from the following table is denoted as $t_{\alpha,df}$ where α refers to the value in the table and df refers to degrees of freedom. Thus, $t_{.05,30}$ is the value of t equal to 2.04, $t_{.1,12}$ = 1.78; etc.

Now, let us make use of the tabulated values of t to construct inter-

Table XI.3. Tabulated values of t.

Degrees of freedom	α = probability that a larger value of $\|t\|$ is obtained.						
	.5	.4	.3	.2	.1	.05	.01
2	.82	1.06	1.39	1.89	2.92	4.30	9.92
3	.76	.98	1.25	1.64	2.35	3.18	5.84
4	.74	.94	1.19	1.53	2.13	2.78	4.60
5	.73	.92	1.16	1.48	2.01	2.57	4.03
6	.72	.91	1.13	1.44	1.94	2.45	3.71
8	.71	.90	1.11	1.40	1.86	2.31	3.36
9	.70	.88	1.10	1.38	1.83	2.26	3.25
12	.69	.87	1.08	1.36	1.78	2.18	3.06
30	.68	.85	1.06	1.31	1.70	2.04	2.75
∞	.67	.84	1.04	1.28	1.64	1.96	2.58

val estimates of the difference between any two treatment means in example X.1. The $\hat{e}_{ij}$ must be normally and independently distributed with population mean zero and the same population variance. The sample variance $s_e^2 = 19/3$ and the variance of a difference of two means, say 1 and 2, was computed as $s^2_{\bar{y}_{1.}-\bar{y}_{2.}} = 2(19/3)/(r=4) = 19/6$. Now we compute a $1 - \alpha = 90\%$ interval estimate of the difference between any two means as follows:

Mean 1 minus mean 2

$$\bar{y}_{1.} - \bar{y}_{2.} \pm t_{.1,6} s_{\bar{y}_{1.} - \bar{y}_{2.}} = 7\text{-}9 \pm 1.94(19/6)^{\frac{1}{2}}$$

$$= 7 - 9 \pm 3.45 = -5.45 \text{ to } +1.45$$

Mean 1 minus mean 3

$$\bar{y}_{1.} - \bar{y}_{3.} \pm t_{.1,6} s_{\bar{y}_{1.} - \bar{y}_{3.}} = 7\text{-}5 \pm 1.94\ (19/6)^{\frac{1}{2}}$$

$$= 7 - 5 \pm 3.45 = -1.45 \text{ to } 5.45$$

Mean 2 minus mean 3

$$\bar{y}_{2.} - \bar{y}_{3.} \pm t_{.1,6} s_{\bar{y}_{2.} - \bar{y}_{3.}} = 9\text{-}5 \pm 1.94\ (19/6)^{\frac{1}{2}}$$

$$= 9 - 5 \pm 3.45 = .55 \text{ to } 7.45$$

The previous two intervals contain the point zero, but the last one does not. Since the last interval does not contain the point zero, this is evidence that the hypothesis of no difference (the null hypothesis) between means for treatments 2 and 3 may not be true.

In the above, we have described a confidence interval for the difference between two treatment means. If it is desired to construct a $(1-\alpha)\%$

confidence interval for a single mean with a sample size of r and with a variance s_e^2 associated with f degrees of freedom, the following formula is used for treatment 1 in the above example:

$$\bar{y}_{1\cdot} \pm t_{\alpha,f}\left(s_e^2/r\right)^{\frac{1}{2}}$$

which for $\alpha = .2$, $f = 6$, $r = 4$, and $s_e^2 = 19/3$ is

$$7 \pm 1.44\left(\frac{19}{3}\left(\frac{1}{4}\right)\right)^{\frac{1}{2}} = 7 \pm 1.44(1.26)$$

$$= 7 \pm 1.81, \text{ or from } 5.19 \text{ to } 8.81 .$$

The interval 5.19 to 8.81 is an 80% = (1-.2)100 confidence interval of the population mean for treatment 1.

The above procedure of constructing confidence intervals is for the normal distribution. The intervals may be constructed for any distribution for which appropriate tables are available. Tables and charts for the binomial distribution are available in many places. Charts for constructing 95% and 99% confidence intervals have been developed and are known as Clopper and Pearson charts (see figures XI.4. and XI.5.).

To illustrate the construction of confidence intervals with binomial variation, let us suppose that $\hat{p} = .4$, that n = 50, and that we are sampling from a binomial distribution. The various $(1-\alpha)\%$ confidence intervals for $\hat{p} = .4$ as read from the charts are:[1]

90%	29	to	52
95%	27	to	55
99%	23	to	59

Here we note that the size of the confidence interval increases as $1-\alpha$ in-

[1] The 90% confidence values were obtained from page 414 of Dixon, W. J. and Massey, F. J., Jr., Introduction to Statistical Analysis, 2nd edition, McGraw-Hill Book Company, Inc., New York, 1957.

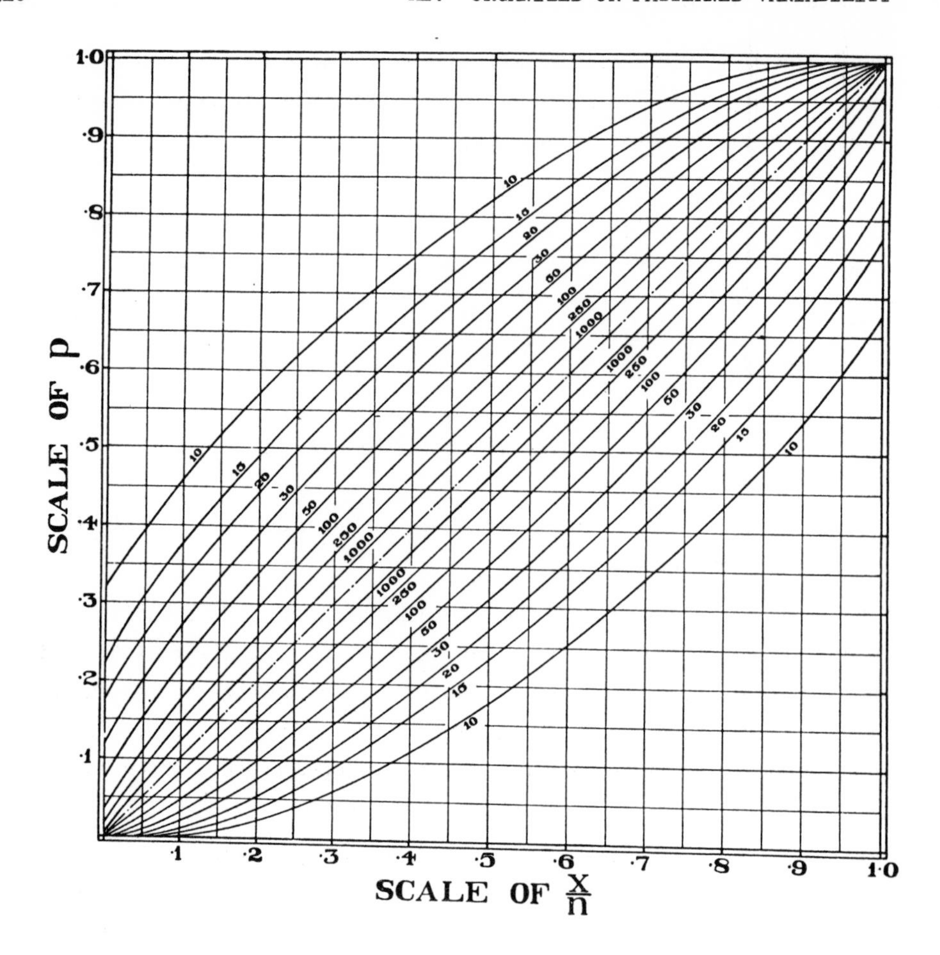

Figure XI.4. Confidence belts for p (Confidence Coefficient = .95).

Reprinted by permission of the authors, C. J. Clopper and E. S. Pearson, and the publishers, the Biometrika Office.

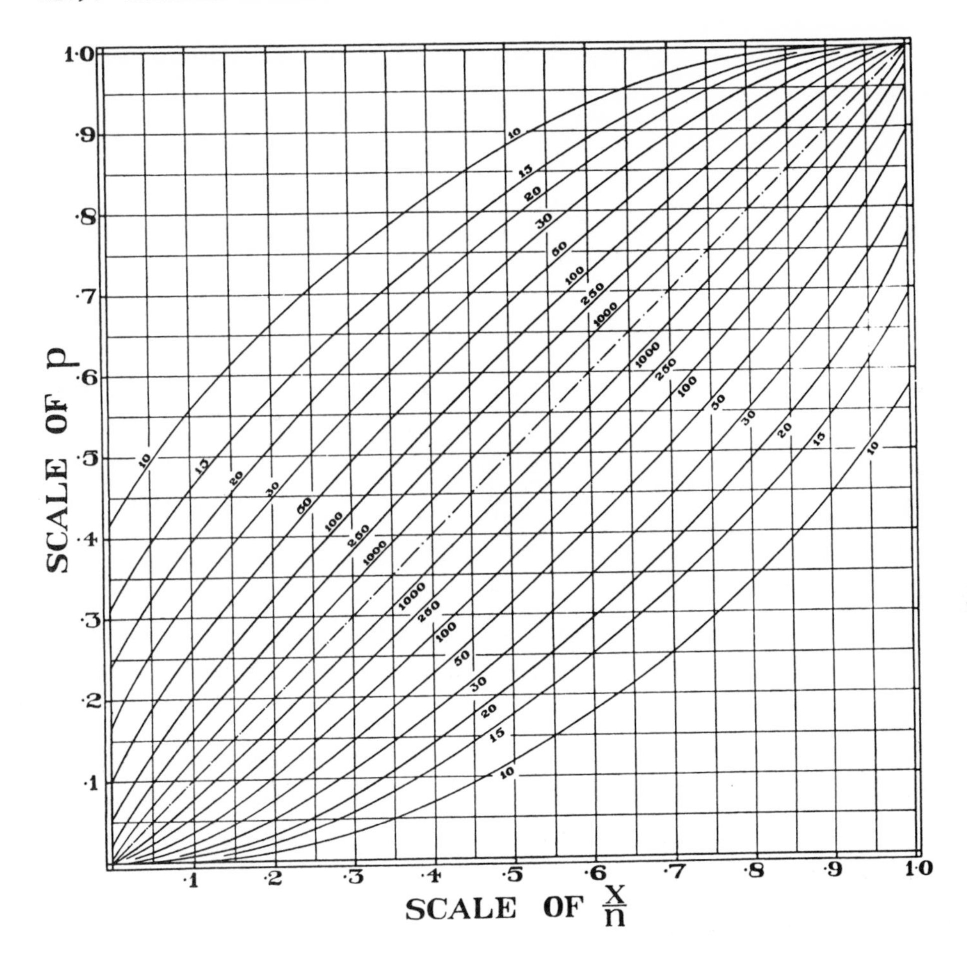

Figure XI.5. Confidence belts for p (Confidence Coefficient = .99)

Reprinted by permission of the authors, C. J. Clopper and E. S. Pearson, and the publishers, the Biometrika Office.

creases. We may use these values to put $(1-\alpha)\%$ confidence intervals on $\hat{q} = .6$ as follows:

90%	48	to	71
95%	45	to	73
99%	41	to	77

The values for $\hat{p} = .4$ are subtracted from 100.

The confidence intervals on p from the binomial distribution may be approximated by using the t-distribution. The approximation becomes better and better as n, the sample size, increases. To illustrate, let us use the above example with a sample size of n = 50. $t_{.10,49\ df} = 1.68$, $(1-\alpha)\% = 90\%$, and $\hat{p} \pm 1.68\sqrt{\hat{p}(1-\hat{p})/n} = 0.4 \pm 1.68\sqrt{0.4(0.6)/50} = 0.4 \pm 1.68(0.069) = 0.4 \pm 0.116$, or from 28.4% to 51.6%. This agrees within rounding errors and within errors in reading the chart. The number 2 used in the example in chapter V corresponds to the tabulated value for $t_{.05,49\ df}$ as tabulated in t-tables.

Thus far we have been concerned with summarizing information from survey data or experimental data. In order to draw inferences about the population parameter, we need to characterize the interval estimate in terms of the parameter being estimated. We have noted the following characteristics about the interval estimate:

1. The average length of the confidence interval of a mean decreases as $1/\sqrt{n}$ where n = sample size.
2. The length of the confidence interval increases as $1-\alpha$ increases.
3. The construction of a confidence interval depends upon the underlying distribution.
4. The point estimate of the parameter does not always fall at the midpoint of the confidence interval.
5. The length of the confidence interval is a statistic and has a distribution. Some confidence interval estimates are relatively short, and some are relatively long for the same sample size and the same $1-\alpha$, depending upon the size of the sample variance.
6. In repeated sampling with the same sample size n and the same α, the

true value of the parameter will be contained in the estimated intervals $1-\alpha$ of the time and will not be α percent of the time.

The last item above is relevant in writing a definition of a confidence interval. A $(1-\alpha)\%$ confidence interval is an interval which is said to bracket the value of the true unknown parameter; in repeated samplings the value of the parameter will be contained in the estimated intervals $(1-\alpha)\%$ of the time and will not be $\alpha\%$ of the time. When we assert that the confidence interval contains the true value of the parameter, we are either right or wrong in making this inference.

We can increase the confidence proportion $1-\alpha$, but this increases the width of the interval; all we have to do to be 100% certain that the true parameter lies within an interval is to make the interval large enough! For example, the interval from minus infinity to plus infinity contains 100% of the possible values of the population mean of normally distributed observations; the interval from zero to plus infinity contains all possible values of the population variances; the interval from zero to unity contains all possible values of a proportion in this interval; and so on. However, we would like to reduce the size of this interval without making $1-\alpha$ too small or the sample size n too large. The selection of α is determined by the investigator; odds of 4:1, 9:1, 19:1, and 99:1 are often used. The latter two odds are popular with statistical textbook writers and editors of scientific journals. The selection of α is, however, made by the investigator. Once he selects an α, then one of the ways to reduce the size of the confidence interval is to increase the size of the sample. (Other ways are to change the experimental procedure, to select an experimental design which controls all assignable variation, etc.).

XI.6. Problems

XI.1. Show how to compute the 80%, 90%, and 95% confidence intervals for the differences between treatment means for the data given in problems X.6, X.7, and X.8.

XI.2. In a genetic experiment, let us suppose that eye color in the common fruit fly segregates in a 3:1 ratio of red-eyed to white-eyed flies. Suppose that we observe 70% red-eyed flies and 30% white-eyed flies out of a sample size of n = 100 flies. What are the values of the 95% and 99% confidence intervals as read from the Clopper and Pearson charts (figures XI.4 and XI.5)? Do these intervals contain the theoretical proportion of red-eyed flies? What would be the confidence intervals for the sample percentage of 70% if the sample sizes had been 50, 250, and 1000? Would these intervals contain the theoretical proportion of red-eyed flies? Would the same statements made about the theoretical proportion of red-eyed flies hold for the theoretical proportion of white-eyed flies? Illustrate.

XI.3. From statistics or other types of textbooks list three examples of data (not listed in this text) which are assumed to follow the distributions: binomial, Poisson, normal. Give complete reference citations, including page number.

XI.4. Show how to compute the 90% confidence intervals for the data of problem XI.2, using the t-statistic and tabulated values of t as follows: $t_{.10,49\ df} = 1.676$, $t_{.10,249\ df} = 1.65$, and $t_{.10,\ 999\ df} = 1.65$.

XI.5. A graduate student performed the following experiment using a simple change-over design. He obtained 10 sets of 15 two-digit numbers from a random number table; these are given below:

10 Sets of Size 15 of Two Digits From Random Digit Table

1	2	3	4	5	6	7	8	9	10
30	86	18	48	40	75	53	56	36	06
20	82	32	36	28	43	55	80	65	61
62	27	67	97	92	77	76	02	56	82
67	04	01	30	13	44	38	66	89	60
00	00	38	17	05	74	49	87	61	66
73	34	71	69	77	58	20	04	85	53
76	65	97	79	12	32	72	31	90	36
30	16	49	24	26	06	22	96	67	56
10	05	67	93	88	45	07	59	69	90
77	83	89	32	28	20	25	18	53	62
76	01	56	07	65	46	02	28	40	24
28	07	51	71	60	67	11	88	99	21
26	80	78	47	56	81	94	93	38	73
31	16	82	26	43	93	17	95	57	60
73	46	66	36	10	23	60	32	05	51

He had two calculating machines, a hand operated Curta and an automatic Monroe. He wished to observe the difference in mean length of time required to compute a sum of squares of 15 two-digit numbers on the two machines. Since there could be a learning process in computing a sum of squares a second time, he decided to compute 5 of the sets on the Curta first and on the Monroe second and the reverse for the other 5 sets. The design used and the time in seconds required to compute each sum of squares are given below where A = Curta and B = Monroe.

1	2	3	4	5	6	7	8	9	10	
A 255	B 115	A 280	B 107	B 105	A 240	A 195	B 110	A 202	B 85	1694
B 113	A 200	B 117	A 238	A 210	B 104	B 90	A 200	B 105	A 180	1557
368	315	397	345	315	344	285	310	307	265	3251

$\Sigma\hat{e}^2_{hij} = 2219.50$ $s^2_e = 277$ A total = 2200

B total = 1051

Compute a 95% confidence interval on the difference between the two means of the machines.

CHAPTER XII. SAMPLE SIZE

XII.1. Introduction

The problem of determining sample size or replicate number arises whenever an experiment or a survey is to be conducted. There are many criteria for determining sample size. The suitability of a given criterion depends upon the objectives and the nature of the investigation, and has to be determined by the investigator or by the administrator. Various criteria will be set forth and discussed individually, and the investigator will need to determine which criterion fits his situation.

The ensuing discussion relates equally well to experiments and to sample surveys. The results for one are applicable to the other type of investigation. In general, the number of samples utilized in an investigation is determined by the following interrelated factors:

1. the degree of precision desired,
2. the amount of variability present in the material under investigation,
3. available resources including personnel and equipment, and
4. size and shape of experimental and sampling units.

The nature of the material under investigation, the characters observed, and/or the expected magnitude of the treatment and category differences determine the degree of precision desired. The very nature of some treatments, strata, and categories may be such that large differences are expected or the observations may be such that low variability is expected. Some characters are more variable than others and therefore require more observations. In connection with determining the number of replicates for an experiment or the sample size for a survey, the characters of interest with their respective standard deviations should be listed. Replicate number or sample size should be determined for the most variable character to be measured. This number will be sufficiently large for all other characters. The relative importance of the characters will also affect the determination of sample size for the investigation in that sample size will be determined for the most important character.

The amount of variability present in experimental material is determined by the experimental conditions, the characters measured, and the treatments tested. The variance of a treatment or category mean relative to the mean or to the difference between means may be small for some characters and large for others.

Larger experimental and sampling units tend to have smaller variation (on a unit basis) than smaller ones. Technique and other errors do not usually have as great an effect on large units as on small ones. An error in weight of 100 pounds on a small plot may have a relatively large effect, whereas an error in weight of 100 pounds on a large plot may be of little consequence. In fact, some of the scales used for weighing sugar cane from experimental plots are only accurate to the nearest 200 pounds.

The shape of the experimental or the sampling unit usually has a relatively small effect on the variance of a treatment mean. In general, long narrow field plots tend to be less variable than square ones. In surveys, geographic nearness acts in the same manner; it has been found that for several characteristics of the population simple geographic stratification into compact areas is as effective as complex stratification.

The following list of criteria for determining sample size is not all-inclusive, but it does illustrate some of the diverse considerations used by investigators in determining sample size. For these criteria it is assumed that the investigation will be conducted in a fixed period of time; this is called a one-stage investigation as opposed to multi-stage or sequential investigation where the results of the first stage determine whether additional stages are to be conducted.

XII.2. The "Head-in-the-Sand" Approach

As indicated by one researcher's statement, "Away with the duplicate plot! Give me one plot and I know where I am!", the use of only one observation could imply no variation since none is observed. Using a sample size of one to avoid variation in observations is a "head-in-the-sand" approach to the problem. Variation remains, whether one or a dozen observations are made. The experimenter is only deceiving himself with such an attitude and such an approach to experimentation.

XII.3. "Favorite Number" Procedure

An unscientific criterion for determining replicate or sample number is to select a number without any other consideration. The experimenter may use 7 replicates because he likes the number 7, because a friend used 7 replicates, or because 7 is the number that he has been using in the past. Some surveyors like to use 100 samples for no specified reason other than that they like to use this sample size. This procedure can be denoted as the "favorite number" procedure. It is a subjective criterion for determining replicate or sample number.

XII.4. Available Resources

Some of the resources that might limit the number of replicates or samples for a given experiment or survey are:

1. amount of material,
2. available personnel or funds,
3. amount of experimental land, pots, equipment, etc. available for an experiment, or number of enumerators available for a survey.

The amount of experimental seed or treatment material may be limited in the early stages of an investigation. This is quite often the situation with new varieties, new chemicals, new drugs, new products, and the like, and the experimenter must decide whether to run the experiment or to wait until more experimental material becomes available. Likewise, limited personnel or other experimental resources may result in insufficient replication. The procedure usually followed is to conduct the experiment with the available number of replicates. Then, based on the results of the first experiment, the experimenter determines whether further testing is required. This "sequential" method of testing has long been used by experimenters.

If the experiment is to be conducted with the number of replicates dictated by available resources with no further testing, serious consideration should be given to the idea of not running the experiment; the limited resources may be allocated to other experiments. The same comment applies to sample surveys.

XII.5. Number of Degrees of Freedom in the Variance

A criterion that has some usefulness, especially for experiments with few treatments, is to use sufficient replication such that the variance is associated with at least 12 to 16, preferably 20, degrees of freedom. The reasoning here is that the variance of the estimated variance will not be so large, relatively, for 20 or more degrees of freedom. The estimated variance of a variance, s^2, is 2 $(\text{variance})^2/(\text{degrees of freedom}) = 2s^4/\text{df}$. From the following table we see that relatively small decreases in the variance of an estimated variance are made with greater than 20 degrees of freedom:

Degrees of freedom = df	2/df
2	1.00
4	.50
8	.25
10	.20
16	.125
20	.10
50	.04
100	.02

The error degrees of freedom may be increased by increasing the number of replicates or the number of treatments in an experiment and the sample size in a survey. Thus, with 20 or more treatments, 2 replicates would be sufficient to satisfy the criterion of at least 20 degrees of freedom for the error variance.

XII.6. Standard Error Equal to a Specified Percentage of the Mean

Some experimenters use sufficient replication to obtain a standard error of a mean which is not greater than a specified percentage of the mean. For this criterion it is required that a relatively good estimate of the ratio of the standard deviation of a single observation to the mean, which is the coefficient of variation, be available. Given this estimate, a sample number is selected which yields the desired percentage. For example, suppose that the coefficient of variation is 10 percent and that it is desired to have a standard error of a mean which is less than 2.5 percent of the mean. Since $10/\sqrt{16} = 10/4 = 2.5$, 16 observations would be required to obtain a standard error which is 2.5% of the treatment mean.

The requirements for this criterion are:

1. the estimated standard deviation, s, and
2. the estimated mean, $\bar{x}$.

With these two statistics the coefficient of variation is computed as $s/\bar{x} = \text{c.v.}$

XII.7. Effect of Sample Size on the Confidence Interval

Suppose that the samples are from a binomial distribution with an unknown p value and that our estimate $\hat{p}$ of p is .4. Then, from figures XI.4 and XI.5 and from McCarthy [1957], pages 201 to 204, the 90%, 95%, and 99% confidence intervals on p are read from the charts as:

sample size = n	$(1 - \alpha)$% confidence intervals					
	90%		95%		99%	
	lower	upper	lower	upper	lower	upper
10	14	70	12	75	7	82
50	29	52	27	55	23	59
100	31	49	30	50	28	53
250	34	45	34	46	32	48
1000	38	42	37	43	36	44

From the above we may note that the length of the interval decreases with sample size for any specified value of $1 - \alpha$. In order to obtain relatively small lengths of intervals, say less than 10% for $\hat{p}$ in the range from .2 to .8, the sample size must be greater than n = 250 even for 90% confidence intervals.

Likewise, for the normal distribution, the sample size must be relatively large and/or the variance must be relatively small in order to have short lengths of confidence intervals on the average. The effect of sample size on length is the reciprocal of the square root of sample size in the estimated confidence interval. Thus, the $(1 - \alpha)$% confidence interval of the population mean is:

$$\bar{y} \pm t_{\alpha,f} \sqrt{s_e^2/n} \ ,$$

where there are f degrees of freedom associated with s_e^2; the confidence interval on the difference of two means with different sample sizes n_1 and n_2 is:

$$\bar{y}_{1\cdot} - \bar{y}_{2\cdot} \pm t_{\alpha,f} \sqrt{s_e^2 \left\{\frac{1}{n_1} + \frac{1}{n_2}\right\}} \ .$$

If $n_1 = n_2 = r$, then the confidence interval is:

$$\bar{y}_{1.} - \bar{y}_{2.} \pm t_{\alpha,f} \sqrt{2s_e^2/r} \; .$$

In the above formulae we note that the sample size is in the denominator of the estimated variance of a mean. Thus the variance of a mean, or of a mean difference, decreases with an increase in sample size; this in turn affects the length of the confidence interval.

XII.8. Specified Probability that the Confidence Interval is Less than or Equal to a Specified Length

Some statisticians prefer a criterion for determining sample size that is based on more of a probability basis than any of those discussed in preceding sections. One such method is described below. Let us consider all possible estimated $(1-\alpha)\%$ confidence intervals; these will have a known distribution for each specified sample size n and error rate base. Now, let us define the one-half length of a confidence interval for the difference of two means as $d = t_{\alpha,f} s_e \sqrt{2/n}$ and then divide the lengths of all one-half confidence intervals into two parts, all those less than or equal to d and all those whose lengths are greater than d. We pick a d such that $(1-\gamma)\%$, for a given γ, of the $(1-\alpha)\%$ confidence intervals is less than or equal to a specified value of d. Computationally, we proceed as follows for a confidence interval on the difference between two means. Let $d = t_{\alpha,f_2} \sqrt{2s_1^2/n}$ $= t_{\alpha,f_2} s_1 \sqrt{2/n}$, where the variance s_1^2 has f_1 degrees of freedom. Squaring both sides of the equation we obtain $d^2 = t^2_{\alpha,f_2} s_1^2(2/n)$; $n = 2t^2_{\alpha,f_2} s_1^2/d^2$. If we multiply the right-hand side by an F-statistic with f_2 and f_1 degrees of freedom (F tables are available in many statistical texts, for instance, Snedecor and Cochran [1967]), we obtain:

$$n \geq 2t^2_{\alpha,f_2} F_{\gamma}(f_2,f_1)(s_1/d)^2 \; ,$$

where s_1^2 is an available estimate of the variance and is associated with f_1 degrees of freedom, and f_2 equals the degrees of freedom to be used in the proposed experiment or survey. Thus, we specify s_1^2, f_1, α, γ, and d and then compute n. Since n specifies f_2, an iterative process must be used to obtain a solution for n.

This procedure for obtaining sample size is presented to illustrate that statistical procedures based on probabilities are available for computing sample sizes for a proposed experiment or survey. Other procedures are available, but many are mathematically more complex than the present one. A statistician is frequently confronted with problems of sample size in the course of his consultation with research workers. Hence, when the reader is faced with a sample size problem, it is suggested that he consult a statistician who has had previous experience with the problem of sample size determination.

XII.9. Problems

XII.1. Describe what your method of determining sample size had been before reading this book. Did your method vary with the characteristic observed?

XII.2. List the 95% and the 99% binomial confidence intervals obtained for the proportion $\hat{p} = .61$ for sample sizes of 10, 15, 20, 30, 50, 100, 250, and 1000. Plot the length of the confidence interval as the ordinate against the sample size as the abscissa and connect the points with straight lines. What conclusions may be drawn from the plots?

XII.3. Given that $s^2 = 25$ with 4 degrees of freedom, that $\bar{y} = 10$, and that $r = 5$ from a completely randomized design, compute $(1 - \alpha)\%$ confidence intervals on the population mean for $(1 - \alpha)\%$ equal to 50%, 60%, 70%, 80%, 90%, 95%, and 99%. Plot the results using $(1 - \alpha)\%$ as the ordinate and the endpoints of the interval as the abscissa. What would be the shape of the curve through the endpoints of the intervals if all $(1 - \alpha)\%$ intervals had been plotted?

XIII.4. Given that $s^2 = 25$ with f degrees of freedom, that $\bar{y}_1 - \bar{y}_2 = 10$, and that $r = 10$, compute the 80% and the 99% confidence intervals for $f = 2, 3, 4, 5, 6, 8, 9, 12$, and 30 degrees of freedom. Plot the results with the length of the interval as the ordinate and degrees of freedom as the abscissa. What conclusions do you draw from the plotted results?

XII.10. References and Suggested Reading

Huff, D. [1954]. *How to Lie with Statistics.* W. W. Norton and Company, Inc., New York, pp. 142.

McCarthy, P. J. [1957]. *Introduction to Statistical Reasoning.* McGraw-Hill Book Company, Inc., New York, Toronto, and London, pp. xiii + 402.

Snedecor, G. W. and Cochran, W. G. [1967]. *Statistical Methods.* 6th edition. The Iowa State University Press, Ames, Iowa, pp. xiv + 593.

Wilson, E. B. [1952]. *An Introduction to Scientific Research.* McGraw-Hill Book Company, Inc., New York, Toronto, and London, pp. x + 373, Sections 3.10 and 4.7.

CHAPTER XIII. STATISTICAL PUBLICATIONS

XIII.1. Introduction

Procedures for obtaining meaningful numbers have been described in the preceding chapters. The reader has been repeatedly cautioned to cast a critical and wary eye on all sets of numbers purported to contain information on particular characteristics of a population. With this attitude still uppermost in our minds, let us now turn our attention to some of the vast accumulations of statistical data available for the United States and for the world. Special publications have been singled out for comment, and brief descriptions are given to illustrate some of the data available on many and varied characteristics of the world's populations. The particular publications to be discussed are the *U.S.D.A. Agricultural Statistics*, the *Statistical Abstracts of the United States*, the *Statistical Yearbook of the United Nations*, the *FAO Production Yearbook*, the *FAO Trade Yearbook*, and the *Yearbook of International Trade Statistics*.

XIII.2. Statistical Publications of the United States

Statistical publications of the United States are many and varied; each department of the United States Government issues many statistical publications on various items. For example, the Department of Labor issues unemployment statistics by month, by group, by section of the country, and so on; the Department of Commerce does likewise for items of interest to them, and this Department, which contains the Bureau of the Census, probably issues more statistical tables and summaries than any other United States Department.

The United States Department of Agriculture has its share of statistical publications with probably the most frequently used one being *U.S.D.A. Agricultural Statistics*, Library of Congress number HD 1751. This is an annual publication of data on agricultural production, supplies, consumption, facilities, costs, and returns. Each annual publication contains the most recent 10 years' data on a particular item. The 1962 issue contains historical tables dating back to 1866 for production figures on the principal crops and to 1867 for livestock numbers. Some international and world statistics are included where needed for comparison. The following major groupings of statistics were shown in the 1968 issue:

1. Grains (food and feed)
2. Cotton, sugar, and tobacco
3. Oilseeds, fats, and oils
4. Vegetables and melons
5. Fruits, tree nuts, and beverage crops
6. Hay, seeds, and minor field crops
7. Cattle, hogs, and sheep
8. Dairy and poultry products
9. Farm resources, income, and expenses
10. Taxes, insurance, cooperatives, and credit
11. Stabilization and price-support programs
12. Agricultural conservation and forestry statistics
13. Consumption and family living
14. Miscellaneous statistics (weather, fishery, refrigeration, etc.)

Three types of data are included: (1) actual counts of items covered, (2) estimates from surveys of various types made by the U.S.D.A, and (3) census enumeration data. The latter was obtained with the cooperation of the Bureau of Census and published in their _United States Census of Agriculture_, Library of Congress number HD 1753.

Many types of information are readily available from these tables. For example, the following 5 states are ranked in order of production of wheat in 1964, with their combined total being almost one-half the production of wheat in the entire United States:

	Total Production (1964)
Kansas	215,460,000 bushels
North Dakota	150,842,000 bushels
Oklahoma	96,623,000 bushels
Montana	90,821,000 bushels
Washington	82,206,000 bushels
Total	635,952,000 bushels
U.S. Total	1,290,468,000 bushels

These results are further broken down into production of spring wheat, winter wheat and durum wheat. Each crop is broken down into similar details.

The Bureau of the Census in the Department of Commerce has been publishing the _Statistical Abstracts of the United States_ annually since 1878. It is a summary of statistics on the social, political, and economic organization of the United States. It is designed to serve as a reference and guide to other statistical publications and sources. Both government and private statistical publications are utilized in compiling each volume. Although the emphasis is on national statistics, many tables present data for regions, individual states, and outlying areas of the United States. It contains 33 sections with many tables in each section as follows: Population; Vital statistics,health, and nutrition; Immigration and naturalization; Education; Law enforcement, federal courts, and prisons; Area, geography, and climate; Public lands, parks, recreation, and travel; Labor force, employment, and earnings; National defense and veterans' affairs; Social insurance and welfare services; Income, expenditures, and wealth; Prices; Elections; Federal government finances and employment; State and local government finances and employment; Banking, finance, and insurance; Business enterprise; Communications; Power; Science; Transportation - land; Transportation - air and water; Agriculture; Forests and forest products; Fisheries; Mining and mineral products; Construction and housing; Manufacturers; Distribution and services; Foreign commerce and aid; Outlying areas under the jurisdiction of the United States; Comparative and international statistics; and Metropolitan area statistics.

In addition to these 33 sections there are three appendices: (1) weights and measures, (2) historical series tables comparisons from colonial times to the present, and (3) a guide to the sources of statistics utilized to compile the Abstracts. A study of the last appendix will indicate the great diversity and the completeness on population characteristics in the United States.

To illustrate one type of data available from these tables, figures XIII.1 and XIII.2 have been prepared from tables 5 and 6 of the 90th annual edition of the _Statistical Abstract of the United States, 1969_. The projections to future years represent extrapolations beyond the data. We were cautioned to be wary of extrapolation in chapter VIII, but sometimes this is necessary; demographers have devised and continue to devise extrapolation and projection procedures necessary for planning for the future welfare of a nation and of the world. Various series of projections are utilized, but we have selected the data for Series A projections simply because they appeared first in the table. Projections for Series B, C, and D may be obtained from this publication. The projections are smaller for the latter series than they are for Series A.

Many other statistical publications of the federal government and of state governments are available in the various libraries in the country.

XIII.3. Statistical Publications of the United Nations

The first issue of the United Nations' *Statistical Yearbook* was published in 1948; the 23rd issue for the year 1971 was distributed in 1972. The Library of Congress number for this series is HC 57 A19. The series continues and expands the work of the *Statistical Yearbook of the League of Nations* in providing a convenient summary of international statistics on a world basis. The latter publication was discontinued in 1945; the 1948 issue of the *Statistical Yearbook* provides for continuity by presenting data for several previous years, generally back to 1928. The official languages of this publication are English and French, with both being presented.

The *Statistical Yearbook* is an annual publication supplemented by several United Nations' publications. Current monthly data for many of the tables and series may be found in the *Monthly Bulletin of Statistics*. Population data in a comprehensive form and breakdown are published in a *Demographic Yearbook*. Only population data of general interest are included in the *Statistical Yearbook*. The order of presenting data for countries is first by continental groups, which are ordered as follows: Africa; America, North; America, South; Asia; Europe; Oceania (Australia, New Zealand, etc.); United Soviet Socialist Republic (both Asian and European parts). The order of countries within each continental group is in alphabetical order of the country's English name. Any changes in territory are described in the book. The reader must be aware of the many territorial changes, especially in the European and United Soviet Socialist Republic groups. The main groups of tables are the following: Population; Manpower; Agriculture; Forestry; Fishing; Industrial production; Mining, quarrying; Manufacturing; Construction; Electricity, gas; Transport; Communications; Internal trade; External trade; Balance of payments; Wages and prices; National income; Finance; Social statistics; Education, culture; Appendices (conversion factors, index, etc.)

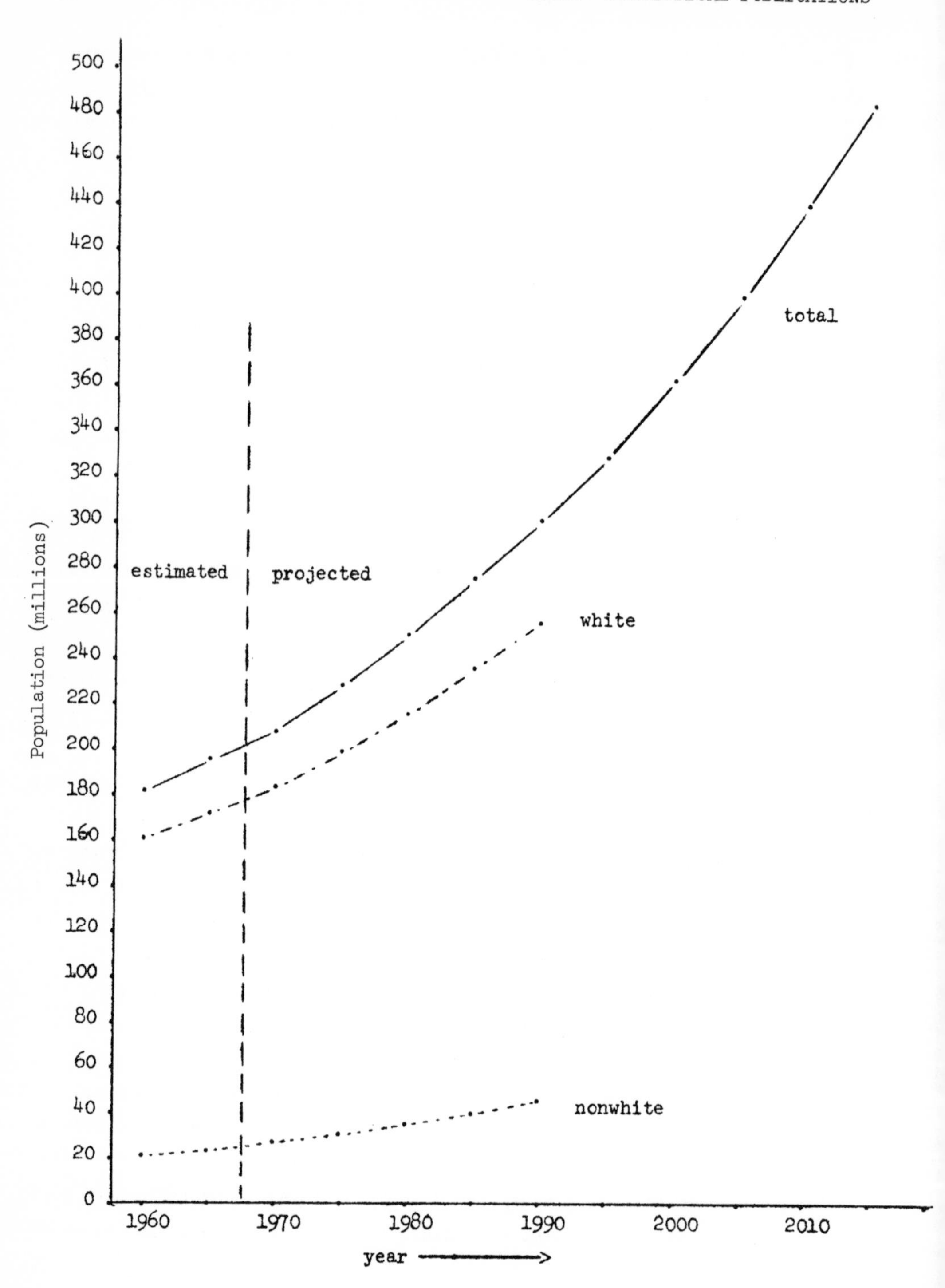

Figure XIII.1. Estimated and projected size (in millions) of white, nonwhite, total population of the United States.

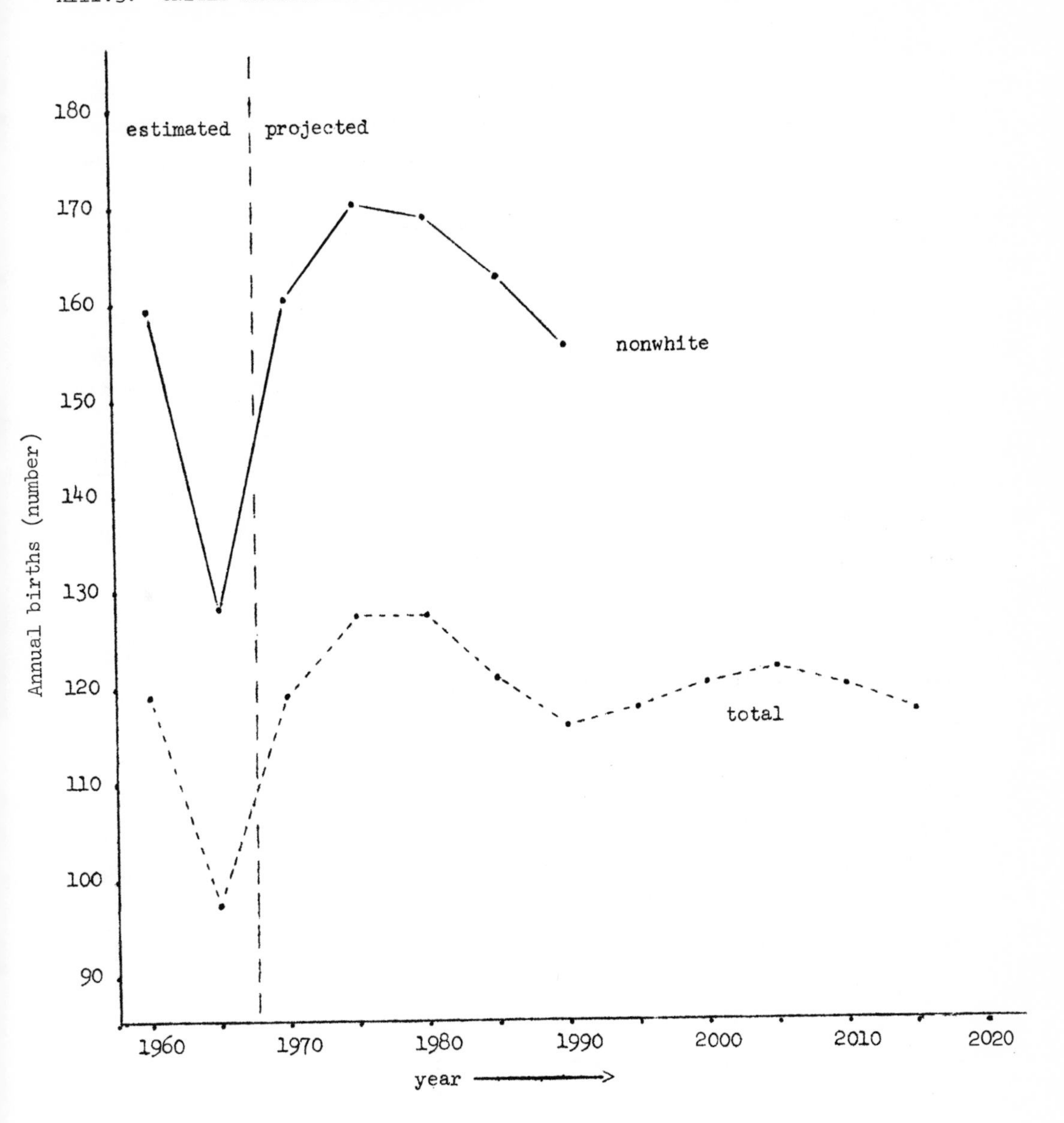

Figure XIII. 2. Estimated and projected fertility rates (annual births per 1000 females 15 to 44 years of age) for total and for non-white populations in the United States.

One may note that there is a vast amount of statistical information available on most countries of the world; however, many of these statistics are very crude, being little if any better than educated guesses, while others are relatively accurate. In the main, these statistics are extremely helpful to the various countries and allow such reports as the World Economic Survey, 1965, to be compiled. In this publication, considerable emphasis is placed on the developing countries of the world.

The Food and Agriculture Organization (FAO) of the United Nations has been active in preparing statistical yearbooks on production and trade. The official languages for these are English, French, and Spanish; all three are presented in the FAO Production Yearbook and the FAO Trade Yearbook. The former was initiated in 1947 and the latter in 1948. The Production Yearbook continues the statistical series on crop acreages and yield and on livestock numbers formerly published by the International Institute of Agriculture at Rome. Current monthly statistics are available in a supplemental bulletin.

In connection with the above, a world census of agriculture is a relatively new concept, the first one having been taken in 1930, the second in 1950, and a third in 1960. FAO took a strong lead in the 1950 census of agriculture and developed the plans for the 1960 census. Sixty-eight countries cooperated in the 1930 census and 106 in the 1950 census; communist country participation was almost lacking in the census for the 1950 program. The first two parts of a three-part publication of the 1960 census of agriculture were published in 1966 and 1967; the 3rd, 4th, and 5th parts were published in 1969, 1968, and 1971, respectively (Library of Congress number HD 1421).

In all these publications, the diversity of material is to be noted. If one wishes to know the number of camels in the United Soviet Socialist Republic or in Israel, for example, numbers are given in the tables. In the FAO Trade Yearbook one can find items such as value of agricultural trade for tea and maté or the commodity trade in groundnuts in the shell and not in the shell. Egg trade is broken down in the same manner as groundnuts, but in addition classifications such as frozen, liquid, and dried egg trade are tabulated.

The Statistical office of the United Nations compiles a Yearbook of International Trade Statistics among many other publications. In the tables for the year 1966 (published in 1968) there are tables of imports and exports for 142 countries by commodity groups in value and in amount. The total imports and exports are given as well as imports and exports to specific countries. The data for figure XIII.3 were obtained from the 1966 Yearbook. We may note from the graph that Cuba became an importing country in 1958 and that the total value of her exports has remained relatively constant since 1946. Russia makes up a sizeable amount of the difference between imports and exports.

It is hoped that the above illustrates the very many types of numbers available for many and sundry characteristics. More and more numerical information is being accumulated for more and more countries of the world. Some researchers spend all their time compiling statistics for a certain characteristic (as for instance, heavy industry and steel production in Russia and satellites in Europe) in order to obtain the most authentic information possible. These statistics are usable and quite valuable for many studies, and are useless for others.

A complete list of publications under the auspices of the Statistical Office of the United Nations has not been compiled. It should be noted that the publications are many and diverse. This active and energetic organization is serving a valuable need.

XIII.4. A Comment

Knowledge of available statistical data is important to politicians, social scientists, students, administrators, businessmen, and others for making decisions in everyday life. Ignorance of what information is available can lead to procurement of duplicate or triplicate sets of data by the survey method, with costly consequences. Also, in politics, social programs, economics, and other fields, ignoring available statistical data may lead to costly blunders.

The method of procuring and summarizing data for all of the nations of the world is a subject in itself. This brief introduction is intended to make the reader aware of this fact and to give him an idea of the sources of statistical information on various population characteristics for a specified country.

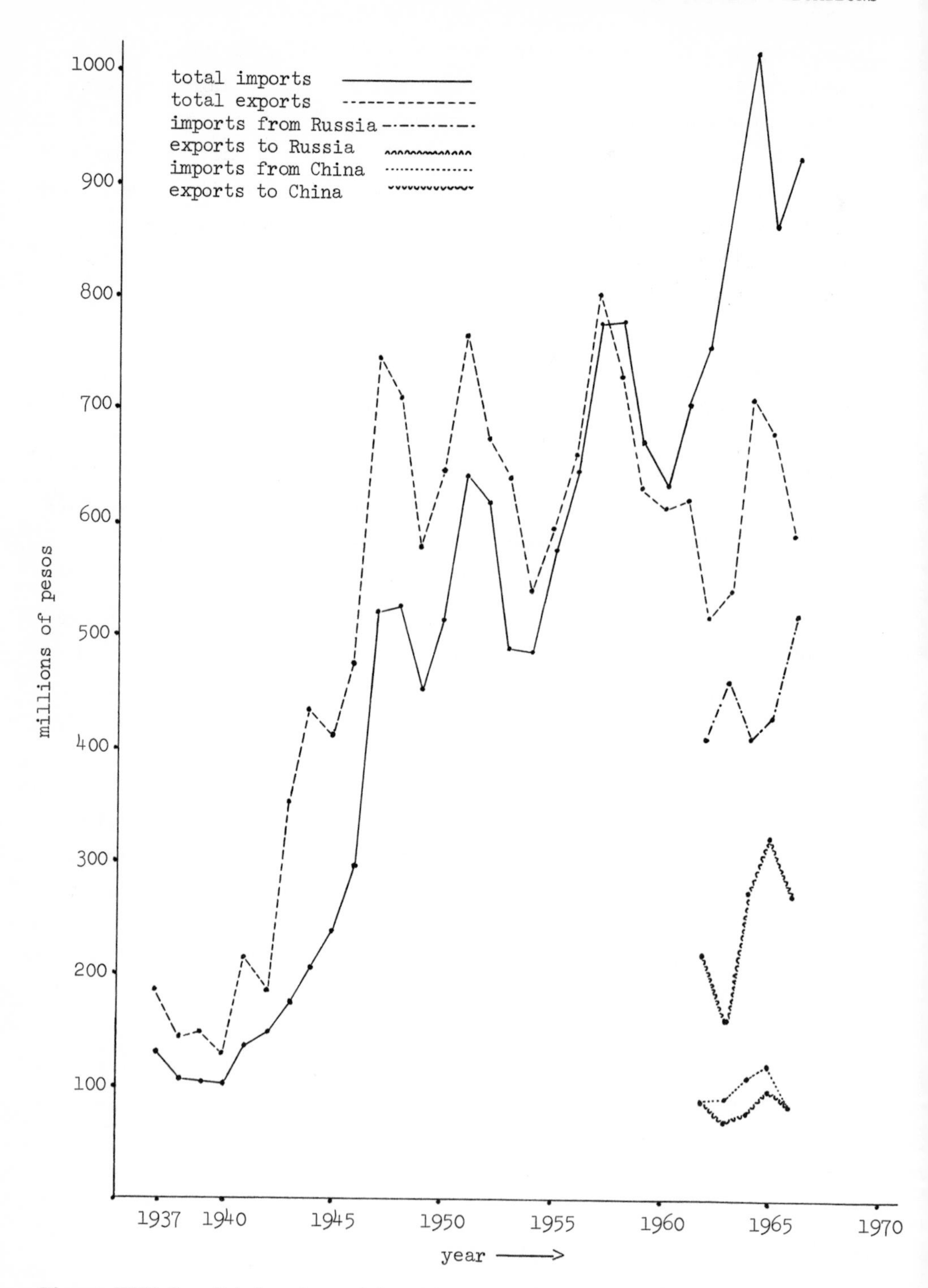

Figure XIII.3. Total value of imports and exports for Cuba and for imports from and exports to Russia and China (mainland).

XIII.5. Problems

XIII.1. Using a linegraph, plot the production of wheat, infant mortality rate, and number of passenger cars for the 15-year period 1949 to 1963 for the country listed on your card. The data are given in the *Statistical Yearbook* of the United Nations. (Note that some of these data may not be available for every country for all years; the countries were selected because they produced wheat. Each student has been given a different country.)

XIII.2. From the United States Department of Agriculture *Agricultural Statistics* for 1968, prepare bargraphs of the characteristic listed on your card for the years given in the table. Note the diversity of information available. (Each student has been given a different characteristic.)

XIII.3. Prepare a graph, pictograph or a bargraph, depicting the number of camels in China, the United Soviet Socialist Republic, Egypt, India, Israel, Iran, and Pakistan for the years 1930, 1950, and 1960 as compiled in the *FAO Trade Yearbook*. How reliable do you think these statistics are? Give reasons for your answers.

XIII.4. Obtain current data for figures XIII.1, XIII.2, and XIII.3, and extend the graphs to include the current data.

XIII.6. Suggested Reading

Statistical Reporting Service [1969]. The story of United States agricultural estimates. United States Department of Agriculture, Miscellaneous Publication No. 1088, April. pp. v + 137.

XIV. MORE ON STATISTICAL METHODOLOGY

XIV.1. Introduction

In chapters VIII, X, and XI, some methods for summarizing data were presented. Graphs, frequency distributions, and two-way contingency tables were discussed in chapter VIII, measures of central tendency and measures of variation were discussed in chapter X; and interval estimates of a parameter were discussed in chapter XI. A vast array of statistical procedures is available for summarizing and extracting the information from a set of data. The theoretical development and the application of statistical procedures form the basis for courses in Statistics. A Statistics course following this text should be either an introductory statistical methods course or an introductory probability and mathematical theory course.

In the following sections, we present a few additional statistical procedures for summarizing the information contained in a set of data. The particular procedures discussed were selected because of their connection with procedures and concepts utilized in the preceding chapters.

XIV.2. Analysis of Variance

In chapter X, a variance was defined to be the average of the squares of the deviations from a mean. The squares of the deviations were summed and divided by the degrees of freedom to obtain this average. The same concept will be used to obtain additional variances for a given survey or experiment.

Suppose that we have a sample survey design consisting of a random selection of v relatively large and equal-sized clusters with a random selection of r observational units (the sampling unit) in each cluster. If the model for a randomly selected observation is:

Y_{ij} = true population mean + true cluster effect + random error

= estimated population mean + estimated cluster effect + estimated random error

$$= \bar{y}_{..} + (\bar{y}_{i.} - \bar{y}_{..}) + (Y_{ij} - \bar{y}_{i.})$$

$$= \bar{y}_{..} + \hat{d}_i + \hat{e}_{ij}$$

where $\bar{y}_{..}$ is the arithmetic mean of the rv observations, $\bar{y}_{i.}$ is the arithmetic mean of the r observations in the ith cluster, $\hat{d}_i = (\bar{y}_{i.} - \bar{y}_{..})$, and $\hat{e}_{ij} = (Y_{ij} - \bar{y}_{i.})$ = the estimated random error. Note that $\hat{e}_{ij}$ is the estimated deviation from the mean and that $s_e^2 = \sum_{i=1}^{v} \sum_{j=1}^{r} (Y_{ij} - \bar{y}_{i.})^2/v(r-1)$ as described in chapter X.

The quantity $(\bar{y}_{i.} - \bar{y}_{..}) = \hat{d}_i$ represents a deviation of the ith cluster sample mean $\bar{y}_{i.}$ from the overall sample mean $\bar{y}_{..}$. If these deviations are squared, summed, and divided by the degrees of freedom, (v-1), a variance among cluster means is obtained as $\sum_{i=1}^{v} (\bar{y}_{i.} - \bar{y}_{..})^2/(v-1)$. The sum of the deviations of the v cluster means from the overall mean $\sum_{i=1}^{v} (\bar{y}_{i.} - \bar{y}_{..})$, is forced to be zero; thus, (v-1) of the deviations $(\bar{y}_{i.} - \bar{y}_{..})$ may vary freely, but the last deviation must be such that $\sum_{i=1}^{v} (\bar{y}_{i.} - \bar{y}_{..}) = 0$.

Since the variance of a mean is s_e^2/r, the preceding variance must be multiplied by r in order to place the squares of the deviations on the same basis as those in s_e^2 as computed above. Thus the sum of squares comparable to $\sum_{i=1}^{v} \sum_{j=1}^{r} \hat{e}_{ij}^2$ is $r \sum_{i=1}^{v} \hat{d}_i^2 = r \sum_{i=1}^{v} (\bar{y}_{i.} - \bar{y}_{..})^2$.

The above results are summarized below in an <u>analysis of variance</u> table:

Source of Variation	Degrees of Freedom	Sum of Squares	Variance
Due to mean	1	$Y_{..}^2/rv = rv\,\bar{y}_{..}^2$	-
Due to clusters	v-1	$r \sum_{i=1}^{v} (\bar{y}_{i.}-\bar{y}_{..})^2$	$r \sum_{i=1}^{v} (\bar{y}_{i.}-\bar{y}_{..})^2/(v-1)=s_d^2$
Due to deviations from cluster means	v(r-1)	$\sum_{i=1}^{v} \sum_{j=1}^{r} (Y_{ij}-\bar{y}_{i.})^2$	s_e^2
Total	vr	$\sum_{i=1}^{v} \sum_{j=1}^{r} Y_{ij}^2$	-

The total sum of squares may be partitioned into the same number of sums of squares as there are kinds of effects in the observational equation. The above is a simple algebraic partitioning of the total sum of squares into a number of parts since the equation

$$\sum_{i=1}^{v}\sum_{j=1}^{r} Y_{ij}^2 = rv\,\bar{y}_{..}^2 + r\sum_{i=1}^{v}(\bar{y}_{i.} - \bar{y}_{..})^2 + \sum_{i=1}^{v}\sum_{j=1}^{r}(Y_{ij} - \bar{y}_{i.})^2$$

is an algebraic identity. For the survey design described, the two variances s_d^2 and s_e^2 are summary statistics for two sources of variation in the survey; s_d^2 summarizes the information about variation among cluster means, and s_e^2 summarizes the information about variation among observations within clusters. The analysis of variance represents a partitioning of the total sum of squares into a number of components and then computing the various variances. This is an analysis of the variation in the survey.

A comparison of the variance s_d^2 with s_e^2 may be of interest in certain situations. For example, the investigator may wish to determine whether or not he should stratify by clusters or take a completely random sample in his next survey on similar material. If the variance among clusters is relatively large compared to the variance within clusters, say $s_d^2/s_e^2 = 1.5$, he may decide to stratify; if $s_d^2/s_e^2 < 1.5$, he may decide to use a simple random sample design. Given that the surveyor has used a cluster simple random sample design, he may compute the efficiency of the sample survey design used relative to a simple random sample design. The estimate of the variation when there is no stratification into clusters is $\left(s_d^2(v-1) + s_e^2\,v(r-1)\right)/\left(v - 1 + v(r-1) = rv - 1\right) = s_e^{2\prime}$. The ratio $s_e^{2\prime}/s_e^2$ estimates the relative efficiency of a simple random sample design to a cluster simple random sample design. The ratio $s_e^{2\prime}/s_e^2 = \left(s_e^2\,v(r-1) + s_d^2\,(v-1)\right)/(rv-1)s_e^2 = (vr-v)/(vr-1) + (s_d^2/s_e^2)\cdot\left((v-1)/(rv-1)\right)$. In this form we note that the ratio of the variances and the values of v and r as they affect the degrees of freedom determine the relative efficiency. For $s_d^2/s_e^2 = 1.5$, the relative efficiency is equal to $v(r-1)/(rv-1) + \left((v-1)/(rv-1)\right)(1 + \frac{1}{2}) = 1 + (v-1)/2(rv-1)$; if $r = 2$, then

$1 + (v-1)/2(rv-1) = 1 + (v-1)/2(2v-1)$ is approximately equal to $1 + \frac{1}{4} = 125\%$ for $v \geq 20$. The selection of the value 1.5 depends upon the cost of sampling under the two schemes; that is, it may cost 25% more to take a simple random sample than to take a random selection of clusters and of individuals within clusters. The selection of the value 1.5 also depends upon the number of degrees of freedom associated with s_d^2 and with s_e^2.

The variance of a cluster mean is s_e^2/r. The overall mean $\bar{y}_{..}$ contains variation among cluster effects as well as among observations within the cluster; hence, the variance of $\bar{y}_{..}$ is s_d^2/rv. The variance of the total $Y_{..}$ is $rv\ s_d^2$. With these variances, the confidence limits may be computed for the overall mean $\bar{y}_{..}$ or for the total $Y_{..}$ (see chapter XI). Also, the population total may be estimated from the sample total, given that a specified proportion of the clusters and of the observations within a cluster have been sampled. This is accomplished by multiplying the sample total by the reciprocal of the proportion sampled. For example, if 1/100 of the population has been sampled, then the sample total $Y_{..}$ multiplied by $1/(1/100) = 100$ is the estimated population total. Likewise, estimated confidence limits on the population total are obtained by multiplying the confidence intervals for the sample total by the reciprocal of the proportion sampled.

The variances of an estimated treatment mean or differences between means from a completely randomized design are described in chapter X. The variances s_e^2 used in chapter X may also be obtained from the above analysis of variance table. Likewise, the computation of confidence intervals follows the description given in chapter XI when s_e^2 is obtained from the above analysis of variance table.

For data obtained from a randomized complete block design, the sources of variation are the mean, the block effects, the treatment effects, and random error effects. Variances due to these sources may be computed as described in the following analysis of variance table:

Source of Variation	Degrees of Freedom	Sum of Squares	Variance
Due to mean	1	$Y_{..}^2/rv = rv\bar{y}_{..}^2$	-
Due to block effects	r-1	$v\sum_{j=1}^{r}(\bar{y}_{.j}-\bar{y}_{..})^2$	$v\sum_{j=1}^{r}(\bar{y}_{.j}-\bar{y}_{..})^2/(r-1)=s_b^2$
Due to treatment effects	v-1	$r\sum_{i=1}^{v}(\bar{y}_{i.}-\bar{y}_{..})^2$	$r\sum_{i=1}^{v}(\bar{y}_{i.}-\bar{y}_{..})^2/(v-1)=s_t^2$
Due to deviations	(r-1)(v-1)	$\sum_{i=1}^{v}\sum_{j=1}^{r}(Y_{ij}-\bar{y}_{i.}-\bar{y}_{.j}+\bar{y}_{..})^2$	s_e^2
Due to total	rv	$\sum_{i=1}^{v}\sum_{j=1}^{r}Y_{ij}^2$	-

In the above table, the symbols are those described in chapters VI and X. The total sum of squares is partitioned as indicated above, and the sum of the four sums of squares is equal to the total sum of squares. The variances s_b^2, s_t^2, and s_e^2 are summary statistics on variation in the experiment. In addition to constructing confidence intervals on treatment means and differences in means, one may compare this randomized complete block design to the completely randomized design using the ratio $s_e^{2\prime}/s_e^2$, where $s_e^{2\prime}$ is computed as:

$$s_e^{2\prime} = \frac{s_b^2(r-1) + s_e^2(r-1)(v-1) + s_e^2(v-1)}{(r-1) + (r-1)(v-1) + (v-1) = rv - 1} .$$

If all treatments had been the same treatment, then the within blocks variance would contain only variation among units unaffected by treatment differences and would have $r(v-1) = (v-1) + (v-1)(r-1)$ degrees of freedom. The term $s_e^2(v-1)$ estimates what the treatment contribution to the within blocks sum of squares would have been had all treatments been the same treatment. Thus, $s_e^{2\prime}$ estimates what the residual error variance would have been for a completely randomized design. If, for example, s_e^2 were only 3/4 as large as $s_e^{2\prime}$, then a completely randomized design with r = four blocks would give the same variance for a treatment mean as a randomized complete block with r = 3 replicates, that is, $s_e^2/3 = s_e^{2\prime}/4$. Thus if an experimenter used a randomized complete block design instead of the completely randomized design, he could use one replicate less to achieve the same value for the standard error of a mean or of a difference between two means. This also means that a randomized complete block design is $33\frac{1}{3}\%$ more efficient than a completely randomized design for material like that in the experiment conducted.

To illustrate the computations of the variances for a randomized complete block design with $v = 4$ treatments and $r = 3$ blocks, the following set of numbers was selected for ease of computation:

Treatment	block 1	block 2	block 3	Total	Mean $\bar{y}_{i.}$	$\bar{y}_{i.}-\bar{y}_{..}$
A	$Y_{A1}=1(0)$	$Y_{A2}=3(0)$	$Y_{A3}=5(0)$	$Y_{A.}=9$	$\bar{y}_{A.}=3$	-1
B	$Y_{B1}=2(-2)$	$Y_{B2}=4(-2)$	$Y_{B3}=12(4)$	$Y_{B.}=18$	$\bar{y}_{B.}=6$	2
C	$Y_{C1}=3(1)$	$Y_{C2}=4(0)$	$Y_{C3}=5(-1)$	$Y_{C.}=12$	$\bar{y}_{C.}=4$	0
D	$Y_{D1}=2(1)$	$Y_{D2}=5(2)$	$Y_{D3}=2(-3)$	$Y_{D.}=9$	$\bar{y}_{D.}=3$	-1
Total	$Y_{.1}=8(0)$	$Y_{.2}=16(0)$	$Y_{.3}=24(0)$	$Y_{..}=48$	-	0
Mean $\bar{y}_{.j}$	$\bar{y}_{.1}=2$	$\bar{y}_{.2}=4$	$\bar{y}_{.3}=6$	-	4	-
$(\bar{y}_{.j}-\bar{y}_{..})$	-2	0	2	0	-	-

The values in the parentheses are obtained as $(Y_{ij}-\bar{y}_{i.}-\bar{y}_{.j}+\bar{y}_{..}) = \hat{e}_{ij}$ = estimated random error deviations. For example, $\hat{e}_{11} = (1-2-3+4) = 0$. The deviations of the block means are obtained as $(\bar{y}_{.1}-\bar{y}_{..}) = (2-4) = -2$, $(\bar{y}_{.2}-\bar{y}_{..}) = (4-4) = 0$, and $(\bar{y}_{.3}-\bar{y}_{..}) = (6-4) = 2$. The deviations of treatment means from the overall mean are similarly obtained. The various sums of squares are computed as:

$$rv\bar{y}_{..}^2 = 3(4)(4^2) = 192 ,$$

$$v\sum_{j=1}^{r}(\bar{y}_{.j}-\bar{y}_{..})^2 = 4\{(-2)^2 + (0)^2 + (2)^2\} = 32 ,$$

$$r\sum_{i=1}^{v}(\bar{y}_{i.}-\bar{y}_{..})^2 = 3\{(-1)^2 + (2)^2 + (0)^2 + (-1)^2\} = 18 \text{ , and}$$

$$\sum_{i=1}^{v}\sum_{j=1}^{r}(Y_{ij}-\bar{y}_{i.}-\bar{y}_{.j}+\bar{y}_{..})^2 = (0)^2 + (-2)^2 + (1)^2 + (1)^2 + (0)^2 + (-2)^2 + (0)^2 + (2)^2 + (0)^2 + (4)^2 + (-1)^2 + (-3)^2 = 40 .$$

The above may be summarized in the following analysis of variance table:

Source of Variation	Degrees of Freedom	Sums of Squares	Variance
Due to mean	1	192	-
Due to blocks	(r-1) = 2	32	$16 = s_b^2$
Due to treatments	(v-1) = 3	18	$6 = s_t^2$
Due to deviations	(r-1)(v-1) = 6	40	$20/3 = s_e^2$
Total	12	282	-

The variance of a difference in treatment means is $2s_e^2/r = 2(20/3)/3 = 40/9$. The estimated variance for a completely randomized design is obtained as:

$$s_e^{2\prime} = \frac{s_b^2(r-1) + s_e^2(r-1)(v-1) + s_e^2(v-1)}{(r-1) + (r-1)(v-1) + (v-1)}$$

$$= \frac{16(2) + (20/3)(6) + (20/3)(3)}{2 + 6 + 3 = 11}$$

$$= \frac{32 + 40 + 20}{11} = \frac{92}{11} .$$

$s_e^{2\prime}/s_e^2 = 92/11/20/3 = 3(92)/11(20) = 69/55 = 125\%$. It is estimated that this randomized complete block design is 25% more efficient than a completely randomized design would have been. This means that 4 replicates of a randomized complete block would result in the same variance of a difference between two treatment means as would have been obtained with 5 replicates of a completely randomized design.

Each of the above sums of squares, except that due to the mean, may be further partitioned if this is desired by the investigator. The total sum of squares for experiments designed as a latin square, a Youden square, a balanced incomplete block, or any other design may be partitioned into various components and the resulting variances computed. The particular partitioning utilized depends upon which summary statistics on variation are desired by the investigator. The partitioning of sums of squares is the same conceptually for all investigations regardless of the orthogonality or nonorthogonality of the effects. The computation of sums of squares is considerably more difficult arithmetically when the effects are nonorthogonal, but the concepts do not change.

The word "analysis" in the analysis of variance does not mean "interpretation." The term "reduction of variance" or "partitioning of variance" may be more meaningful to some than the term "analysis of variance." The last term is the one used in published literature utilizing this concept.

XIV.3. Regression

A 19th century investigator, F. Galton, studied the heights of fathers and sons to determine the usefulness of a father's height in predicting his son's height. He noted that very tall fathers tended to have tall sons who tended to be somewhat shorter than their fathers. Likewise, short fathers tended to have short sons who were somewhat taller than their fathers. There was a tendency for heights to "regress" toward the mean height of all fathers, and hence the term regression. Of course, some sons were taller than their tall fathers but in general they tended to be shorter.

A father's height, say X, may be used to predict a son's height, say Y, by establishing a functional relation between these two variables X and Y. As described in chapter VII, this functional relation may take many forms. One form which appears to describe the above situation is the simple linear regression equation $Y_i = \mu + b(X_i - \bar{x}_.) + e_i$, where X_i is the height of the ith father and Y_i is the height of his son, $\bar{x}_.$ is the mean height of all fathers, μ is the population mean height of all sons, b is the slope of the line and is called the regression coefficient, and e_i is the deviation of a son's height from the mean value of all sons' heights, $\mu + b(X_i - \bar{x}_.)$, for a specified height of a father (for example, $X_i = 70''$). Thus, around each mean value $= \mu + b(X_i - \bar{x}_.)$, for X_i specified, there is a distribution of heights of sons. If the population variance in the distribution for each X_i

is equal to some constant σ^2 , if the X_i are measured without error, and if a Y_i (or a set of Y_i) is randomly selected for each X_i , then we may estimate the parameters μ and b in the above linear regression equation by the equations:

$$\hat{\mu} = \bar{y}_. = \sum_{i=1}^{n} Y_i/n \quad \text{and}$$

$$\hat{b} = \sum_{i=1}^{n} (Y_i - \bar{y}_.)(X_i - \bar{x}_.) / \sum_{i=1}^{n} (X_i - \bar{x}_.)^2 ,$$

where there are n pairs of observations (X_i, Y_i) in the sample. That is, for each father's height, his son's height is obtained. In the equation for the estimated linear regression coefficient $\hat{b}$, the numerator divided by n-1 is called the estimated <u>covariance</u>, and the denominator divided by n-1 is called the variance of the X_i . Therefore an estimated regression coefficient is the ratio of the estimated covariance of X_i and Y_i to the variance of the X_i . The variable Y is called the <u>dependent variable</u> since it depends upon the <u>independent variable</u> X.

To illustrate the computations in linear regression, consider the following example:

Speed records, Y_i, attained in the

Indianapolis Memorial Day auto races

	X_i=Year	Y_i	$(X_i-\bar{x}_.)$	$(Y_i-\bar{y}_.)$	$(X_i-\bar{x}_.)^2$	$(X_i-\bar{x}_.)(Y_i-\bar{y}_.)$	$\hat{Y}_i$
	1935	106	-3	-7	9	+ 21	108.5
	1936	109	-2	-4	4	+ 8	110.0
	1937	114	-1	1	1	- 1	111.5
	1938	117	0	4	0	0	113.0
	1939	115	1	2	1	+ 2	114.5
	1940	114	2	1	4	+ 2	115.0
	1941	115	3	3	9	+ 9	117.5
Total	13566	790	0	0	28	41	791
Mean	1938	113	-	-	-	-	113

From the above

$$\hat{b} = \sum_{i=1}^{7}(X_i-\bar{x}_.)(Y_i-\bar{y}_.)/\sum_{i=1}^{7}(X_i-\bar{x}_.)^2 = 41/28 = 1.5$$

and

$$\hat{Y}_i = \bar{y}_. + \hat{b}(X_i-\bar{x}_.) = 113 + 1.5(X_i-1938) \ .$$

The mean $\bar{y}_.$ was calculated to the nearest whole integer. From the equation for $\hat{Y}_i$, the regression line may be computed for the data in figure XIV.1. Computation of two values for $\hat{Y}_i$ is sufficient to draw the line of the equation in figure XIV.1, but three points are computed as an additional check on the computations. The line must pass through all three points. The $\hat{Y}_i$ values for X_i = 1936, 1938, and 1940 are:

$$\hat{Y}_2 = 113 + 1.5(1936 - 1938) = 113 - 3 = 110 \ ,$$

$$\hat{Y}_4 = 113 + 1.5(1938 - 1938) = 113 + 0 = 113, \text{ and}$$

$$\hat{Y}_6 = 113 + 1.5(1940 - 1938) = 113 + 3 = 116 \ .$$

The estimated regression coefficient $\hat{b} = 1.5$ indicates that speed records tend on the average to increase by 1.5 miles per hour each year. For each unit increase in X, Y is estimated to increase by 1.5 miles per hour. The estimated regression coefficient $\hat{b} = 1.5$ is a summary statistic of the average slope or average increase in miles per hour (Y) per unit increase in X. If there were only two pairs of values (X_1, Y_1) and (X_2, Y_2), the slope would be $(Y_1 - Y_2)/(X_1 - X_2)$ which is a result used in high school geometry books. The formula for $\hat{b}$ given above is used when there are more than n = 2 pairs of values.

From the formula for $\hat{b}$, two facts should be noted. The first is that $\sum_{i=1}^{n}(X_i - \bar{x}_.)(Y_i - \bar{y}_.)$ may be negative or positive and that its value is limited only by the values of $(X_i - \bar{x}_.)$ and $(Y_i - \bar{y}_.)$. This means that the covariance may be negative or positive, whereas the variance is never negative. The second fact to note is that $\hat{b}$ may be positive, negative, or zero. Thus Y may tend to increase, decrease, or vary independently with X.

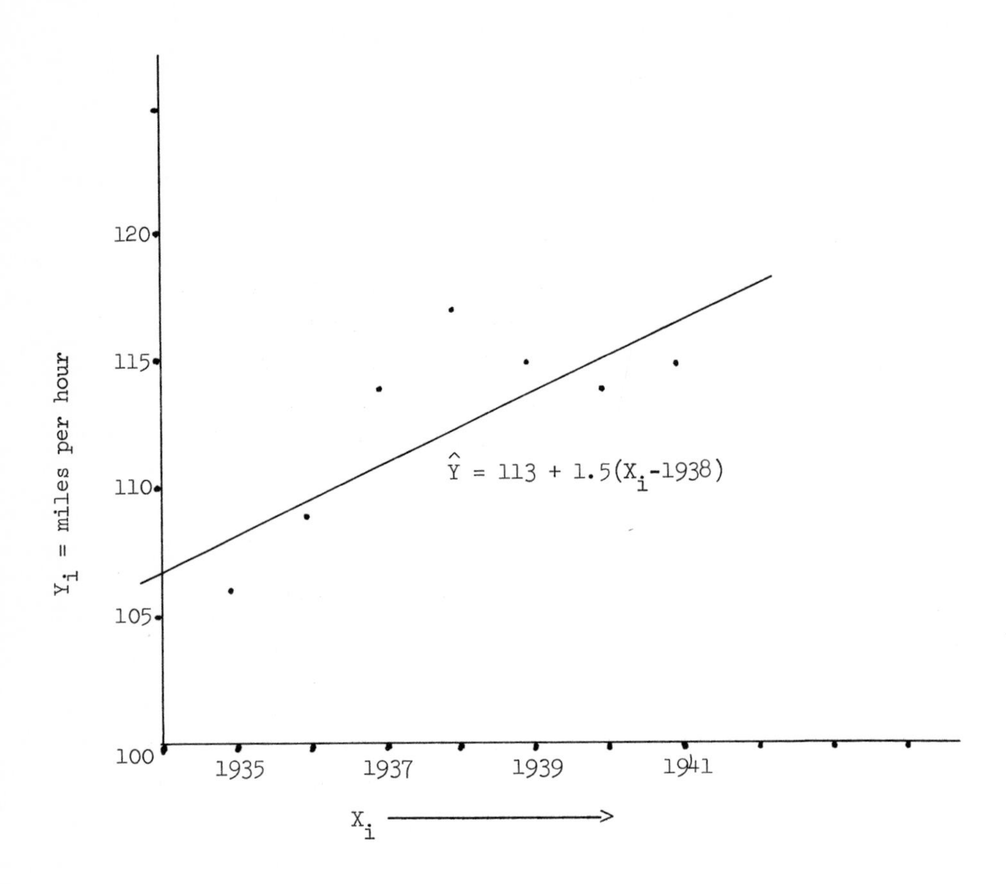

Figure XIV.1. Speed records Y_i in miles per hour in the Indianapolis Memorial Day automobile races for 1935 to 1941.

The deviations from the regression line may be obtained as $(Y_i - \hat{Y}_i)$ = observation minus its computed value on the regression line. The total deviation $(Y_i - \bar{y}_.) = (Y_i - \hat{Y}_i) + (\hat{Y}_i - \bar{y}_.)$, that is, the total deviation is partitioned into two parts. The first is the deviation of the observation from the computed value on the regression line, and the second is the deviation of the value on the regression line from the overall mean. The values of the deviation from the regression line are -2.5, -1.0, 2.5, 4.0, 0.5, -2.0, and -2.5. (The deviations do not add to zero as they should because of rounding errors. At least one decimal in $\bar{y}_.$ and two in $\hat{b}$ should have been retained.) The sum of squares is equal to 40.00 with n - 2 = 5 degrees of freedom. Therefore, $s_e^2 = 40.00/5 = 8.00$ = the average squared deviation.

Additional independent variables, for example, the height of the mother and the height of a grandfather[1], may be useful in predicting the height of the son. A multiple regression equation of the form $Y_i = \mu + b_1(X_{1i} - \bar{x}_{1.})$ $b_2(X_{2i} - \bar{x}_{2.}) + b_3(X_{3i} - \bar{x}_{3.}) + e_i$ may be useful to describe the height of a son, where X_{1i} = height of father, $\bar{x}_{1.}$ = mean height of fathers, X_{2i} = height of mother, $\bar{x}_{2.}$ = mean heights of mothers, X_{3i} = height of grandfather, $\bar{x}_{3.}$ = mean height of grandfathers, Y_i = height of son of ith father, mother, and grandfather, b_1, b_2, and b_3 are the respective population regression coefficients, and e_i is the deviation of height of a son from the mean height for specified values of X_{1i}, X_{2i}, and X_{3i} .

Many other examples resulting in regression equations of the above form are available in published literature. In studying performance in college = Y , several variables such as X_1 = high school grade average, X_2 = amount of outside activities, X_3 = Scholastic Aptitude Test score, and others are studied. Based on these studies, a students' performance in college is predicted, and based on this prediction, he may or may not be admitted to college.

[1]Experience has shown that little is gained from using a grandfather's height to estimate a grandson's height if the heights of the father and of the mother are available, and that b_3 is small relative to b_1 and b_2 .

We may summarize the information about sources of variation in linear regression in an analysis of variance table as follows:

Source of Variation	Degrees of Freedom	Sum of Squares	Variance
Due to mean	1	$n\bar{y}_{.}^2$	-
Due to linear regression	1	$\hat{b}^2 \sum_{i=1}^{n} (X_i-\bar{x}_{.})^2$	-
Deviations from regression	n-2	$\sum_{i=1}^{n} (Y_i-\bar{y}_{.}-\hat{b}(X_i-\bar{x}_{.}))^2$	s_e^2
Total	n	$\sum_{i=1}^{n} Y_i^2$	-

In the above analysis of variance table, the total sum of squares is partitioned into three parts in the algebraic equation:

$$\sum_{i=1}^{n} Y_i^2 = n\bar{y}_{.}^2 + \hat{b}^2 \sum_{i=1}^{n} (X_i-\bar{x}_{.})^2 + \sum_{i=1}^{n} (Y_i-\bar{y}_{.}-\hat{b}(X_i-\bar{x}_{.}))^2 \quad .$$

The variance $\hat{b}^2 \Sigma(X_i-\bar{x}_{.})^2$ may be compared with the residual variance s_e^2 to determine relatively how much of the variation is attributable to regression. In the sum of squares due to the mean, $n\bar{y}_{.}^2$, n is the sample size. In the sum of squares due to regression, $\hat{b}^2 \Sigma(X_i-\bar{x}_{.})^2$, the quantity $\Sigma(X_i-\bar{x}_{.})^2$ plays the same role as n in the preceding sum of squares. Thus, in selecting the X_i for regression, consideration should be given to making $\Sigma(X_i-\bar{x}_{.})^2$ as large as possible. In chapter VII it was suggested that n/2 of the observations should be placed at the lowest value of X_i, and n/2 should be placed at the highest value of X_i among all X_i being considered. If we do this, then $\sum_{i=1}^{n} (X_i-\bar{x}_{.})^2$ attains its maximum value. This is also of importance in the estimated variance of $\hat{b}$ which is $s_e^2/\sum_{i=1}^{n} (X_i-\bar{x}_{.})^2$, where it may be noted that a maximum value of $\sum_{i=1}^{n} (X_i-\bar{x}_{.})^2$ minimizes this variance.

Other more complex forms of regression equations may be treated in a similar manner in an analysis of variance table. Since we showed that the variation in a regression situation may be treated in an analysis of variance table in the same manner as a set of data from an experiment designed as a completely randomized design, we shall now illustrate how the sums of squares in an analysis of variance table for the latter case may be treated from the standpoint of regression. The resulting sums of squares will be identical. Suppose that there are $v = 3$ treatments replicated r times in a completely randomized design. From simple geometric considerations, the slopes of the three nonhorizontal lines in figure XIV.2 are computed as follows:

$$\text{top line} \quad - \text{ slope} = (\bar{y}_{1.} - \bar{y}_{..})/1 = (\bar{y}_{1.} - \bar{y}_{..})$$

$$\text{middle line} - \text{ slope} = (\bar{y}_{2.} - \bar{y}_{..})/1 = (\bar{y}_{2.} - \bar{y}_{..})$$

$$\text{bottom line} - \text{ slope} = (\bar{y}_{3.} - \bar{y}_{..})/1 = (\bar{y}_{3.} - \bar{y}_{..})$$

The slope is defined to be the increase in Y for each unit increase in X. Since all regression lines are required to go through the point $(0, \bar{y}_{..})$, the sum of squares $\sum_{j=1}^{r} (X_j - 0)^2$ is equal to $\sum_{j=1}^{r} X_j^2 = \sum_{i=1}^{r} 1^2 = r$. Therefore the sum of squares due to regression for each of the $v = 3$ regression lines is

$$r(\bar{y}_{1.} - \bar{y}_{..})^2 + r(\bar{y}_{2.} - \bar{y}_{..})^2 + r(\bar{y}_{3.} - \bar{y}_{..})^2 = r \sum_{i=1}^{3} (\bar{y}_{1.} - \bar{y}_{..})^2 .$$

This is the same sum of squares obtained previously for this source of variation.

If the deviations e_{ij} follow a normal distribution, we may set confidence limits on various quantities in linear regression. The various variances are summarized below:

variance of $\bar{y}_.$: $\quad s_e^2/n$

variance of $\hat{b}$: $\quad s_e^2 \Big/ \sum_{i=1}^{n} (X_i - \bar{x}_.)^2$

variance of $\hat{Y}_i = \bar{y}_. + \hat{b}(X_i - \bar{x}_.)$: $\quad s_e^2 \left(\frac{1}{n} + \frac{(X_i - \bar{x}_.)^2}{\Sigma(X_i - \bar{x}_.)^2} \right)$

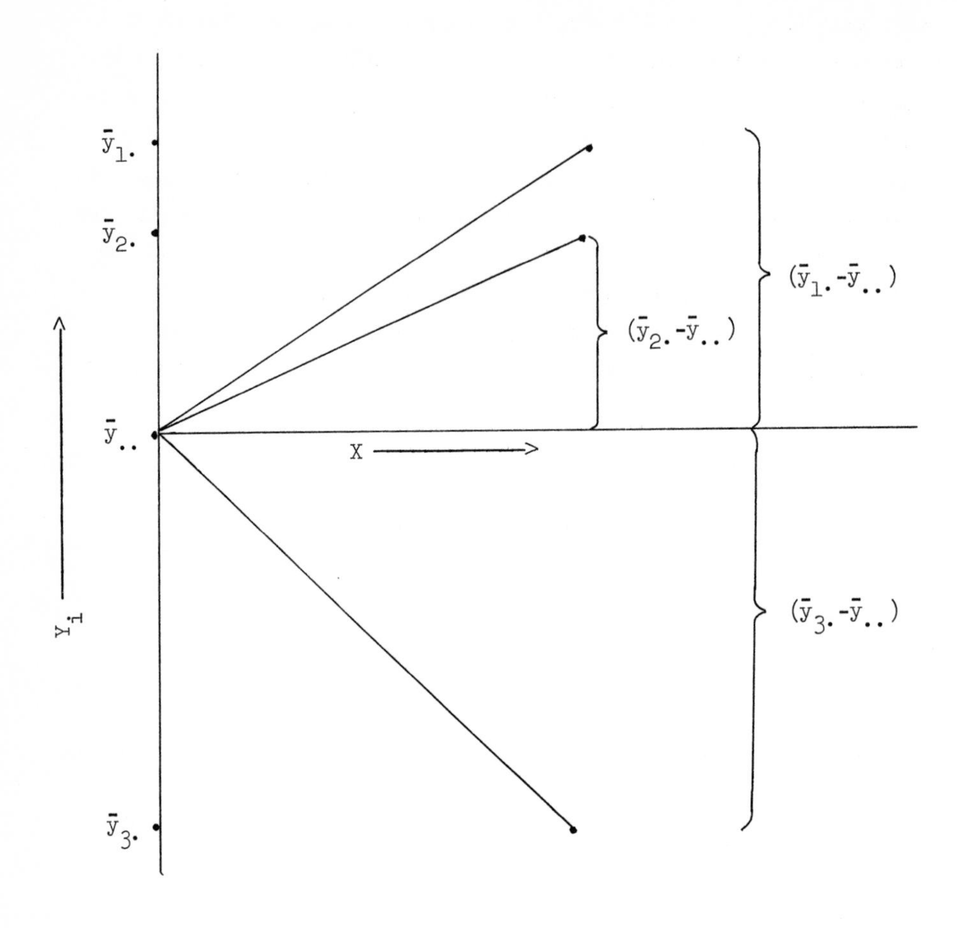

Figure XIV.2. Graphical representation of means $\bar{y}_{i.}$ of three treatments relative to the overall mean $\bar{y}_{..}$.

$$\text{variance of } Y_i = \bar{y}_. + \hat{b}(X_i - \bar{x}_.) + \hat{e}_i : \quad s_e^2 \left(\frac{1}{n} + \frac{(X_i - \bar{x}_.)^2}{\sum_{i=1}^{n} (X_i - \bar{x}_.)^2} + 1 \right) .$$

The standard deviation is obtained as the square root of the corresponding variance. Then a standard error is multiplied by $t_{\alpha,n-2}$ from the t table in chapter XI, where n-2 = degrees of freedom and α is the selected level for the $(1-\alpha)\%$ confidence interval.

XIV.4. Correlation

In the case of regression, an X_i was considered to be measured without error and to be the independent variate. Suppose that neither Y_i nor X_i are independent variates, that both are random variates, and that we wish to measure the relationship between the n pairs of measurements in a sample.

The interpretation of the measured relationship must be considered carefully and thoughtfully before conclusions are drawn. For example, the percentage of people speaking English in country i = Y_i and the number of cars owned per individual = X_i may be highly related variates but this does not mean that increasing the percentage of people speaking English in a country will increase the number of cars owned per individual, nor does it mean that an increase in the number of cars owned per individual will increase the percentage of people speaking English. The variates Y_i and X_i are related through some common factors and tend to increase together.

The correlation coefficient is a measure of relationship and is defined as a covariance divided by a geometric mean of the variances, that is covariance$/\sqrt{(\text{variance of Y})(\text{variance of X})}$. The estimated correlation coefficient is designated as $r_{xy} = \left(\sum_{i=1}^{n} (X_i - \bar{x}_.)(Y_i - \bar{y}_.)/(n-1) \right) \div$

$$\sqrt{\sum_{i=1}^{n} \frac{(X_i - \bar{x}_.)^2}{n-1} \sum_{i=1}^{n} \frac{(Y_i - \bar{y}_.)^2}{n-1}} = \sum_{i=1}^{n} (X_i - \bar{x}_.)(Y_i - \bar{y}_.) \Big/ \sqrt{\sum_{i=1}^{n} (X_i - \bar{x}_.)^2 \sum_{i=1}^{n} (Y_i - \bar{y}_.)^2} .$$

The correlation coefficient varies between plus and minus one. Correlation coefficients near zero indicate little or no relationship, and values near unity indicate a high relationship.

In order to set confidence limits on r and to use the t table in chapter XI, it is necessary to use the following function of r_{xy} :

$$Z = (\log_e(1+r_{xy}) - \log_e(1-r_{xy}))/2 .$$

The variable Z is normally distributed with variance $1/(n-3)$. Hence the $(1-\alpha)\%$ confidence limits of Z are computed as $Z \pm t_{\alpha,\infty} / \sqrt{(n-3)}$. The, Z is transformed back to r_{xy} to obtain the confidence limits for r_{xy} .

To illustrate the arithmetic associated with computing the correlation coefficient, consider the first five values for own height and own weight in table VIII.1. These are:

	X_i = own ht.	Y_i = own wt.	$(X_i-\bar{x}_.)$	$(Y_i-\bar{y}_.)$	$(X_i-\bar{x}_.)^2$	$(Y_i-\bar{y}_.)^2$	$(X_i-\bar{x}_.)(Y_i-\bar{y}_.)$
	71	148	2	- 5.6	4	31.36	- 11.2
	65	120	-4	-33.6	16	1128.96	134.4
	72	175	3	21.4	9	457.96	64.2
	70	173	1	19.4	1	376.36	19.4
	67	152	-2	- 1.6	4	2.56	3.2
Total	345	768	0	0	34	1997.20	210.0
Mean	69	153.6	0	0	-	-	-

$$r_{xy} = 210 / \sqrt{34(1997.20)} = 210/260.6 = 0.81 .$$

In all investigations involving small samples, interpretations should be made with caution. The example above and the one on regression contain only n=5 and n=7, paired observations, respectively. These are small sample sizes, and conclusions drawn from them are risky. The construction of confidence limits will aid in interpretation, but this does not overcome the vagaries of sampling associated with small sample sizes.

XIV.5. Tests of Significance

In the manner used herein, a test of significance is a measure of the standardized distance or deviation of a sample value from a hypothetical or hypothesized value. To illustrate, let us consider a difference between the two sample means, $\bar{y}_{1.}$ and $\bar{y}_{2.}$ obtained from two randomly selected samples, and let us consider the test of significance known as a t-test which is $t = (\bar{y}_{1.}-\bar{y}_{2.})/s_e\sqrt{2/r}$, where $s_e\sqrt{2/r}$ is the estimated standard error of the difference between the sample means $\bar{y}_{1.}$ and $\bar{y}_{2.}$. Suppose that the randomly selected sample values used to compute $\bar{y}_{1.}$ came from a population with mean μ_1 and that the randomly selected sample values used to compute $\bar{y}_{2.}$ came from a population with mean μ_2 . Then, $\bar{y}_{1.} - \bar{y}_{2.}$ is an estimate of $\mu_1-\mu_2$. If the two samples were from the same population, then $\mu_1-\mu_2 = 0$. Suppose one hypothesizes that $\mu_1-\mu_2 = 0$, then the larger the value of $|t|$, the less likely it is that $\mu_1-\mu_2 = 0$. In other words, the size of the value of the t-statistic measures the "strength of evidence" against the hypothesis that $\mu_1-\mu_2 = 0$ (or some other specified value). The larger the value of $|t|$, the greater is the strength of evidence against the hypothesis.

In order to ascertain when a value of the t-statistic is large enough to be significant or large enough for the investigator to attach some significance to the result, one may compare the value of the t-statistic with the values computed from the distribution of t, such as those given in chapter XI. As an illustration, let $\bar{y}_{1.} = 10$, $\bar{y}_{2.} = 5$, $s_e^2 = 16$, and $r = 8$, then $t = (10-5)/4\sqrt{2/8} = 5/4/2 = 5/2$. In figure XIV.3, we have constructed a "likelihood function", given that the error deviations are normally distributed. The $(1-\alpha)\%$ confidence intervals are computed for the various values of α, and then a smooth curve (dotted line) is fitted through the points. The confidence intervals computed are $\bar{y}_{1.} - \bar{y}_{2.} \pm t_{\alpha,d.f.}s_e\sqrt{2/r} = 5 \pm t_{\alpha,28}4\sqrt{2/8}$ $= 5 \pm t_{\alpha,28}(2)$ where $s_e^2 = 16$ is associated with 28 degrees of freedom; the $t_{\alpha,28}$ values may be obtained from various sources (e.g., Fisher, R. A., Statistical Methods for Research Workers). The $(1-\alpha)\%$ confidence interval calculations for figure XIV.3 are:

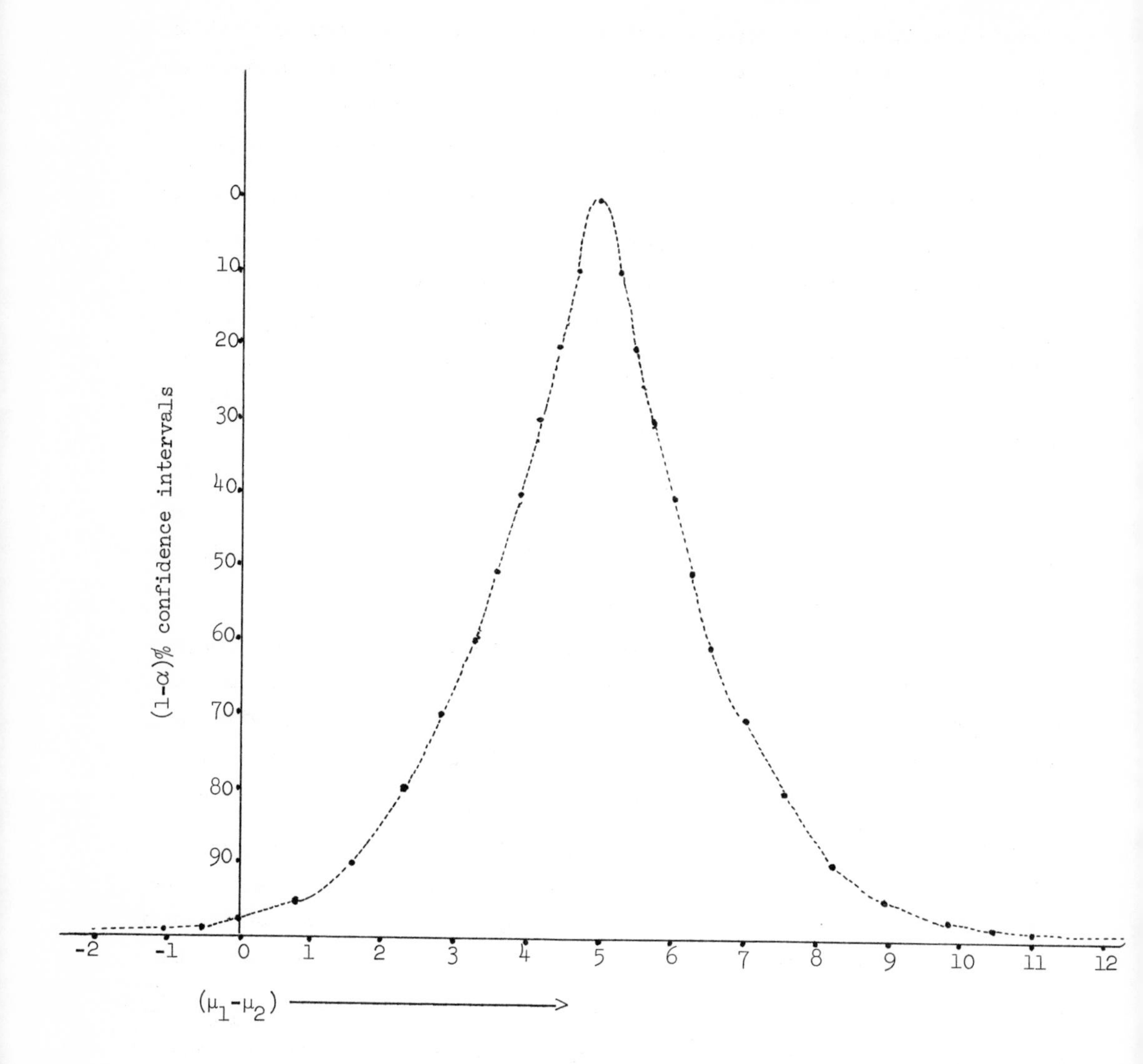

Figure XIV.3. Likelihood function for $(\mu_1-\mu_2)$ given that $(\bar{y}_{1.}-\bar{y}_{2.}) = 5$, $s_e\sqrt{2/r} = 2$, and degrees of freedom = 28.

α	$\bar{y}_{1.}-\bar{y}_{2.} \pm (s_e\sqrt{2/r})(t_{\alpha,28})$	confidence limits lower	confidence limits upper
P = .9	5 ± 2(.127)	4.746	5.254
P = .8	5 ± 2(.256)	4.488	5.512
P = .7	5 ± 2(.389)	4.222	5.778
P = .6	5 ± 2(.530)	3.940	6.060
P = .5	5 ± 2(.683)	3.634	6.366
P = .4	5 ± 2(.855)	3.290	6.710
P = .3	5 ± 2(1.056)	2.888	7.112
P = .2	5 ± 2(1.313)	2.374	7.626
P = .1	5 ± 2(1.701)	1.598	8.402
P = .05	5 ± 2(2.048)	0.904	9.096
P = .02	5 ± 2(2.467)	0.066	9.934
P = .01	5 ± 2(2.763)	-0.526	10.526
P = .005	5 ± 2(3.047)	-1.094	11.094

In figure XIV.3, the proportion of the area under the curve between the values 3 to 7 indicates that these are likely values for $\mu_1 - \mu_2$. Values from -2 to 1 or 9 to 12 would represent unlikely values for $\mu_1 - \mu_2$, because the area under the curve is relatively small. The proportion of the area under the curve between two values gives the "likelihood" of $\mu_1 - \mu_2$ being in the interval in repeated samplings. The "strength of the evidence" that the hypothesized value for $\mu_1 - \mu_2$ may or may not be true, may be obtained by noting where it lies on the abscissa of a graph such as figure XIV.3.

A second test of significance used extensively is the chi-square test computed as $\chi^2(k - 1 \text{ degrees of freedom}) = \sum_{i=1}^{k} (Y_i - \hat{Y}_i)^2/\hat{Y}_i$ where Y_i is the number observed in a particular category, and $\hat{Y}_i$ is the number computed for that category according to some expectation. In table VIII.1, the heights of recording team A were compared with the heights of recording team B, and the differences A - B were computed. Of these 85 differences, 51 were negative, 3 were zero, and 31 were positive. Based on a hypothesis of no difference, one would expect $85/2 = 42.5 = \hat{Y}_1$ negative and $42.5 = \hat{Y}_2$ positive differences. Place $1\frac{1}{2}$ of the 3 zero differences into the negative group, making $52.5 = Y_1$, and $1\frac{1}{2}$ into the positive group, making $32.5 = Y_2$. Since $k = 2$,

$\chi^2(1 \text{ d.f.}) = \frac{[52.5 - 42.5]^2}{42.5} + \frac{[32.5 - 42.5]^2}{42.5} = \frac{200}{42.5} = 4.71$. A comparison of 4.71 with tabulated values of χ^2 for one degree of freedom gives an indication of how much significance to attach to this value of χ^2 . The following table has been reproduced for this purpose:

Degrees of freedom	Probability of a larger value of χ^2 [1]						
	.90	.70	.50	.30	.10	.05	.01
1	0.02	0.15	0.46	1.07	2.71	3.84	6.64
2	0.21	0.71	1.39	2.41	4.60	5.99	9.21
3	0.58	1.42	2.37	3.67	6.25	7.82	11.34
4	1.06	2.20	3.36	4.88	7.78	9.49	13.28
5	1.61	3.00	4.35	6.06	9.24	11.07	15.09
6	2.20	3.83	5.35	7.23	10.64	12.59	16.81
7	2.83	4.67	6.35	8.38	12.02	14.07	18.48
8	3.49	5.53	7.34	9.52	13.36	15.51	20.09
9	4.17	6.39	8.34	10.66	14.68	16.92	21.67
10	4.86	7.27	9.34	11.78	15.99	18.31	23.21

From the above table we note that the probability of exceeding a chi-square value equal to 4.71 with one degree of freedom is between 5% and 1%. Hence it would appear that a 50:50 split of the negative and positive difference is not a plausible hypothesis. The strength of the evidence indicates that team A records lower heights than does team B.

As another illustration, consider the number of students with brown hair in the first 20, the second 20, the third 20, and the fourth 20 reported under own measurements. The following results were obtained:

[1] Taken from table III of Fisher: Statistical Methods for Research Workers, published by Oliver and Boyd, Edinburgh, and by permission of the author and publishers.

brown hair	Own hair color (no.) 1st 20	2nd 20	3rd 20	4th 20	total
observed Y_i	11	19	16	17	63
expected $\hat{Y}_i$	15.75	15.75	15.75	15.75	63
$(Y_i-\hat{Y}_i)$	-4.75	3.25	0.25	1.25	0
$(Y_i-\hat{Y}_i)^2/\hat{Y}_i$	1.43	0.67	0.00	0.10	2.20

Since the listing of names is alphabetical and since the number having brown hair should not be influenced by the first letter of one's name, one would expect one-fourth of the total, 63/4 = 15.75 , to be in each category. There are three degrees of freedom associated with this table, since three of the values $(Y_i-\hat{Y}_i)$ may vary, but the fourth is required to be the negative sum of the previous ones. This is required because the sum of the observed values must equal the sum of the expected or computed values. Since

$\chi^2(3d.f.) = \sum_{i=1}^{4} (Y_i-\hat{Y}_i)^2/\hat{Y}_i = 2.20$, we would consider these data to be a good fit to a 1:1:1:1 hypothesis. From the table of tabulated values of $\chi^2(3d.f.)$, we note that a value of 2.20 would be exceeded in more than 50% of random samplings from a chi-square distribution.

Let us consider the same set of data on hair color but from another viewpoint. Suppose we wish to obtain evidence on the independence of the two classifications listed below:

classification 1 = hair color	group = classification 2 1st 20	2nd 20	3rd 20	4th 20	total
brown	11	19	16	17	$63 = Y_{1.}$
not brown	9	1	4	3	$17 = Y_{2.}$
total	$20 = Y_{.1}$	$20 = Y_{.2}$	$20 = Y_{.3}$	$20 = Y_{.4}$	$80 = Y_{..}$

The above is a 2 by 4 contingency table with hair color as one classification and with groups of 20 names as the other classification. The observed number is Y_{ij}, and the computed or expected number is $\hat{Y}_{ij} = Y_{i.}Y_{.j}/Y_{..}$. The border totals are regarded as fixed, and hence only (r-1)(c-1) (where r = number of rows and c = number of columns) quantities in the table may vary; the remaining quantities are obtained by subtraction from the border totals. This results in (r-1)(c-1) degrees of freedom for the resulting chi-square test computed as:

$$\chi^2\left((r-1)(c-1)\text{d.f.}\right) = \sum_{i=1}^{r}\sum_{j=1}^{c}(Y_{ij}-\hat{Y}_{ij})^2/\hat{Y}_{ij} .$$

The computed values for the above table are:

hair color	Group: 1st 20	2nd 20	3rd 20	4th 20	total
brown	15.75	15.75	15.75	15.75	63
not brown	4.25	4.25	4.25	4.25	17
total	20	20	20	20	80

where 63(20)/80 = 15.75 and 17(20)/80 = 4.25 . The deviations $Y_{ij}-\hat{Y}_{ij}$ are:

hair color	Group: 1st 20	2nd 20	3rd 20	4th 20	total
brown	-4.75	3.25	0.25	1.25	0
not brown	4.75	-3.25	-0.25	-1.25	0
total	0	0	0	0	0

The quantities $(Y_{ij}-\hat{Y}_{ij})^2/\hat{Y}_{ij}$ are:

	Group				
hair color	1st 20	2nd 20	3rd 20	4th 20	total
brown	1.43	0.67	0.00	0.10	2.20
not brown	5.31	2.49	0.01	0.37	8.18
total	6.74	3.16	0.01	0.47	10.38

A chi-square value of 10.38 with three degrees of freedom has a probability between one and five percent of being exceeded in random sampling. Hence this is evidence that the grouping and hair color are not independent, even though we might believe that they should be. A chi-square test used in this manner results in a test of independence, whereas in the previous example the test was a goodness of fit test. In chapter XI, observed frequencies were compared with computed or expected frequencies utilizing the chi-square test for the binomial and Poisson distributions. Many forms of a chi-square test are used in statistical methodology.

XIV.6. Problems

XIV.1. Show how to compute the analysis of variance for example X.1, page 273. (Remember that to show how to compute does not mean to compute.)

XIV.2. Show how to compute the analysis of variance for example X.2, page 275.

XIV.3. Show how to compute the analysis of variance for example X.3, page 277.

XIV.4. Show how to compute the analysis of variance for problem X.6.

XIV.5. Show how to compute the analysis of variance for problem X.7.

XIV.6. Show how to compute the analysis of variance for problem X.8.

XIV.7. Show how to compute the analysis of variance for problem XI.5.

XIV.8. Show how to compute the analyses of variance for the first three examples of factorial treatment designs given in section VII.6, page 161.

XIV.9. Show how to compute the linear regression coefficients for Y = grade-point average on X_1 = high school grade and on X_2 = aptitude test score for the data given in section VII.6, page 165.

XIV.10. Show how to compute the correlation coefficient for X_1 and X_2 in the preceding problem.

XIV.11. The following data were obtained from table VIII.3 by grouping all ages less than or equal to 20.5 years and all those older than 20.5 years; the other classification was omitted. The regrouped data are:

age class	class standing					total
	freshman	sophomore	junior	senior	graduate	
less than 20.5 years	13	32	8	4	0	57
more than 20.5 years	0	1	10	19	4	34
total	13	33	18	23	4	91

From what we know about students, the two classifications should not be independent. Consequently, a large value of chi-square should be obtained. Show how to compute the chi-square test for independence for this contingency table.

XIV.12. Show how to compute the goodness of fit chi-square test for the data in tables XI.1 and XI.2.

XIV.13. Show how to compute a goodness of fit chi-square test for the data given by Rao and Chakravarti in section XI.3 when the zero class is included and when it is excluded.

XIV.14. Show how to compute a goodness of fit chi-square test to the 1:1, the 3:1, and 7:1 ratios given that 80 red-eyed flies and 20 white-eyed flies were obtained in the experiment described in problem XI.2.

XIV.7. References and Suggested Readings

(See the references listed in chapter X.)

APPENDIX

Examinations for one semester

The sequence of chapters taught for the following set of examinations was chapters I, II, III, IV, VI, VII, V, VIII, X, XI, and XIII. Some of the material in chapters IX and XII was included, but no formal discussion of probability or of sample size was given. None of the material in chapter XIV was presented. The ordering of the chapters in the text is considered to be better pedagogically since sample survey design concepts are easier to grasp than are experiment design concepts. If the order given in the text is followed, the instructor may simply move the questions on sampling from the third examination to the front part of the second examination and then may include some of the material from the back part of the second examination on the third examination. For example, for the 1968 set of examinations, take questions 7 (15 points) and 10 (9 points) from the second examination and place them at the beginning of the third examination; then, take questions 17 to 28 (24 points) from the third examination and place them at the beginning of the second examination. Each of the revised examinations has 100 points.

The examinations were given in the fourth, eighth, and twelfth weeks of the semester with the final examination covering the material in the last chapters as well as the material in the earlier chapters.

The instructor is urged to use questions which employ reasoning more than memory. In this way, the examination can be more of an educational tool. It is felt that the first examination relied too heavily on memory. Because of the nature of the material, it was not possible to write an examination relying more on the powers of reasoning and less on the powers of memory for the first examination. Both the preparation for an examination and the taking of an examination should provide an educational experience for the student.

It is highly recommended that an instructor using this text write the examination questions together with the answers prior to the time that the formal course work begins. With such a set of examinations available, the course will have a well-defined direction and pace. The presentation will be more comprehensible and less ambiguous to the student since they were written at a time when the instructor was not under the strain of presenting the course and being forced to prepare an examination "the night before".

Also, it is suggested that the examination be given at a time when the student has sufficient time to complete the examination, such as at night. This procedure was followed for the 1968 to 1971 examinations, but the 1966 and 1967 examinations were given during a class period.

Additional problems for the individual chapters may be obtained from the examinations. If discussion periods are organized as was done for the course taught by the author, the questions on the examinations from previous years form an excellent medium for discussion. This has been found to be an effective means to aid the student in understanding the concepts of the course.

Name ______________________________

Statistics — Spring, 1968

FIRST PRELIM (Problems I to VII)

Points

10 — I. Define STATISTICS used as the name of a subject matter field.

Distinguish between the subject of Statistics and "statistics" used as a noun.

10 — II. Match the most appropriate term, by using the assigned letter, with the following definitions:

Terms:

(a) datum	(i) treatment
(b) statistic	(j) absolute experiment
(c) hypothesis	(k) comparative experiment
(d) number	(l) single phenomenon experiment
(e) research	(m) re-search
(f) experimental design	(n) law
(g) experiment	(o) parameter
(h) theory	(p) postulation

Definitions:

i) A __________ is one designed specifically to compare two or more treatments.

ii) The quest or pursuit of new knowledge is defined to be __________.

Points (1st Prelim continued)

iii) An __________ is defined to be the planning and collection of measurements or observations according to a prearranged plan for the purpose of obtaining factual evidence on the plausibility of hypotheses.

iv) The arrangement of the treatments in an experiment is known as __________.

v) A number that is used to describe a characteristic of a population and that is derived from a consideration of all members in the population is called a __________.

vi) A number derived from a sample and used to estimate the value of a characteristic of the population is known as a __________.

vii) A fact from which a conclusion is to be drawn is known as a __________.

viii) A tentative or postulated explanation of a phenomenon is known as a __________.

ix) A relatively well verified postulation which possesses some degree of generality is known as a __________.

x) A __________ is a single entity or phenomenon under study in an experiment.

30 III. Complete the following statements (2 points each).

1. ______________ is the term applied to the procurement and systematization of scientific knowledge.

2. The method whereby scientific knowledge is acquired is known as ______________________________.

3. An explanation of a phenomenon which has been verified beyond all reasonable doubt is known as a ______________.

4. An experiment containing but one treatment or phenomenon is ____.

5. __________________ inference is the process of determining the implications contained in a set of propositions.

Points (1st Prelim continued)

6. A 100% sample from a population or universe of individuals is a ________________.

7. An error in measurement is composed of ________________ plus ________________.

8. A ________________ yields information on what is present in the universe.

9. The type of reasoning involved when we judge the temperature of the water of a lake by sticking our toe in the water at one spot in the lake is ________________.

10. The process of reaching a conclusion from the data given in the following example is known as ________________. Ex. "If all students who studied pass this test and if you studied, then you will pass this test."

11. Statistics is more concerned with ________________ inference while mathematics is more concerned with ________________ inference.

12. ________________ inference is used to a considerable extent in formulating hypotheses.

13. A figure or number which pertains to some clear-cut characteristic and which is impossible to obtain is known as ________________. (E.g. Dognapping is big business, grossing $1,000,000 per year.)

14. A figure or number used in connection with an undefined term or an ambiguous thought in such a way that it is unclear what is meant is ________________. (E.g. The National Safety Council estimates that traffic accidents caused 1.9 million disabling injuries and $11 billion in economic loss in 1967.)

15. In terms describing variability the variation among the true means of populations is defined as ________________.

20 IV. Problem III.2 was given as follows: An experimenter used 1000 ripe peaches (from Georgia) in an experiment. He wished to determine which quarter of a peach fruit was the most tender

Points (1st Prelim continued)

when squeezed. He designated the peach fruit by quarters starting with the suture, or indentation, on the fruit as follows: (1) left front, (2) left back, (3) right back, and (4) right front. (He did not cut the peach fruit.) A device was utilized which measured the amount of pressure required to penetrate the peach skin; this was used to measure tenderness. On each peach the experimenter measured the quarters in the order as numbered above. He reached the conclusion that the least pressure was required to penetrate the right front quarter, the next lowest pressure was required to puncture the right back quarter, the next in amount of pressure required for penetration was the left back, and the left front required the greatest pressure for penetration. He published in a scientific journal.

Show how each of the eight principles of scientific experimentation as listed in chapter IV of the text apply to this investigation.

1. ____________________

2. ____________________

3. ____________________

4. ____________________

5. ____________________

6. ____________________

7. ____________________

12. V. With arrows indicate the relationship between the steps discussed relative to data collection in chapter III of the text with the steps in scientific experimentation as listed in chapter IV:

Points (1st Prelim continued)

Steps in Chapter III	Steps in Chapter IV (fill in blanks)
Why collect data? ——→	1. A clear and precise statement of the problem
What data are to be collected?	2. ________
How are data to be collected?	3. ________
Where and when are the data to be collected?	4. ________
Who collects the data?	5. ________
Complete description of data.	6. ________
Disposal of data.	7. ________
Conclusions.	8. ________

3 VI. What is the difference between precision and accuracy? Illustrate with an example.

5 VII. A characteristic is described for three populations (I, II, and III). The following numbers a, b, c, d, and e, as designated on the real line below, are defined as follows:

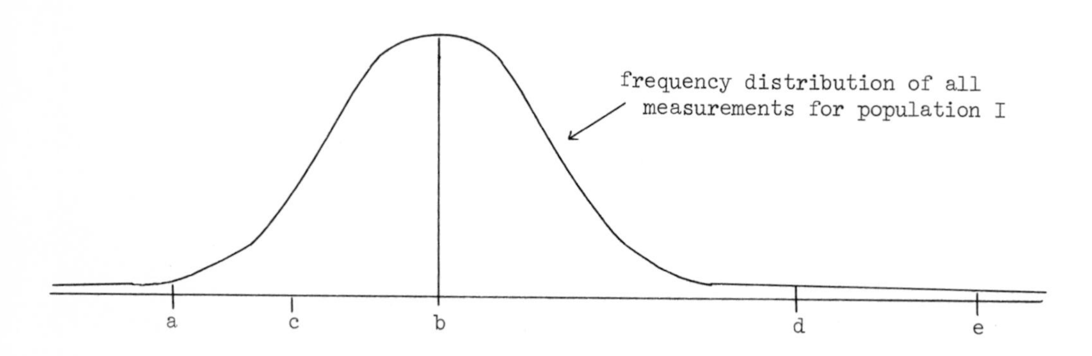

a = true mean of population I

d = true mean of population II

e = true mean of population III

b = mean obtained from all the measurements for population I

c = a single measurement from population I

Points (1st Prelim continued)

In the terms used for describing variability, the line segment ab is a measure of ______________________________,
the line segments ad, ae, and de are measures of variation due to
____,
the line segment ac is a measure of ____________________,
the line segment bc is a measure of ____________________,
and the line segment ce is a measure of ____________________
____.

Name____________________

Statistics Spring, 1968

SECOND PRELIM (problems 1-10, 100 points)

Points

6 1. Miss I. C. Food was conducting an experiment on different kinds of pie thickeners to measure their effect on consistency as defined and measured by her. She was investigating $v=4$ kinds of thickening agents (the treatments) and decided to use $r=3$ repetitions of each treatment. She could prepare 12 pies at one time and then add the treatments. She could also bake 12 pies at one time in an oven which exhibited no gradients. Therefore, she decided to use a completely randomized design for the experiment. Describe the yield equation for this experiment when the effects are additive and enter the equation in linear form.

Are the assignable effects orthogonal to each other? Why or why not?

Construct an experimental plan properly randomized using the attached random number table and starting at the top of column 1 and proceeding down the column. (Write down the sequence of random numbers used.)

8 2. Miss I. M. Fashion was conducting an experiment on length of time different colors of eye shadow stayed presentable which was a measurement made by her. She used $v=5$ preparations of eye shadow which only differed in the color component. She was using five different time periods for which she knew that there were only random variations and no known gradients. Therefore, she decided to use $r=4$ different girls as the blocks and to use a randomized complete block design.

Describe the yield equation for this experiment when the effects are additive and enter the equation in linear form.

Which of the assignable effects are orthogonal to each other?

Construct the experimental plan using the enclosed random number table and starting at the top of column 2 and proceeding down the column. (Write down the sequence of random numbers used.)

Points (2nd Prelim continued)

12 3. Mr. Bunny Buck wishes to compare v=5 feeding treatments on the growth of rabbits from three to six weeks of age. Since he has available to him only litters of 3 rabbits each he decides to use a balanced incomplete block design.

Construct a balanced incomplete block design for v=5 treatments in blocks of size k=3 rabbits each by completing the following:

Litter number = Block number

1	2	3	4	5	6	7	8	9	10
1	1	1	1	1					
2	2	2	3	3					
3	4	5	4	5					

The number of replicates is r = ______________________.

The number of times any pair of treatments occur together in the b incomplete blocks is λ = ______________________.

Which of the assignable effects are orthogonal to each other?

Describe and illustrate the randomization procedure for this design using the attached random number table starting at the top of column 3 and proceeding down the column. (Write down the sequence of random numbers used.)

Random Digits (9001 - 13,750)

(Published in Jour. Amer. Stat. Assoc., 48:931-934 (1953). Taken from A Million Random Digits, published by the Rand Corporation, Santa Monica, Calif.)

3780	28391	05940	55583	81256	38175	38422	64677	80358	52629
5325	05490	65974	11186	15357	21805	10371	95812	84665	74511
8240	92457	89200	94696	11370	75517	82119	09199	30322	33352
2789	69758	79701	29511	55968	19195	92261	44757	98628	57916
7523	17264	82840	59556	37119	77869	08582	63168	21043	17409
8853	59083	95137	76538	44155	79419	22359	65206	54941	95992
0274	79932	44236	10089	44373	59914	04146	01419	48575	77822
2805	21149	03425	17594	31427	43374	25473	60982	27119	16060
4971	49055	95091	08367	28381	22199	11865	26201	18570	72803
3606	46497	28626	87297	36568	13786	27475	31254	36050	73736
7286	28749	81905	15038	38338	45445	41059	55142	55585	39829
5670	72111	91884	66762	11428	21067	57238	35352	57741	98761
4262	09513	25728	52539	86806	30302	95327	12849	15795	97479
7375	85062	89178	08791	39342	70040	91385	96436	58982	91281
9483	62469	30935	79270	91986	13351	48321	28357	88526	74396
1206	65749	11885	49789	97081	15564	04716	14594	22363	85700
0908	21506	16269	54558	18395	30987	57657	33398	63053	46792
9944	65036	63213	56631	88862	79172	72764	66446	78864	96004
4963	22581	17882	83558	31960	57875	45228	49211	69755	27896
9286	45236	47427	74321	67351	57146	64665	31159	06980	64709
6075	20517	69980	18293	44047	42826	06974	61063	97640	13433
3375	62251	58871	70174	52372	93929	01836	36590	75052	89475
0487	38794	36079	23362	24902	83585	00414	62851	48787	28447
9473	45950	18225	09899	87377	27548	37516	24343	63046	02081
7703	83717	18913	66371	53629	32982	56455	53129	77693	25022
7612	72738	26995	50933	92936	30104	67126	76656	29347	28492
9042	37595	04931	73622	69902	35240	00818	09136	01952	48442
9609	35653	15970	37681	96326	94031	62209	43740	54102	76895
5354	65770	15365	41422	29451	99321	11331	06838	03818	77063
9452	71674	30260	97303	31002	78236	71732	04704	61384	57343
9867	89294	59232	31776	54919	43108	56592	42467	88801	91280
6719	06144	82041	38332	64452	91058	60958	20706	31929	57422
6970	45907	99238	74547	19704	98172	44346	60430	59627	26471
5747	78956	11478	41195	58135	12523	57345	41246	98416	08669
1838	07526	07985	60714	88627	66682	82517	33150	27368	53375
4361	34534	70169	24805	63215	01056	27534	23085	49602	74391
9278	17082	26997	32295	10894	18730	96197	64483	40364	90913
8124	84721	23544	88548	65626	07794	60475	49666	17578	12830
2025	16908	82841	24060	40285	48883	77154	74973	42096	34934
0326	86370	91949	19017	83846	70171	59431	76033	40076	20292
4855	27029	01542	72443	72302	48830	55029	10371	09963	85857
5434	12124	91087	87800	34870	73151	64463	50058	11468	93553
6800	16781	65977	65946	65728	06571	95934	09132	13746	82514
1233	81409	46773	69135	36170	76609	52553	47508	25775	91309
2933	77341	20839	36126	18311	32138	61197	95476	69442	54574

Points (2nd Prelim continued)

15 4. Mr. John Henry is studying v=3 teaching methods (A, B, and C) with four different teachers in four different class periods. He wishes to include treatment A twice as frequently as treatments B and C. He knows that there are two sources of variation, i.e., teachers and class periods, so he selects a latin square design.

Construct a latin square design for Mr. Henry.

Describe the yield equation for this experiment and demonstrate which effects are orthogonal to each other.

The number of replicates on treatment A equals_______________.

The number of replicates on treatments B and C equals _______.

Construct an experimental plan, properly randomized, using the attached random number table starting at the top of column 4 and proceeding down the column. (Write down the sequence of random numbers used.)

10 5. Suppose that Miss Fashion (problem 2) knew that there was a gradient from the first period to the last period of the experiment. In this case she would have used a 5 row (the time periods) by 4 column (the girls) Youden square design.

Construct the 5 row by 4 column Youden square for v=5 treatments.

Explain why each of various assignable sources of variation are or are not orthogonal to each other.

Construct a properly randomized experimental plan for the experiment starting with the top of column 5 and proceeding down the column. (Write down the sequence of random numbers used.)

10 6. An experimenter used the following blocked design for v=3 treatments, (A,B,C) replicated r=3 times in b=3 blocks of size k=3 experimental units each.

Block		
1	2	3
A	C	C
B	B	C
A	B	A

Points (2nd Prelim continued)

Are the occurrences of pairs of treatments within the blocks balanced? Why or why not?

Illustrate which of the assignable causes of variation are orthogonal to each other.

What design would have been more efficient than the one used above?

15 7. In chapter VI we studied five properties of experimental designs after introducing the subject by stating three characteristics that we desired for our experimental arrangements. List the five properties in the upper boxes below and list the three characteristics in the lowest three boxes. Interrelate the properties and characteristics with each other by use of arrows.

Valid estimate of experimental error variation

Reduction of experimental error variation

characteristic (1) characteristic (2) characteristic (3)

Points (2nd Prelim continued)

6 8. List the two purposes of randomization.

i)

ii)

3 9. One of the statements in each set is correct, circle the correct one. Given the following experimental arrangement without any additional information,

C	B	A
A	C	B
B	A	C

, which of the following statements are correct:

(a) The experimental design is a latin square design.

(b) The experimental design is a randomized complete block design.

(c) Nothing can be said about the experimental design until we know what randomization procedure was used.

3 An experimental unit is

(a) The unit consisting of one comparative experiment.

(b) The smallest unit to which one treatment is applied

(c) The smallest unit on which an observation is made.

3 Sources of variation in an experiment refer to

(a) The different assignable causes of variation and the random error variation with each item being a source of variation.

(b) The measurement errors that arise in experimentation which are denoted as sources of variation.

(c) The variation caused by the experimenter and denoted as sources of variation.

10. In the problems below, 3 of the 6 statements in each problem are correct. Circle the letter of the correct statement.

Points (2nd Prelim continued)

3 A placebo is:

(a) A remedy for any kind of headache.

(b) A control in an experiment.

(c) Or may be, a pharmacologically inactive treatment.

(d) Sometimes used to divide participants in a study into a group that responds and a group that does not respond to the stimulus.

(e) A necessity in all medical experiments.

(f) Not of any value in experimentation.

3 With regard to the conditions under which an experiment is conducted, which of the following statements are correct?

(a) The conditions of the experiment must be such as to allow an expression of differences of treatment responses if in fact they exist.

(b) The conditions of the experiment must conform to ease with which the experiment can be conducted.

(c) For practical inferences, the experiment must be conducted under the same conditions as found in practice.

(d) For theoretical (not practical) inferences only, it is necessary to consider only statement (a) above.

(e) All experiments must be conducted in the same manner.

(f) All experiments must be conducted in the laboratory.

3 Relative to the inclusion of a control in the treatment design

(a) A control is included to have a point of reference.

(b) Often more than one control is required.

(c) A control is included to enlarge the experiment.

(d) A control is included to obtain estimates of differences in effects between non-control treatments.

(e) A control is necessary to complete a factorial experiment.

(f) A placebo is a form of a control treatment.

Name________________________________

Statistics Spring, 1968

THIRD PRELIM
(100 points)

Points

Miss I. M. Fashion wishes to use 3 colors (red, yellow, and blue) with two shades (light and dark) of each color; she also wishes to observe the effect of two oils (lanolin and olive oil) in two different amounts (12% and 14% by weight). She wishes to use all possible combinations of these ingredients as listed above in eye shadow preparations. The response to be measured is number of hours the eye shadow stays presentable (a measurement she makes reliably) on Cornell coeds. The experiment is to be conducted during the academic year using r=20 girls such that all v treatments are used on each girl. The differences between treatment periods or experimental units is known to represent only random sampling variation. Therefore, the v treatments are randomly allotted to the v time periods for each girl. There is a real difference between girls with respect to the length of time an eye shadow preparation stays presentable.

3 1. What is the treatment design?

3 2. What is the experimental design?

6 3. Write out the treatment combinations that Miss Fashion will use.

3 4. Suppose that the two treatments involving 14% olive oil in combination with the light shades of blue and yellow were omitted from the experiment because it was known that they gave unsatisfactory results. What would the resulting treatment design be?

3 5. Suppose that 8 experimental periods are available for each of 12 girls and that there is a real (assignable) effect of period as well as for girls. What experimental design should Miss Fashion use?

5 6. Given the following treatment design

$a_0b_1c_1d_0$ $a_0b_2c_1d_0$ $a_0b_3c_1d_0$

$a_0b_1c_2d_0$ $a_0b_2c_2d_0$ $a_0b_3c_2d_0$

$a_0b_1c_3d_0$ $a_0b_2c_3d_0$ $a_0b_3c_3d_0$

can the experimenter estimate effects for a and d?

Points (3rd Prelim continued)

Why or why not?

What is an alternative way to write the above treatment design?

10 7. Given the following data and solutions for effects

level of factor a	level of factor b 0	1	effect of factor a
	estimated interaction effects		
1	$\widehat{a_1b_0} = 2$	$\widehat{a_1b_1} = -2$	$\hat{a}_1 = 3$
2	$\widehat{a_2b_0} = -2$	$\widehat{a_2b_1} = 2$	$\hat{a}_2 = -3$
effect of factor b	$\hat{b}_0 = 4$	$\hat{b}_1 = -4$	$\hat{\mu} = 10$

Show how to compute the responses Y_{ij} given the above effects

Y_{10} = ______________ Y_{11} = ______________

Y_{20} = ______________ Y_{21} = ______________

Note that the sum of the estimates of main effect levels in the above 2x2 factorial add to zero and that the estimated interaction effects add to zero across rows or across columns. Is this true for any pxq factorial? yes no (circle one).

10 8. Show how to compute the remaining main effects and remaining interactions for the following data for a 2x3 factorial:

Level of factor a	Level of factor b b_0	b_1	b_2	sum	mean
a_0	3	4	5	12	4
a_1	1	8	15	24	8
sum	4	12	20	36	-
mean	2	6	10	-	6

Points (3rd Prelim continued)

$\hat{a}_0 = -2$ $\hat{a}_1 =$ ______________________

$\hat{b}_0 = -4$ $\hat{b}_1 =$ ____________ $\hat{b}_2 =$ ____________

$\hat{ab}_{00} = 3$, $\hat{ab}_{01} = 0$, $\hat{ab}_{02} = -3$, $\hat{ab}_{10} = -3$

$\hat{ab}_{11} =$ ______________________

$\hat{ab}_{12} =$ ______________________

Given the same means in the above table what numbers would need to replace numbers 3, 1, 5, and 15 in the above table in order to have all interaction effects equal to zero?

6 9. The following plan involves 8 treatments with r=6 blocks in which the treatments have been randomly allotted to the 8 experimental units within each block. Treatments 1, 2, 3, and 4 represent one level of a factor, say color, and treatments 5, 6, 7, and 8 represent the second level of this factor. For the second factor treatments 1 and 8 represent one level, treatments 2 and 5 represent a second level, treatments 3 and 7 represent a third level, and treatments 4 and 6 represent a fourth level.

What is the treatment design?

What is the experimental design?

In the problems below one-half of the statements are correct and one-half are incorrect. Check the correct statements in problems 10-12.

3 10. Which of the following statements pertain to a factorial treatment design?

a) Involves two or more factors.

b) Involves one or more factors.

c) Involves a single factor with equally spaced levels.

d) Involves two or more levels of each factor being studied.

e) Is useful only for discussion purposes in statistics classes.

f) May have the levels of one factor equally spaced and the levels of the remaining factors unequally spaced.

Points (3rd Prelim continued)

3 11. Relative to a biological assay or bioassay treatment design check the correct statements:

a) It relates to all laboratory experiments.

b) It relates to experiments involving reactions on living material.

c) It relates to estimation of potency of materials and relates to identification of constitution of materials.

d) It does not include analytical assays which estimate the potency of a test preparation relative to a standard.

e) It includes slope ratio assays and parallel line assays as two special types.

f) It is used only in human drug experiments.

5 12. In selecting a treatment design involving one or more factors which of the following conditions are pertinent?

a) The nature of the experimenter.

b) The level of factors not varied in the experiment.

c) The specification of factors to be varied.

d) The range over which the factors are to be varied.

e) The experimental design.

f) The form of the material to be used in varying the levels of the factors.

g) The sample survey design.

h) The number and spacing of the levels of each factor studied.

i) The ease with which the results can be explained.

j) Whether or not the levels of all factors can be ordered.

4 13. Define bioassay.

5 14. Define relative potency and illustrate how it is estimated in slope ratio and a parallel-line treatment design.

Points (3rd Prelim continued)

5 15. Define a simple random sampling procedure and illustrate your definition with a population of N=5 objects A,B,C,D,E and with a sample size of n=2 objects.

A student researcher, Mr. I. C. E. K. Facts, wishes information on several characteristics, say Y_1, Y_2, Y_3, Y_4, Y_5, on the population of coeds at Cornell University during the Spring term 1968. The population of undergraduate coeds at Cornell University live in one of the following residence units (no girl will be classified as living in two of the following residence units) within each of two groups:

Non-sorority (group 1)	Sorority (group 2)
1. Comstock - A	13. Alpha Epsilon Phi
2. Comstock - B	14. Alpha Phi
3. Clara Dickson - V	15. Delta Delta Delta
4. Clara Dickson - VI	16. Delta Gamma
5. Balch - 1	17. Delta Phi Epsilon
6. Balch - 2	18. Kappa Delta
7. Balch - 3	19. Kappa Kappa Gamma
8. Balch - 4	20. Phi Sigma Sigma
9. Risley	21. Pi Beta Phi
10. Mary Donlon - 1	22. Sigma Delta Tau
11. Mary Donlon - 2	
12. Other	

Let character Y_1 be the proportion of outside area of stone (brick, mortar, and stone) to other materials in the residence unit.

Let character Y_2 be the number of girls living in the residence unit.

Let Y_3 be the hair color of a coed.

Let Y_4 be the cumulative grade point average of a coed.

Let Y_5 be the attitude of girls toward dating boys who had waxed, handle-barred, red mustaches like the one Mr. Facts was wearing.

Mr. Facts does not have the time nor the resources to perform a complete census. Consequently, he decides to sample the population. He randomly selects 6 non-sorority and 5 sorority residence units. He then randomly selects 10 girls from each of the 11 selected residence units.

Points (3rd Prelim continued)

For character Y_1 Mr. Facts photographs the structures and then measures the area of stone to non-stone materials making up the outside area.

For character Y_2 he counts the number of girls living in each room of the residence unit.

For characters Y_3 and Y_4 he determines hair color and cumulative grade point average only on his sample of 10 girls from each of 11 residence units. He does likewise for character Y_5.

2 17. What is the sampling design for character Y_1?

2 18. How should he have obtained information on character Y_2?

2 19. What is the sampling design for character Y_3?

2 20. What is the sampling design for character Y_4?

2 21. What type of allocation of sample size was utilized within each of the 11 selected residence units?

2 22. What is the sampling design for characters Y_3 and Y_4 within each stratum?

2 23. What type of allocation of sample sizes was utilized for characters Y_1 and Y_2?

2 24. How should the sampling design as described above be changed in order to obtain proportional allocation of sample size for characters Y_3, Y_4, and Y_5?

2 25. Suppose that the ten girls selected in each unit were not randomly selected, but were selected because Mr. Facts had dated them. He believed that they were representative of the population sampled. What type of sampling design would this be for characters Y_3, Y_4, and Y_5?

2 26. Suppose that Mr. Facts selected the girls to be interviewed for characters Y_3, Y_4, and Y_5 as in question 25 but he did this because it simplified his procedure. What type of sampling design would this be?

Points (3rd Prelim continued)

2 27. Suppose that Mr. Facts did not select the 10 girls within each residence unit as described above, but selected them as random except with the proviso that there be 2 freshmen, 2 sophomores, 2 juniors, 2 seniors, and 2 others to make up the sample within each residence unit. What type of sampling procedure would this be?

2 28. Suppose that an alphabetical list of the girls names were available in each of the 11 selected residence units. Mr. Facts decided to number the names from 1 to 1500 (there were 1500 girls in the 11 selected residence units). He selected a random number between 1 and 12, which was 3. He then took the 3rd, 15th, 27th, 39th, 51st, 63rd, etc., names on the list. What type of sampling design would this be?

2 29. State one question that you think should have been on this prelim.

Name______________________

Statistics Spring, 1968

FINAL EXAMINATION (250 points)

An experiment was conducted to compare the relative speeds of three calculating "machines" for computing the sum of squares of a set of numbers. To do this 21 sets of 5 numbers each were obtained. The 21 sets were divided into 7 groups of three sets each. The three sets in each group possessed a common characteristic. For example, one group of numbers involved only numbers between 1 and 14, a second group involved numbers between 15 and 28, a third group involved numbers between 29 and 42, a fourth group involved numbers between 43 and 57, a fifth group involved numbers between 58 and 71, a sixth group involved numbers between 72 and 85, and the seventh group involved numbers between 86 and 99. Three calculating machines, the three treatments, were to be compared for relative time in seconds for computing sums of squares. The three calculating machines were C = the Curta (hand operated), M = Monroe (electrically operated), and H = human (self operated). The observation or response was the number of seconds required to obtain the sum of the squares of five numbers. The three sets of numbers in each group were randomly allocated to the three treatments. The experiment as conducted and the observations (time in seconds) are given below:

	Set 1	Set 2	Set 3	Set 4	Set 5	Set 6	Set 7
	C - 23	M - 16	C - 28	H - 52	H - 56	M - 33	H - 38
	H - 4	H - 16	M - 15	C - 29	C - 38	H - 62	C - 31
	M - 15	C - 28	H - 38	M - 21	M - 26	C - 40	M - 21
Set Means	14	20	27	34	40	45	30

Treatment means: H = 38 seconds = 266/7

C = 31 seconds = 217/7

M = 21 seconds = 147/7

Mean of all 21 observations: 30 seconds = 630/21

Sum of squares of all error deviations = 1100 = $\sum_{i=1}^{3}\sum_{j=1}^{7}\hat{e}_{ij}^2$

Points (Final continued)

For the experiment described above answer the following questions (1-24):

3 1. The experimental design is ______

3 2. The number of degrees of freedom for error is ______

3 3. Show how to obtain the first error deviation in terms of the numbers above ______

______

3 4. Show how to compute s_e^2 = estimated variance of a single observation in terms of the numbers above ______

______

3 5. For the above data the estimated variance of a treatment mean equals ______

3 6. For the above data the estimated variance of a difference between any two treatment means is ______

3 7. Show how to compute the 90% confidence interval on the difference between any two treatment means above. The tabulated t value required is in the following table:

Degrees of freedom	α = proportion of time a larger value of $\|t\|$ is observed				
	.3	.2	.1	.05	.01
2	1.39	1.89	2.92	4.30	9.92
8	1.11	1.40	1.86	2.31	3.36
9	1.10	1.38	1.83	2.26	3.25
10	1.09	1.37	1.81	2.23	3.17
12	1.08	1.36	1.78	2.18	3.06
20	1.06	1.32	1.72	2.09	2.84

The 90% confidence interval or interval estimate for the above data is from ______ to ______.

3 8. How do the computations in (7) change in computing the 70% confidence interval? ______

______

Points (Final continued)

3 9. The sources of variation in the above experiment are__________

____

3 10. What effects are orthogonal to each other in the above design?

____

3 11. The experimental unit is______________________________

____

3 12. The type of experiment described above is one of the following (circle appropriate one):

factorial, comparative, basic, absolute, technical

3 13. In order for the variance of a difference between any two means as computed above to be appropriate the error deviations must be

____

3 14. The sources of assignable error in the above experiment are

____

4 15. Blocking or "local control" is one of the four properties of an experimental design discussed in this course. Describe how blocking was used in the above experiment.

4 16. Randomization is a second principle of experimental design. Describe how it was used in a design of the above type, and why.

4 17. Replication is a third principle of experimental design which was discussed. Describe how it was used in the above design, and why.

4 18. Orthogonality is a fourth principle of experimental design. Describe how this property applies for the above experiment and tell why it is useful.

The seven principles of scientific investigation may be given as:

1. A clear statement of the problem requiring solution.
2. Formulation of a trial hypothesis.
3. A careful, logical, and critical evaluation of the hypothesis.
4. The planning or design of the experimental investigation.

Points (Final continued)

5. Selection of measuring instruments and control of bias and other errors.

6. Complete and critical summarization and interpretation of results in terms of the hypotheses.

7. Preparation of a complete, accurate, and readable report of the investigation.

Describe briefly how each principle was utilized in the above experiment.

Points	No.	Question
4	19.	(1st principle)
4	20.	(2nd principle)
4	21.	(3rd principle)
4	22.	(4th principle)
4	23.	(5th principle)
4	24.	(6th principle)

An experimenter was considering the following two experimental designs for conducting an experiment (questions 25-26):

Design I (schematic arrangement)

Block 1	Block 2	Block 3	Block 4	Block 5	Block 6
A	A	A	A	A	A
B	B	B	B	B	B
C	C	C	C	C	C
D	D	D	D	D	D

Design II (schematic arrangement)

Block 1	Block 2	Block 3	Block 4	Block 5	Block 6
A	A	A	B	B	C
A	A	A	B	B	C
B	C	D	C	D	D
B	C	C	C	D	D

Points (Final continued)

5 25. What properties are possessed by the above two designs?

5 26. What design would you recommend to the experimenter, and why?

An experimenter used eight experimental units (for instance, 8 rats) in his experiment. He grouped the eight experimental units into two groups of four each. He wished to compare three treatments A, B, and C (for example, three nutritional levels) to measure their effectiveness on yield (for example, increase in weight). He randomly allotted two of the experimental units to treatment A and one each to the other two treatments B and C in each group. Thus treatment A was applied to four experimental units and treatment B and C were applied to two experimental units each. (Questions 27-29):

5 27. What is the experimental design ______________________________

____

10 28. Show whether or not the treatments are balanced with respect to occurrence with each other in the groups.

10 29. Show whether or not treatment and group effects are orthogonal.

Given the following set of numbers for questions nos. 30 to 34:

1 2 8 9

3 30. The arithmetic mean is computed as

3 31. The median is computed as

3 32. The mode is

3 33. The harmonic mean is computed as

3 34. The geometric mean is computed as

3 35. Given the following frequency distribution locate the relative positions of the mean, median, and mode.

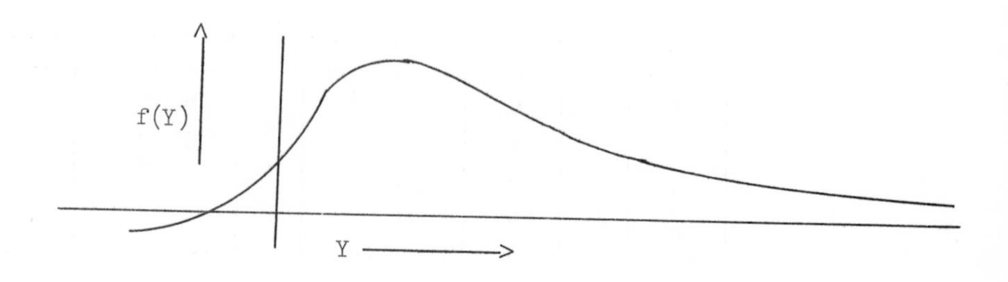

Points (Final continued)

Complete the following statements:

3 36. The ____________________ is an appropriate average when dealing with rates or prices.

3 37. The ____________________ is an appropriate average when dealing with observations which increase in a geometric or exponential manner.

3 38. Suppose that we have the population of a city in 1950 and 1960, that we wish to estimate the population size in 1955, and that birth and death rates stay constant (not equal) and that immigration equals emigration. The appropriate average for estimating the 1955 population would be ____________________ ____________________.

3 39. The difference between the largest member in a sample and the smallest member is the ____________________.

3 40. Another measure of variation which averages the squared deviations between the observations and the arithmetic mean is ____________________.

3 41. If an event either occurs or does not occur in n trials and if the two events are independent and neither includes the other in any way, then the random variable is said to have a ____________________ distribution.

3 42. In the above distribution we are concerned with a success or failure of the event whereas in the ____________________ distribution we are concerned with the number of times an event occurs and not at all with the number of times it fails.

10 43. Using the 99% confidence interval for the binomial distribution as given on the following figure, construct the 99% confidence intervals for the following pairs of values.

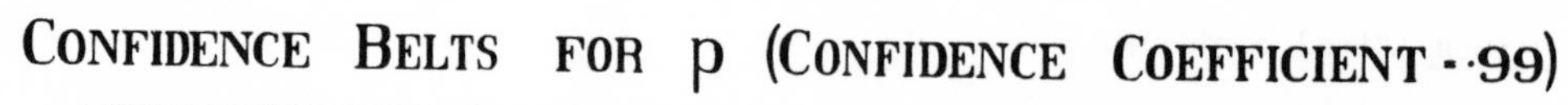

CONFIDENCE BELTS FOR p (CONFIDENCE COEFFICIENT ·99)

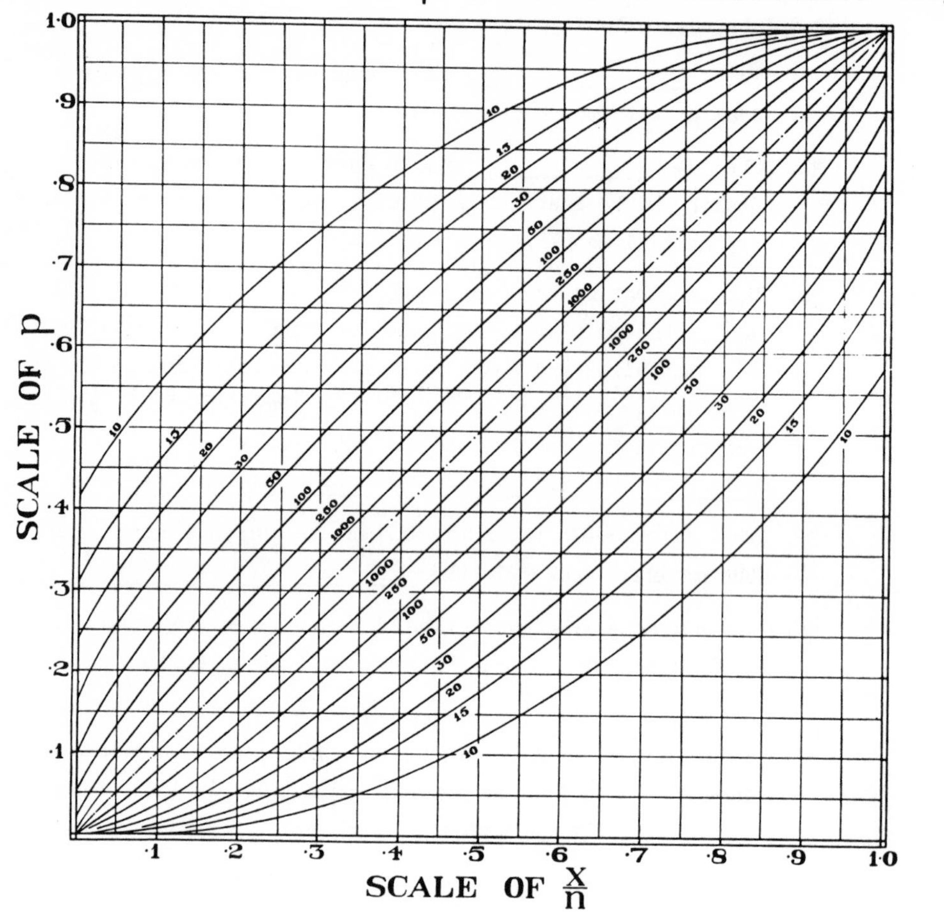

Reprinted by permission of the authors, C. J. Clopper and E. S. Pearson, and the publishers, the Biometrika Office.

Points (Final continued)

Estimated proportion equals x/n	Sample size	lower limit	upper limit
$\hat{p}$ = .2	n = 50		
$\hat{p}$ = .8	n = 50		
$\hat{p}$ = .4	n = 10		
$\hat{p}$ = .4	n = 100		
$\hat{p}$ = .4	n = 1000		

3 44. Circle the correct statement below. One is correct and the other is incorrect.

The probability is 99% that the true proportion p falls in the above 99% confidence intervals.

On the average, 99% of confidence intervals such as those above will include the true proportion p.

5 45. Suppose that the following results were obtained from a 3X2 factorial experiment:

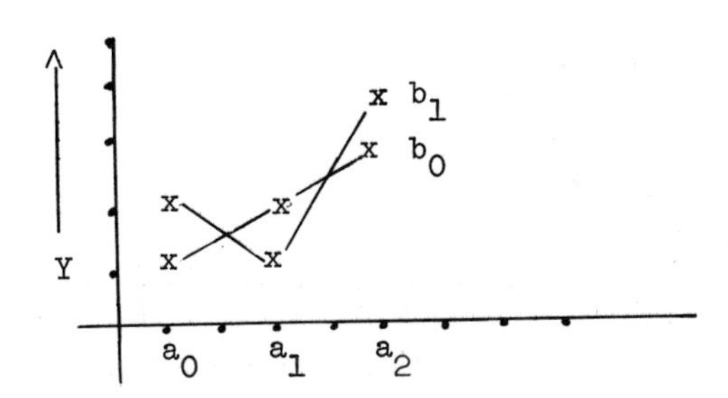

Show with a dashed line what responses Y would be necessary in order to have zero interaction effects.

5 46. Compute the main effects and interaction effects for the following 2x2 factorial treatment design.

Level of factor a	Level of factor b: b_0	b_1	Mean
a_0	12	16	14
a_1	16	8	12
mean	14	12	13

Points

Give the names for the following graphs (questions 47-51):

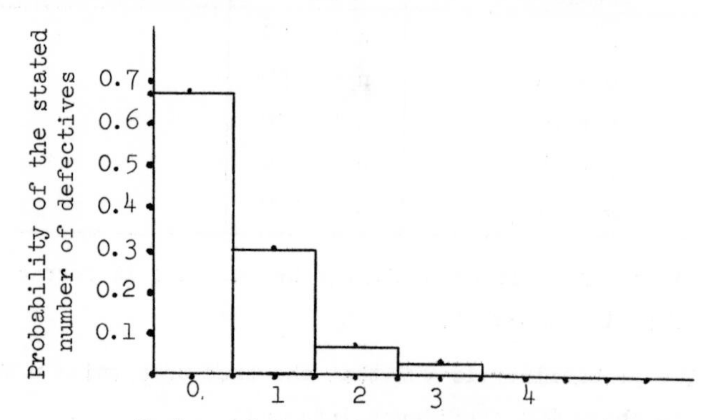

Number of defective in random samples of four items drawn from a population containing 10% defectives (P = 0.1).

2 47. ______________________________

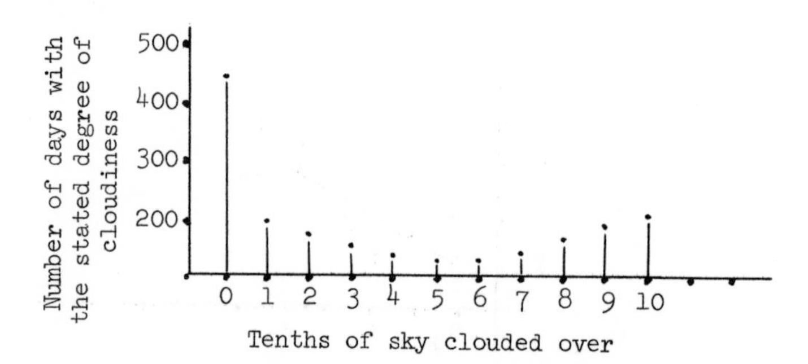

Distribution of cloudiness at Greenwich. Based on data given by Gertrude Pearse (1928) for month of July 1890-1904 excluding 1901) and quoted in M. G. Kendall, Advanced Statistics Vol. 1. Note tendency for sky to be either very clear or very cloudy.

2 48. ______________________________

Points

Author Clinic	Resection Mortality
Finsterer Vienna	5% of 202 cases (1929 report)
Balfour Rochester	5% of 200 cases (1934 report)
Truesdale Fall River	6% of 17 cases (1922 report)
Hanssen Christiana	7% of 51 cases (1923 report)

Lowest reported resection mortality percentages taken from a statistical report by Liningston and Pack: End Results in the Treatment of Gastric Cancer (Paul B. Hoeber, Inc., New York).

2 49. ____________________

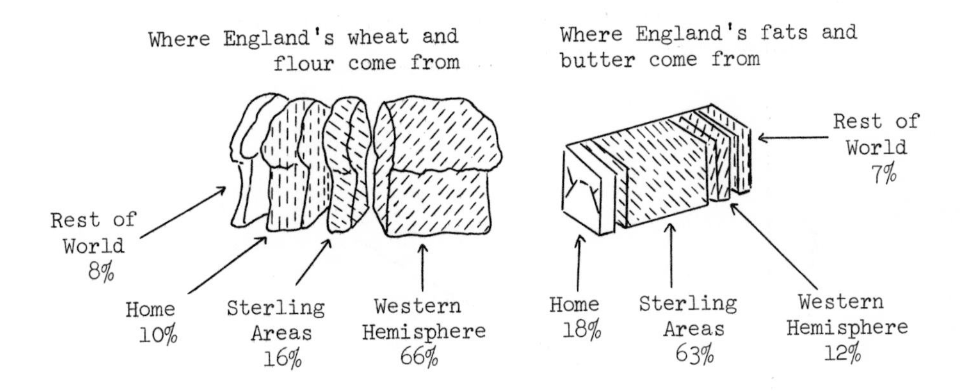

2 50. ____________________

Points (Final continued)

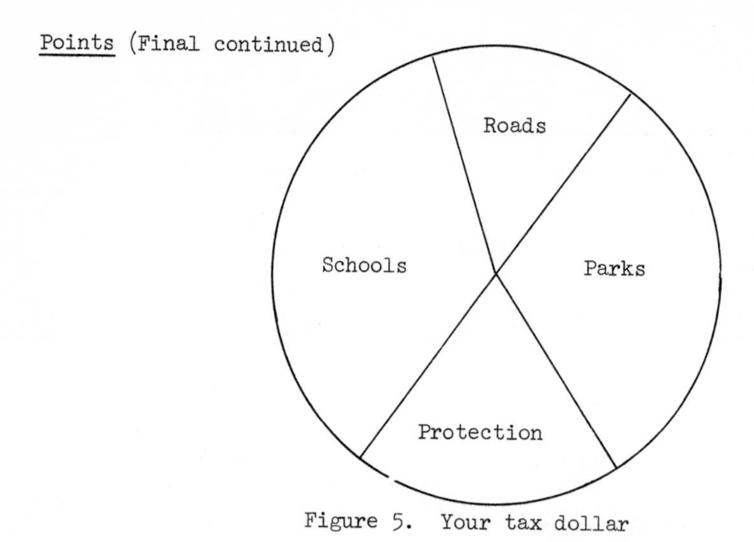

Figure 5. Your tax dollar

2 51. ______________________________

2 52. The following graph appeared in a paper. What comments are relevant to the graph? ______________________________

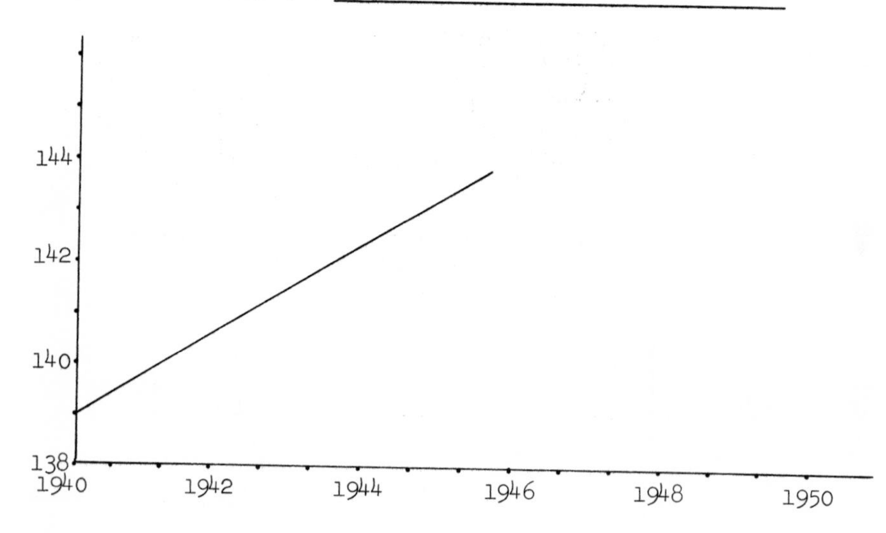

Points (Final continued)

2 53. What comment is relevant for the following graph? ______________

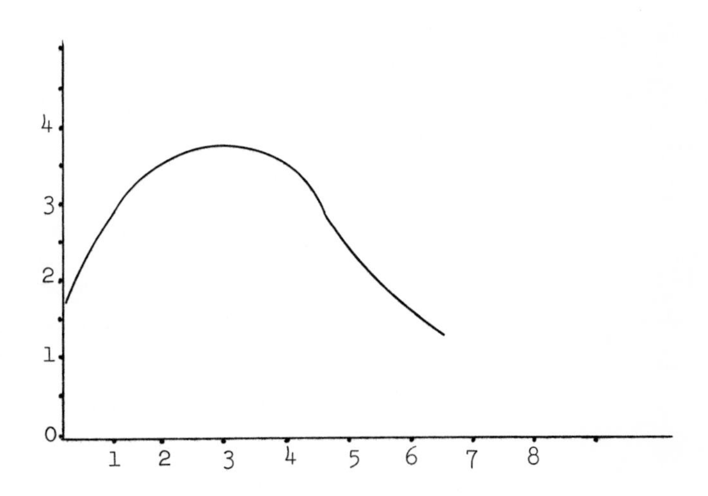

2 54. Circle the appropriate graph for presentation given that Y and X are defined:

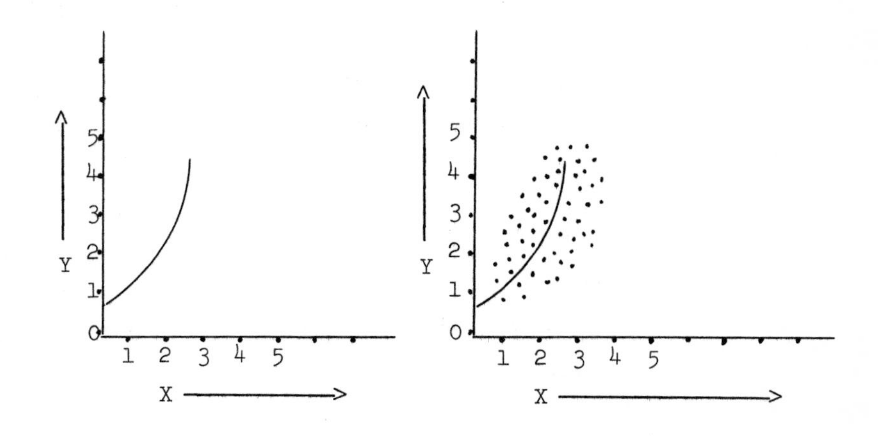

Points (Final continued)

A list of N households in an area is available to a marketing researcher, Mr. John Doe, for the purpose of conducting a survey. The measurement on a household, say Y_1, for the first character has no known assignable error other than a common effect and a random error. Mr. Doe wishes to draw a random sample of n households and to obtain the measurement Y_1 on each household. (Complete the following statements including type of allocation for questions 55-59):

6 55. What sample design is appropriate for the above situation?

____________________________________.

6 56. If Mr. Doe wishes to obtain measurements Y_2 for a second character on all persons living in the n selected households, what is the name of the sample survey design for the second character? ________________________.

6 57. If Mr. Doe randomly selects one individual from among those living in a household to obtain the Y_2 measurements, what is the name of the sample survey design? ____________

____________________________________.

6 58. For a third character, the measurement Y_3 on a household is to be obtained on each of the n households; it is known that the area containing the N households can be subdivided into four subregions which are relatively homogeneous. Mr. Doe selects n_1 observations from subregion one, n_2 observations from subregion two, n_3 observations from subregion three, and n_4 observations from subregion four such that $n_1 + n_2 + n_3 + n_4 = n$. Then, the households are randomly selected within each subregion. What is the name of this design for the third character?

____________________________________.

6 59. For the fourth character, measurements Y_4 are to be made on a random sample of individuals residing in the selected households in 58 above. What is the name of this sample survey design?

____________________________________.

5 60. Define a simple random sample.

Points (Final continued)

Miss I. M. Fashion wishes to use 4 materials (nylon, rayon, cotton, silk) with two gauges (light and heavy) of each material; she also wishes to observe the effect of two colorings (flesh and white) in two different amounts (12% and 14% by weight). She wishes to use all possible combinations of these items as listed above in stocking samples. The response to be measured is number of hours the stocking stays "wearable" (a measurement she makes reliably). The experiment is to be conducted during the academic year using $r = 20$ girls such that all v treatments are used on each girl. The difference between treatment periods or experimental units is known to represent only random sampling variation. Therefore, the v treatments are randomly allotted to the v time periods for each girl. There is a real difference between girls with respect to the length of time a sample stocking stays wearable.

2 61. What is the treatment design?

2 62. What is the experimental design?

4 63. Write out the treatment combinations that Miss Fashion will use.

2 64. Suppose that the four treatments involving light gauge silk were omitted from the experiment because it was known that they gave unsatisfactory results. What would the resulting treatment design be?

2 65. Suppose that only 8 experimental periods are available for each of 32 girls and that there is a real (assignable) effect of periods as well as for girls. What experimental design should Miss Fashion use?

8 66. Miss I. M. Fashion was conducting an experiment on length of time different colors of eye shadow stayed presentable which was a measurement made by her. She used $v = 5$ preparations of eye shadow which only differed in the color component. She was using five different time periods for which she knew that there were only random variations and no known gradients. Therefore, she decided to use $r = 4$ different girls as the blocks and to use a randomized complete block design.

Points (Final continued)

Describe the yield equation for this experiment when the effects are additive and enter the equation in linear form.

Which of the assignable effects are orthogonal to each other?

Construct the experimental plan using the following set of random numbers and describe how you obtained the plan.

3780	28391	05940	55583	81256	38175	38422	64677	80358	52629
5325	05490	65974	11186	15357	21805	10371	95812	84665	74511
8240	92457	89200	94696	11370	75517	82119	09199	30322	33352
2789	69758	79701	29511	55968	19195	92261	44757	98628	57916
7523	17264	82840	59556	37119	77869	08582	63168	21043	17409

INDEX